# Springer Series in Design and Innovation

Volume 29

**Springer Series in Design and Innovation** (SSDI) publishes books on innovation and the latest developments in the fields of Product Design, Interior Design and Communication Design, with particular emphasis on technological and formal innovation, and on the application of digital technologies and new materials. The series explores all aspects of design, e.g. Human-Centered Design/User Experience, Service Design, and Design Thinking, which provide transversal and innovative approaches oriented on the involvement of people throughout the design development process. In addition, it covers emerging areas of research that may represent essential opportunities for economic and social development.

In fields ranging from the humanities to engineering and architecture, design is increasingly being recognized as a key means of bringing ideas to the market by transforming them into user-friendly and appealing products or services. Moreover, it provides a variety of methodologies, tools and techniques that can be used at different stages of the innovation process to enhance the value of new products and services.

The series' scope includes monographs, professional books, advanced textbooks, selected contributions from specialized conferences and workshops, and outstanding Ph.D. theses.

The volumes of the series are single-blind peer-reviewed.

**Keywords:** Product and System Innovation; Product design; Interior design; Communication Design; Human-Centered Design/User Experience; Service Design; Design Thinking; Digital Innovation; Innovation of Materials.

**How to submit proposals**

Proposals must include: title, keywords, presentation (max 10,000 characters), table of contents, chapter abstracts, editors'/authors' CV.

In case of proceedings, chairmen/editors are requested to submit the link to conference website (incl. relevant information such as committee members, topics, key dates, keynote speakers, information about the reviewing process, etc.), and approx. number of papers.

Proposals must be sent to: series editor Prof. Francesca Tosi (francesca.tosi@unifi.it) and/or publishing editor Mr. Pierpaolo Riva (pierpaolo.riva@springer.com).

Umar Zakir Abdul Hamid • Mari Suoheimo
Editors

# Service Design for Emerging Technologies Product Development

## Bridging the Interdisciplinary Knowledge Gap

Springer

*Editors*
Umar Zakir Abdul Hamid
China Euro Vehicle Technology AB,
Sweden
Göteborg, Sweden

Mari Suoheimo
The Oslo School of Architecture and Design
Oslo, Norway

ISSN 2661-8184 ISSN 2661-8192 (electronic)
Springer Series in Design and Innovation
ISBN 978-3-031-29305-4 ISBN 978-3-031-29306-1 (eBook)
https://doi.org/10.1007/978-3-031-29306-1



This Springer imprint is published by the registered company Springer Nature Switzerland AG
The registered company address is: Gewerbestrasse 11, 6330 Cham, Switzerland

# Foreword 1

The design field has gone through many iterations over the years but I can think of no better direction for the field of design and designers than that of service design. Designers will still incorporate important aspects, such as aesthetics, experience design, human factors, sustainability, and production, into their design problems and opportunities, and rightly so as the idea of making the world a better place has always been at the core of design.

Service design encompasses many aspects of the user and consumer experience, whether they are using an online software program to pay their taxes or using visual cues and technology to navigate a museum. The goal of the service design team is to provide the best possible experience. Is the design easy to use and navigate? Is the design easily understandable? Does it include diversity and equity in its offering and is inclusive? Does it provide a space for human emotions to flourish; amusement, caring, enjoyment, and learning are only a few of the positive experiences service design can provide.

With the rise and increasing use of technology, designers and design teams need to understand the tenants of service design now more than ever. Artificial Intelligence, augmented reality, virtual reality, and a growing dependence on data collection are all areas in great need of human touch to keep technology ethical and usable. Service design can serve as an ethical practice for future technologies. Complex experiences with technologies need service design to help demystify their use.

As designers and design teams engage in service design, it is also imperative to incorporate knowledge from many disciplines. Service design problems and opportunities are often complex and can benefit greatly from the fields such as business, economics, engineering, health, psychology, and sociology. Understanding the user is key for service design, and designers need knowledge from these fields from active team members of a cross-disciplinary team.

This book, *Service Design for Emerging Technologies Product Development: Bridging the Interdisciplinary Knowledge Gap*, will benefit those engaged in service design and user experience and those who want to know more about the field. This book offers years of experimentation, research, concerns, and advice to design and provide the optimum user experience. Good service design in every area will enhance people's lives for the better.

Dean Emerita, Professor of Industrial Design,
Rochester Institute of Technology,
Rochester, NY, USA

Lorraine Justice, PhD, FIDSA

# Foreword 2

Twenty years designing medical devices, cars, consumer products and heavy trucks has shown me that many hardware products – however useful – can only be purchased, insured and maintained in certain regions, by certain *clientele*. Having worn virtual "horse-blinders" to obscure a hard product's end-of-life impact on natural resources, I have also gradually come to understand the often untapped promise of *Service Design*.

Shifting focus from Product Development to Service Design – across industry – sounds at first like an aggressive departure from today. But when you consider the "pull" of two simple categories – (1) that *prosperity* should increase among those in the world's emerging markets, and (2) that a more *circular* model should drive down resource use in the global economy, backed increasingly by financiers and policymakers – the crucial role of service design, and the collaboration it demands, is clear.

*Prosperity* Reducing poverty and improving outcomes worldwide motivate a vast *market need* for breakthroughs in transport, housing and food production. Far from a zero-sum game, there is a *growth opportunity* in more prosperity everywhere. The challenge faced by industry – not only in advanced countries – is daunting: why invest to develop new hardware for communities who cannot afford it, and whose infrastructure cannot support it? Service design, however, offers a path for business models to grow, even those enabled by advanced technology. A robust business driven by human behavior in, say, urban goods mobility, can *scale* – without users themselves clamoring over a high entry barrier, or increasing their investment, forced to purchase or lease revision A, B or C of a given technology. The business can grow and gain efficiencies – often in the background – by ever-improving hardware and software. Servitization enables affordability and continuous improvement. As the back-end of a service, the sharing of hardware and logistics is fundamental.

*Circularity* Selling kilometers, hours, temperature or uptime – at first sounding too intangible to be the basis for a business – are often fundamental measurements for practitioners and theorists within the Circular Economy field, in which manufacturers and their partners actually keep ownership of the hardware. This way, the incentives are thought to encourage producers to maximize the value of produced goods. The service *is* the product. Those within industry who develop new technologies will see the path to profitability more clearly when they can squeeze more value out of their R&D investment, and many people from different economic strata can participate, providing a broad user base. Sharing is built-in to "circular" systems, and technology can continuously be tried and brought aboard a service platform.

*Collaborative and Iterative* Design is about defining a purpose, *then* planning, *then* building and *then* using. *Service design thinking* processes simply repeat this flow, but in small steps, to gain new insights – and increase participation – in each loop. Often wrongly seen as the domain of highly-specialized artisans, design is rather, by its nature, collaborative and iterative.

Servitization is the most promising path to the *productization* of new technology. But it cannot be delivered to organizations turnkey by digitalization consultants alone. By its nature, service design requires all functions including finance, logistics, R&D and operations to share knowledge and create value. Sales functions play a vital role, as transitioning from hardware to services changes nearly all of *The Commercial Offer*, redefining *The Customer*.

While certain San Francisco startups come to mind, consider for a moment how service design can transform your own business. The research implications and advances in technology that have most excited you and your colleagues – or communities – showcase their benefit because of what they point to: a tangible human need, based on our shared humanity and a desire for well-being, good health, transportation and community, and to flourish. Working to *serve* these most fundamental aspects of life invites solutions both niche and global, and has far greater promise than merely their enabling technologies.

I believe this book will be of great benefit to you – and the industry – with its technical depth, societal focus and attention to leveraging diverse experience across disciplines.

Founder & Principle, Atlantiska Byrån AB
Gothenburg Sweden
January 2023

Chris Hillman

# Preface and Acknowledgment

## About the Book

Industry 4.0 has seen progress in a lot of emerging technologies development. For example, field robotics, blockchain, vertical take-off-landing vehicles, autonomous robots and robotaxis. However, most of the works currently done are still on the research level with limited productization. Apart from that, there are also myriad of discussions in different topics such as sustainability, circular economy as well as climate change which require interdisciplinary collaborations for the productization efforts. For mass productization of these technologies, there are necessities of paradigm shift in different industries which demand a more human- and customer-centric design philosophy that service design offers.

There are evident gaps of knowledge on the service design topic between practitioners of different areas (business, product, technical, marketing). Furthermore, to output reliable emerging technologies products, Service Design should be understood. It is not enough just to focus on the technology for emerging technologies product development; instead, both external and internal stakeholders service design should be improved. This requires a centralized discussion on the theme.

*Service Design for Emerging Technologies Product Development: Bridging the Interdisciplinary Knowledge Gap* is written to bring a comprehensive discussion on the service design for emerging technologies product development topics. The editors consist of an industrial practitioner and a service design professor who have each worked extensively in the field of emerging technologies productization and service design. The editors believe that the combination of different perspectives in their experiences will be valuable for the editorial process of this book, which aimed to provide comprehensive service design discussions for practitioners in different fields and sectors. The book received contributions by reputable researchers and experts in Service Design field from the United States, China, Finland, Sweden, Norway, Brazil, Denmark and South Africa, among others.

The 16 chapters of this book are divided into four parts. The first part, "Introduction", aims to provide the introduction to the book organization as well as providing overview on the service design for emerging technologies productization for general readers. The second part, "Frameworks and Research Lenses for Service Design in Emerging Technologies Product Development", discusses in detail the service design framework, including its historical background and related-ecosystem. Furthermore, in the same part, discussions are also made on different research lenses in utilizing service design. Part III, "Organizational Transformations and Management for Practicing Service Design", highlights the required transformation needed for organizations to practice service design for their emerging technologies productization, and in Part IV, "Case Studies on the Service Design for Emerging Technologies Applications", several examples on utilizing service design for the product development of emerging technologies are discussed, with examples from different fields.

Summarizing, to enable viable emerging technologies productization, collaboration between different stakeholders are required (internally and externally). And this requires a good understanding on service design. We believe the chapters written for this book will provide good introductory understanding for interested readers from different fields.

## Acknowledgments

### *Dr. Umar Zakir Abdul Hamid*

I have been in the automotive and mobility industry for eight years now and have witnessed the significant transformations taking place in this sector. Based on my experience, I believe that these industrial transformations are larger and more impactful than we might initially expect. The emerging innovations that are being developed will not only affect the automotive and mobility industry but will also disrupt how we live our lives. To thrive in this transformation and deliver sustainable and profitable products, interdisciplinary collaboration is essential. That is why I approached Associate Professor Dr. Mari Suoheimo with the idea of editing a service design book that would benefit practitioners and academics working in emerging technologies product development. Fast forward, after more than a year, we have managed to finalize this book, which includes contributions from distinguished Service Design experts. I would like to thank my co-editor, Associate Professor Dr. Mari Suoheimo, for her efforts throughout the process. Finally, I would like to express my appreciation to SN for the serenity that you always bring.

## *Associate Professor. Dr. Mari Suoheimo*

Associate Professor Mari Suoheimo would firstly like to thank her family for supporting her research and the Oslo School of Architecture and Design for providing the time needed to make this work. Secondly, she would like to thank professor Jonna Häkkilä and docent Ashley Colley for taking her to work previously as a postdoc at the LUX research group at the University of Lapland where emerging technologies were part of the daily exciting research projects. Professor Satu Miettinen for the opportunity in making Ph.D. in service design and for working at the Co-Stars service design research group, where Suoheimo learned more of industry and service design collaborations. These experiences of work and research have shaped the initial road of editing this book. Thirdly, all the special colleagues at the University of Lapland and at the Oslo School of Architecture and Design and other institutes for the fruitful discussion regarding service design, systems, design and technologies, Assoc. Prof. Paola Trapani, Assoc. Prof. Josina Vink, Asst. Prof. Martina Čaić, Assoc. Prof. Jakob Trischler, Assoc. Prof. Ted Matthews, Ma. Harri Pyykkö, Dr. Stefan Walter, Dr. Mira Alhonsuo, Ma. Siiri Paananen, Ma. Krista Korpikoski, Ma. Amna Qureshi. Fourthly, Suoheimo wishes to show gratitude for all the chapter authors of the book, for the lessons that you are providing for us all of combining service design and emerging technologies. Finally, all the peer-reviewers and the editorial team of Springer Series in Design and Innovation.

## *Editors' Joint Appreciation*

The editors would like to express their gratitude to the reviewers and authors for their contributions to this project.

# List of Reviewers

The editors would like to extend their appreciation and gratitude to the following reviewers of the book:

- Toini Harra, Metropolia University, Finland
- Satu Miettinen, University of Lapland, Finland
- Niina Turtola, Edinburgh Napier University, United Kingdom
- Aulikki Laitinen-Tolonen, Lapland University of Applied Sciences, Finland
- Jonna Häkkilä, University of Lapland, Finland
- Aguinaldo dos Santos, Universidade Federal do Paraná, Brazil
- Ashley Colley, University of Lapland, Finland
- Josina Vink, Oslo School of Architecture and Design, Norway
- Danielle Zsifkovits, Hochschule Macromedia, Germany
- Krista Korpikoski, University of Lapland, Finland

# Contents

# About the Editors

**Umar Zakir Abdul Hamid** PhD, has been working in the future mobility (connected and autonomous vehicle) field since 2014 with various teams in different countries and continents. Previously, he led a team of 12 engineers (of ten different nationalities) working in the Autonomous Vehicle Software Product Development with Sensible 4, Finland. Umar is one of the recipients of the Finnish Engineering Award 2020 for his contributions to the development of all-weather autonomous driving solutions with the said firm. He is currently the senior lead strategist for CEVT AB (Geely Group) in Sweden. He is also the secretary of the SAE Cooperative Driving Automation (CDA) Committee and IEEE Sweden. A lifelong learner, Umar Zakir is currently pursuing his Executive MBA at the Gothenburg School of Business, Economics and Law (University of Gothenburg), Sweden. He has authored and edited books and scientific articles on the connected and automated mobility theme, in addition to being the associate editor for the *IEEE Transactions on Intelligent Transportation Systems*, *SAE International Journal of Connected and Automated Vehicles* and *SAE Technical Papers*, among many others. His latest book *Autonomous, Connected, Electric and Shared Vehicles: Disrupting the Automotive and Mobility Sectors* was published in October 2022 by SAE International. As an industrial practitioner, Umar Zakir believes that the productization of emerging technologies can be expedited through cross-sector collaborations.

**Mari Suoheimo** is working as an associate professor of service design at the Oslo School of Architecture and design. Her research concentrates on how to approach wicked problems in service design. Wicked problems are diverse and can range from sustainability to the use of technology. Suoheimo's research interest is in the field of mobility and logistics, emerging technologies, public services, participatory design, design for policy, circular economy, design theories, decolonizing design and sustainability. She is a Doctor of Service Design (2020), Master (2010) and Bachelor of Industrial Design (2008) – University of Lapland; has a degree in art education (2015) – Centro Universitario Claretiano. Suoheimo has working experience in service, UX and product design, marketing and branding. She was able to defend her Ph.D. thesis in three years and during her studies to publish eight peer-reviewed scientific publications. In 2018 she received one-year grant from the Lapland Cultural Foundation for her thesis research. She has worked as a service design researcher and project manager at the University of Lapland for both LUX research group and Co-Stars Service Design research group. Suoheimo has experience in research projects funded by EU (H2020, ESR, EAKR), Academy of Finland, Interreg Nord and Business Finland. She has publications in the following journals: *The Design Journal*, *Processes*, *The International Journal of Design Management and Professional Practice* and *Revista Educação Gráfica*.

Google Scholar: https://scholar.google.com/citations?user=xJsIPf0AAAAJ&hl=en

ORCID: https://orcid.org/0000-0001-6847-0314

LinkedIn: https://www.linkedin.com/in/mari-suoheimo-270b4a28/

# Part I
# Introduction

# Chapter 1
# Introductory Chapter: Service Design for Emerging Technologies Product Development

**Umar Zakir Abdul Hamid and Mari Suoheimo**

## Introduction

This book was born from the idea of taking a comprehensive look at the current state of service design for product development in relation to emerging technologies. Such work requires holistic, multidisciplinary collaboration, particularly between service designers, marketers, businesspeople, managers, and engineers. New technologies trigger innovation and they have and will continue to shape our daily lives.

Making services itself is not an old research theme inside science, but the design of services has actually been recognized as a research topic in the early 1990s (Sun, 2020). Sun (2020) recognizes that the flourishing of service design literature as we know it today in the design field is more from the early 2000. This all seems quite surprising how service design is still so new in the history of science and practice if we think that the services sector is considered being the main contributor to growth and employment in the EU (2016). In addition, the European Union has seen that it is essential to turn from Industry 4.0 to the 5.0 where competency is seen no longer as solely technology but greater focus on human centered strategies (European Commission, 2022, n.d.).

Kimbell (2011) has investigated and written more about the differences between service design adopted in the design field in relation to engineering or marketing fields. We as editors of this book have looked at service design with a broad lens,

U. Z. A. Hamid (✉)
CEVT AB, Gothenburg, Sweden

School of Business, Economics and Law, University of Gothenburg, Gothenburg, Sweden
e-mail: umar.hamid@cevt.se

M. Suoheimo
Oslo School of Architecture and Design, Oslo, Norway



U. Z. A. Hamid, M. Suoheimo (eds.), *Service Design for Emerging Technologies Product Development*, Springer Series in Design and Innovation 29,
https://doi.org/10.1007/978-3-031-29306-1_1

and it can be seen as service design, product service system, design for service, or service ecosystem design to name a few. In this chapter, the aim is to provide brief idea on the service design concept as well as provide the organization of the book.

## Definition of Product

Definition of a "product" by the Merriam-Webster (n.d.) dictionary is explained as follows:

> **1:** the number or expression resulting from the multiplication together of two or more numbers or expressions
> **2: a**: (**1**): something produced *especially*: COMMODITY sense
> (**2**): something (such as a service) that is marketed or sold as a commodity
> **b**: something resulting from or necessarily following from a set of conditions // a *product* of his environment
> **3:** the amount, quantity, or total produced

We see that the product can be understood as a physical object but also as a service as defined in point 2. It is often the amount or quantity that comes into question as well if we discuss industrial products and services. Products can be touched and stored, while services are mainly considered intangible. Often due to the intangible nature, some even claim that services cannot be described as products (Shostack, 1977; Secomandi & Snelders, 2011). Often the two intangible and tangible work in interaction with each other as physical products make part of a holistic service experience. Meroni and Sangiorgi (2016, p. 15) write about how "products" can be interpreted as physical artifacts, experiences, activities, or services.

## Service Design: Brief Concept Overview

Thinking of an example of an experience in a theater will have several touchpoints or service design materials such as buying a ticket, finding a seat through printed signs, seating on a seat, buying refreshments during the break, or having a sensory experience by assisting an audio-visual play. According to Shostack (1977, p. 77), service evidencing is everything that "the consumer can comprehend with their five senses," and it is through these, often tangible, evidences that the service becomes real for the customer. Secomandi and Snelders (2011) also discuss the importance of recognizing the embodied human interactions. Interactions include the experience of being in a theater, such as the feeling of a seat or the music, as well as interactions between people or with digital interfaces, such as purchasing a ticket through an app. Secomandi and Snelders (2011, p. 32) aptly point out that "The distinctive characteristic that stands out in these cases is not intangibility, but the material *heterogeneity* of the service interface." The authors continue describing:

> (...) the material heterogeneity of such an interface, we present a way for letting services *be* on their own, neither equating them with the kind of artifacts associated with goods, nor abstracting them into processes, nor resorting to their socioeconomic circumstances of co-production for a final explanation. (Secomandi & Snelders, 2011, p. 33)

As Secomandi and Snelders (2011, p. 33) write, interactions have connections with other design disciplines, typically "devoted to phenomena in the interface domain of services (e.g., product design, interaction design, graphic design, and many others)." They argue that these disciplines can benefit from grounding themselves in service design research. We agree that taking a more holistic view through service design can also improve the design of physical, tangible products and the interactions and experiences they offer. Service design can provide a perspective on the larger system involved in these contexts of applying tangible interfaces to a service.

Applying interaction and user experience design (UX) in making new product development has had a longer history than service design. Some note that service design was seeing to get birth from cognitive psychology and interaction design (Rytilahti et al., 2015). The coming of service design in the industrial field is more novel. For example, a Google Scholar search with words "industrial service design" made in Google in July 2022 resulted only in 142 results. Industrial service design itself has had less focus in the field service design. Research wise there has been more focus on designing customer experiences (Reason et al., 2015), complexities (Sangiorgi, 2009; Suoheimo et al., 2020), health care (Vink et al., 2019), or tools used in service design (Stickdorn et al., 2018) as some current main streams.

In 2017, a book called *An Introduction to Industrial Service* was edited by Professor Satu Miettinen. The field of "industrial service design" as a research topic has been slowly growing, and the editors hope that this book will serve as a landmark to guide research and practice in the field, especially when applying and productizing emerging technologies. User-oriented research and practice is gaining recognition in other fields and industries, and the coming chapters will explore further into these perspectives. The book is intended for a wide audience, including engineers, designers, marketers, and those in management fields.

## Challenges for Emerging Technologies Product Development

A lot of innovation-related prototypes have been designed and constructed across different industries. However, the actual challenge is to productize the disruptive innovations and turn it into profitable outputs that can be marketed. For example, in the context of the automotive and mobility industry, we have seen for quite a number of years now the prototype of autonomous vehicles roaming in the closed environment in cities globally. However, to turn these driverless vehicles prototypes into actual products, it requires a wider context of discussions, which include understanding different stakeholders and changing management in organizations related to the legacy processes as well as the management of new generation of talents. We believe understanding service design will facilitate better communication between different stakeholders which will then subsequently help to improve the product development activities for emerging technologies. As has been mentioned, this book is expected to provide understanding for the general audience about service design for emerging technologies' productization.

## Summary and Expectations for This Book

This book contains 16 chapters (including this introductory chapter), contributed by service design experts from different countries and continents. It is divided into two main themes, i.e., where the first theme focuses on theory and setting up the scene of making service design in the context of emerging technologies product development (Part I, *Introduction* and Part II, *Frameworks and Research Lenses for Service Design in Emerging Technologies Product Development*). The second theme consists of more detailed discussions and case studies that will shed light on the current research made in the field through practice (Part III *Organizational Transformations and Management for Practicing Service Design* and Part IV *Case Studies on the Service Design for Emerging Technologies Applications*).

We hope that you will find several articles in this book that will be beneficial for your company, research, or public institution as you seek new innovations and services that incorporate emerging technologies.

**Acknowledgments** The editors would like to express their gratitude to all of the authors and reviewers who contributed to the publication of this book. Special thanks go to Mary James and Cynthia Pushparaj from the Springer team for their invaluable support throughout the editorial process.

## References

European Commission. (2016). *European thematic factsheet on services*. Available at https://ec.europa.eu/info/sites/default/files/european-semester_thematic-factsheet_services_en.pdf

European Commission. (n.d.). *Industry 5.0. Research and innovation*. Retrieved October 19, 2022, from https://research-and-innovation.ec.europa.eu/research-area/industrial-research-and-innovation/industry-50_en

Kimbell, L. (2011). Designing for service as one way of designing services. *International Journal of Design, 5*(2).

Meroni, A., & Sangiorgi, D. (2016). *Design for services*. Routledge.

Merriam-Webster. (n.d.). *Product definition & meaning*. Merriam-Webster. Retrieved November 30, 2022, from https://www.merriam-webster.com/dictionary/product

Miettinen, S. (2017). *An introduction to industrial service design*. Routledge.

Reason, B., Løvlie, L., & Flu, M. B. (2015). *Service design for business: A practical guide to optimizing the customer experience*. Wiley.
Rytilahti, P., Miettinen, S., & Vuontisjärvi, H. R. (2015). The theoretical landscape of service design. In *Design, user experience, and usability: Design discourse* (pp. 86–97). Springer.
Sangiorgi, D. (2009, April). Building up a framework for service design research. In *8th European academy of design conference* (pp. 415–420).
Secomandi, F., & Snelders, D. (2011). The object of service design. *Design Issues, 27*(3), 20–34.
Shostack, G. L. (1977). Breaking free from product marketing. *Journal of Marketing, 41*(2), 73–80.
Stickdorn, M., Hormess, M. E., Lawrence, A., & Schneider, J. (2018). *This is service design doing: Applying service design thinking in the real world*. O'Reilly Media.
Sun, Q. (2020). Towards a new agenda for service design research. *The Design Journal, 23*(1), 49–70.
Suoheimo, M., Vasques, R., & Rytilahti, P. (2020). Deep diving into service design problems: Visualizing the iceberg model of design problems through a literature review on the relation and role of service design with wicked problems. *The Design Journal, 24*(2), 231–251.
Vink, J., Joly, M. P., Wetter-Edman, K., Tronvoll, B., & Edvardsson, B. (2019). Changing the rules of the game in healthcare through service design. In *Service design and service thinking in healthcare and hospital management* (pp. 19–37). Springer.

# Chapter 2
# Product Development Challenges for Emerging Technologies and Service Design Roles in Addressing the Issues

**Umar Zakir Abdul Hamid and Mari Suoheimo**

## Introduction

Compared to the last century, general public lives have significantly changed due to the rapid technological development in various sectors. For example, if one takes a look at the telecommunication sector, most older people were not even acquainted with mobile phones at the end of the 1990s. However, especially post-2020, most of the world's population, even in developing countries, is now equipped with knowledge of using smartphones. As can be seen, the usage of smartphones and applications such as WhatsApp and Facebook Messenger is widespread today (Nuryana et al., 2021; Susilawati & Supriyatno, 2020).

This continuous change in user behaviour and industrial transformation has been spurred by the global pandemic that happened circa 2019–2021, that is, COVID-19 (Pratama et al., 2020). The global quarantines have stimulated digitalization in a lot of sectors. Simultaneously, we are also seeing progress which has stimulated a lot of innovations for more advanced emerging technologies, for example, autonomous vehicles, autonomous mobile robotics and blockchain-related technologies.

However, most of these developments are still limited in the early stage of productization, that is, in the form of prototypes or proof of concepts. For example, even though getting a lot of funding, the autonomous vehicle industry is actually still in the infancy stage in the context of product maturity. The proof is even

U. Z. A. Hamid (✉)
CEVT AB, Gothenburg, Sweden

School of Business, Economics and Law, University of Gothenburg, Gothenburg, Sweden
e-mail: umar.hamid@cevt.se

M. Suoheimo
Oslo School of Architecture and Design, Oslo, Norway



U. Z. A. Hamid, M. Suoheimo (eds.), *Service Design for Emerging Technologies Product Development*, Springer Series in Design and Innovation 29,
https://doi.org/10.1007/978-3-031-29306-1_2

after billions of funding, some of the giants in the area have been shut down (https://techcrunch.com/2022/10/26/ford-vw-backed-argo-ai-is-shutting-down/). However, it is not because the technology is not valuable; instead, the proper deduction in the productization of emerging technologies is not as easy as making the prototypes. If the early development process has the privilege to abandon the perspective of customer experience and market, for productization, however, to allow market-and-user acceptance, those factors could not be abandoned.

## The Objective and Contributions of This Chapter

Service design is one of the things that is said to help improve product development, therefore yielding reliable productization. The knowledge will improve the understanding between the desired product and service by customers, internally or externally. However, the knowledge gap between technical and design experts on the topic is still broad. Even though some books have been written, not many books are written with the perspective of bridging the knowledge gap between practitioners in those fields from different perspectives. Therefore, this book is aimed to address the aforementioned proficiency void.

As such, this chapter is written to provide the first flavour of the book, following the previous introductory chapter. It is aimed to provide an overview of emerging technologies' product development challenges as well as the service design potential roles in improving the productization effort. This chapter is expected to benefit the readers before going into individual chapters, to understand the ongoing transformation in the multi-sector industry as well as the internal and external challenges involving the topics. This chapter can also be read as a stand-alone chapter, as such it is not intended to be exhaustive writing, but instead, it is written to be indicative towards the said matter. This chapter is not intended to provide theoretical discussions, but instead the aim is to relate the connection between emerging technologies and service design.

## Emerging Technologies

Etymologically, emerging technologies could be oversimplified as new technologies that are gaining progress in development, but are yet to gain market penetration or traction (https://www.lawinsider.com/dictionary/emerging-technologies). The advance and progress in the computing field have facilitated an ongoing rapid industrial transformation across different sectors, which is also called the Fourth Industrial Revolution (FIR). FIR sees a lot of discussions revolving around the potential of emerging technologies such as autonomous vehicles, 3D printing, nanotechnology and quantum computing (The Fourth Industrial Revolution, n.d.; Walthall & Dixit, 2022).

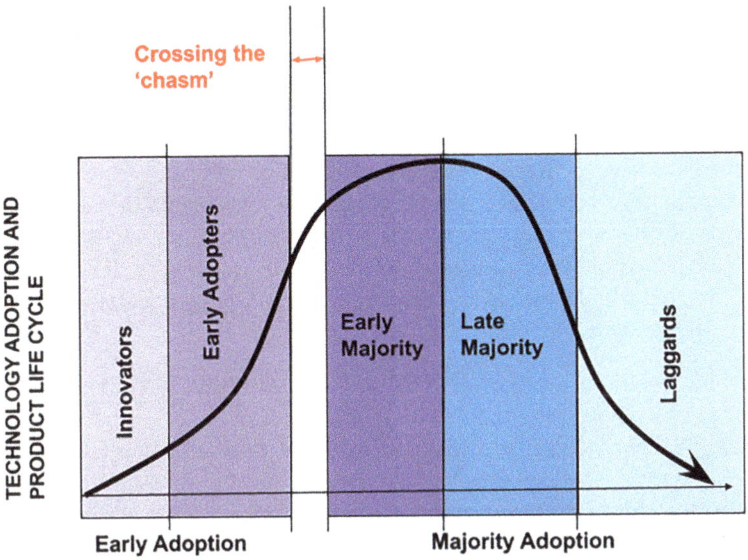


**Fig. 2.1** New technology adoption and product life cycle, re-illustrated from (Abdul Hamid et al., 2021)

The aforementioned emerging technologies are claimed to have disruptive effects, that is, the ability not only to arise as a new product but also to disrupt the corresponding industry. For example, autonomous vehicles, together with vehicle connectivity and electrification as well as shared mobility, is expected to disrupt and alter the legacy ecosystem of the automotive and mobility industry. However, as the technology is still pristine, it is still not getting traction in the market. This is due to the nonmature technology stage of the technology readiness level (TRL) that it possesses. As can be seen in Fig. 2.1, usually, in the technology adoption and product life cycle, new technologies are embraced by innovators and early adopters. These groups are usually coming from technical backgrounds. In reality, if a technology wants to be well accepted by the majority, it should be customer-facing in the design, allowing it to cross the chasm (Abdul Hamid et al., 2021).

'Chasm' here refers to 'conditions' that an early prototype or innovation needs to traverse to be adopted by the majority of consumers. A classic example of this can be represented by the adoption of smartphones. In the early stage of its introduction to the market, most of the people using smartphones were coming from a technology-savvy demographic. However, the booming of smartphones in the market actually transpired when it managed to cross the 'chasm', that is, attractive and stable enough to be bought by the subsequent type of adopters, that is, the early majority, late majority and laggards. This shows that to be profitable, the adoption of emerging technology by the majority of consumers requires the technology to have a consumer-centric design in its nature.

However, as can be imagined, the productization of emerging technologies arrives with challenges. For example, as it arose from the technical perspective, usually it lacks the discussion of customer-facing user experience. This is because the focus of a lot of innovation activities is to prove that certain technology can 'work', thereby abandoning the consumer experience in the initial stage (e.g. the requirement to have a friendly user interface). This then slows down the time for market-and-user acceptance. Furthermore, to be acceptable in the market, the product should focus on delivering value to the customer, assuring improvements to the value added for the customers compared to the previous technology. In the next section, we will denote several challenges regarding product development for emerging technologies.

## Challenges for Emerging Technologies' Product Development

### *Lack of End-User Perspective*

Engineering-led innovation usually faces a prominent concern, that is, absence of customer perspective and end-user perspective in the design of the system (Chiesa & Frattini, 2011; Slater & Mohr, 2006). Consequently, we frequently see a lot of high-tech innovations by start-ups or laboratories that failed to be productized. However, on the other hand, new technologies that integrate the customer perspective, even though they are simple technologically, managed to gain massive traction (Zhang et al., 2009; Görgens, 2019; Bough et al., 2020).

For example, Uber as an application, in the context of technology, is not a sophisticated technology. However, due to the inclusion of end-user perspective, that is, bringing new value to add to the customers, they managed to disrupt the mobility industry. The word *uberization* has been coined to refer to massive growth and disruption. It refers to the act or process of changing the market for a service by introducing a different way of buying or using it, especially using mobile technology (Daidj, 2018).

### *Failure to Understand the Market Demands*

In deep-tech start-ups, especially those working on emerging technologies, one of the top reasons why some of the companies collapsed even after receiving a significant amount of investments is because of the failure to understand the market demands. According to Investopedia, 'productize' expresses 'the process of developing or altering a process, idea, skill, or service to make it marketable for sale to the public' (https://www.sae.org/blog/gp-umar-zakir-abdul-hamid-next-generation-mobility-productization-industrialization). Therefore, building a working prototype is not the same as making it acceptable for the market. An understanding of the market includes doing the study of what the competitors are doing as well as understanding the full current ecosystem of one's industry.

## *Rise of Software*

Compared to the previous industrial revolutions, the fourth industrial revolution witnessed a lot of emerging technologies which has software as their cornerstone. In the automotive industry, for example, the term software-defined vehicle has been coined to indicate that future mobility will revolve around software, instead of software being patched around hardware (https://www2.deloitte.com/cn/en/pages/consumer-business/articles/software-defined-cars-industrial-revolution-on-the-arrow.html; Aust, 2022). With the rise of software, in legacy traditional corporations, the failure to adopt software-related requirements into the organizational ethos has managed to hinder and stall a lot of productization efforts. The adoption of software-based notions and philosophy into the production of software-centred technologies in a company previously heavily laden with mechanical and hardware sees the shift that is not only about technology. But instead, it requires a new process, a new business model and the new generation of talents, among many others (Humble & Kim, 2018).

## *Attracting and Retaining the New Generation of Talents*

If we are looking at the previous subsections, the rise of software brings the need to bring a new generation of talents, that is, software-oriented experts. In the automotive industry, for example, a lot of companies are starting to hire people from computer-science backgrounds to support the future mobility transformation (Hamid, 2022).

However, the implementations are not as easy as that. The productization of emerging technologies requires hiring the right person for the right topics with the right quantity with the right merit. Adam Smith in *The Wealth of Nations* has mentioned the ability to differentiate the specialization and the division of labour as one of the keys to achieving success and prosperity (Smith, 1937; https://www.econlib.org/library/Enc/DivisionofLabor.html; https://www.econlib.org/library/topics/highschool/divisionoflaborspecialization.html). This notion remains true even in this context. To elaborate more, the productization of emerging technologies not only requires building the right tech but also constructing an organization that will support the ambition. Figure 2.2 illustrates some of the organizations that are needed to be hired in the product development life cycle for emerging technologies. The challenge is to assure every department moves with the same vision and why. Furthermore, as can be seen from Fig. 2.2, for early-stage start-ups, it is important

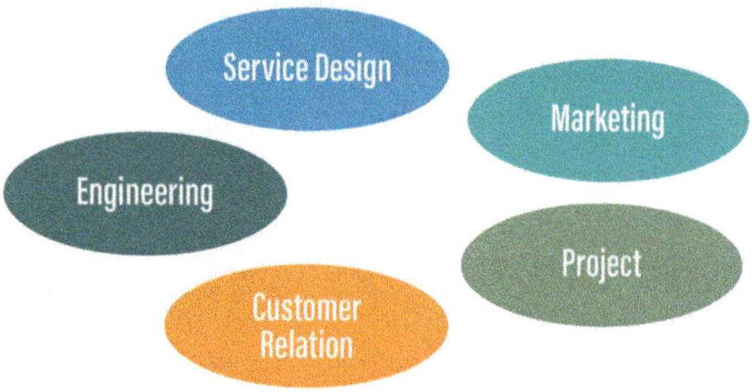


**Fig. 2.2** Typical organizations and talents that are needed to productize emerging technologies. (The photo is meant to be indicative)

to have the right balance between different type of talents, which then require a management with sufficient prioritization skills in hiring and structuring the organization.

For these emerging areas, different talents that are required are not only coming from a technical background, but companies also need business experts with software mindsets (as an example). *The Great Resignation* that started amid the global COVID-19 pandemic, where a mass exodus of employees occurred, is not only contributed by the pandemic (Hopkins & Figaro, 2021). But instead, the new generation of talents are facing new kinds of challenges including moral injury, a situation where employees are no longer working for the 'payslips', but instead they are also looking for security and safety from emotional and mental distress (https://www.ptsd.va.gov/professional/treat/cooccurring/moral_injury.asp; Veenema et al., 2022; Linzer et al., 2022).

As these experts are coming from not only different fields but also new generations, their concerns are also different (Pelliccione, 2020). As such a lot of efforts have been done to discuss and propose new and improved working cultures. For example, the NextGen Project, a think-tank organization in Helsinki discussing the future of leadership, was formed for this related purpose (https://www.thenextgenproject.io/). This discussion is important as bringing the wrong people to lead the productization of emerging technologies not only will delay the productization, but it potentially will create stress and high employee turnover.

The right question is how can we identify, retain and attract these new kinds of talents to develop emerging technologies? This is because it is not only the external customers that we need to attract, but we also need to handle internal 'customers and stakeholders' for the product development of emerging technologies.

## *Unclear Visions and Unclarified 'Why'*

Building a product with no good understanding not only will cause building the wrong products, but instead, it will also contribute to the unclear vision from the top

**Fig. 2.3** Among the most important questions when building the products deriving from emerging technologies are 'Why?', that is, 'Why do we need this?' 'Why do markets and customers want this?'

management. With the wrong input regarding the market and customer perception, the vision that will be created by the top management will also be distorted. This then will affect the 'why' for the product development team, which then will affect the product definition and consequently the manufacturing of the product, therefore potentially causing building a product that is not required by the customers and markets (Fig. 2.3).

## *Overhyped and Misleading Marketing*

Continuing our discussions in the previous section, unclear vision will create unclear requirements for the product. Furthermore, on top of that, without clear communications and unified understanding on what kind of products the organization is building, this might lead to overhyping the products. This eventually might bring risks, or even fatalities due to the wrong customer understanding. According to the latest research by the Thatcham Research, more than 50% of drivers in the United Kingdom mistakenly assume that they can purchase a fully autonomous vehicle today. The same situation also happens in the United States, but the number is higher, that is, 72% (On the road to an automated future, n.d.). This is actually very dangerous as customers might also overestimate the capabilities of the technology. Therefore, it is important to have a clear 'why' during the development of emerging technologies to prevent misleading marketing efforts.

As noted from this section, it is obvious that product development for emerging technologies does not only require the perspective of technical but also human-centric design. And this perspective should not only consist of external, but it should also include internal stakeholders' feedback. This involves designing the right communication between departments and having the right design structure of the organization to achieve the latest business objective. This will then help in reducing the silo and miscommunications. One of the topics that can help to address these is service design.

It is important to note that the factors mentioned above are only indicative, and there are a lot more challenges for emerging technologies that have not been mentioned, but the authors believe it will provide the picture to the readers.

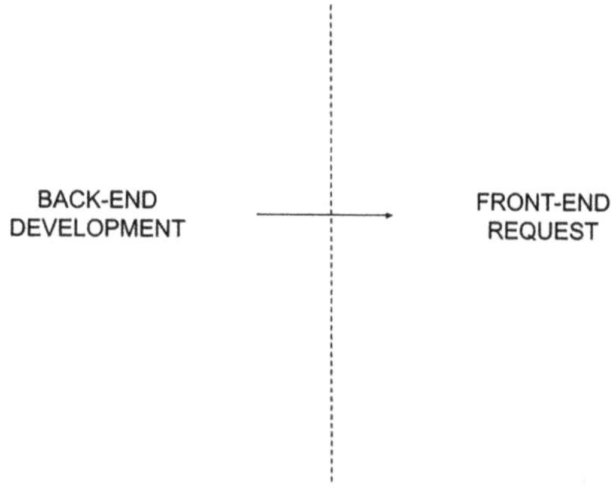


**Fig. 2.4** Service design can help to improve the transparency between front-end customer requests with the back-end development challenges (Glushko & Tabas, 2009)

## Service Design

Service design is a framework and philosophy which emphasizes the customer experience throughout the journey of product utilization. It underscores the importance of delivering the best quality of service to the customer in the whole user journey. The discussions are comprehensive and not only from the technical but also from the entire journey.

In the context of product development for emerging technologies, for example, for autonomous vehicles, not only the service of a robotaxi should be designed to fit the front-end user that will eventually ride the vehicle, but the need of the front-end user will simultaneously affect what will happen in the back-end development. This means any changes on the front-end requirements will also affect the back-end process, structure and organization. The complexity of emerging technologies' development should not be underestimated. The figure below illustrates more in detail. In Glushko and Tabas (2009) mentioned designing integrated service systems by blurring the difference between the 'front stage' and 'back stage' (Fig. 2.4). This is true as a transparent process and service in the whole journey will help to yield a high-quality product.

## How Service Design Can Improve the Product Development for Emerging Technologies

Korper et al. (2020) listed six principle themes of service design which include the following:

- Human- and meaning oriented
- Co-creation and inclusion
- Transformative and betterment oriented

- Emergent and experimental
- Explicative and experientially explicit
- Holistic and contextual

As can be seen, the principles heavily accentuate the need for human perspectives. According to Sangiorgi and Prendiville (2014), service design is about designing the customer experience, and it concerns several parties and items which include end-users, organizations and technology-related topics (Korper et al., 2020; Sangiorgi & Prendiville, 2014).

Stickdorn et al. (2011) have promoted similar service design principles as user-centredness, co-creation, sequencing, evidencing and holism (Stickdorn et al., 2011). Two different terms to Korper et al. (2020) are sequencing and evidencing (Korper et al., 2020). It is essential through sequencing and evidencing to picture or visualize, for example, through service, what the current pain points of a service are and how they could be improved. This is one way of making the invisible parts visible. Once they are visible, then the 'pain points' encountered are possible to tackle or to enhance the parts that already work well. These ideas are developed further through often 'quick and dirty' prototyping, which means that the ideas are tested with low-level cost as paper or cardboard to get user feedback as early on as possible. Through different iterations, the prototypes are enhanced until a product or a service takes form. Using service design in making digital services has been seen in costs of re-coding a software, thus reducing costs, since the engineers participating in the service design process knew what the software was about (Suoheimo et al., 2022).

In the next subsections, we are going to describe and highlight how the adoption of service design principles into the product development life cycle can improve the process as well as add value not only to end customers but also to the employees that are developing it (e.g. improving the communications in the organizations).

## *Fusion of Service Design Output in the PDLC Help to Early Rectification*

As mentioned earlier in this paper, one of the challenges with product development is the lack of communication and silos between departments. In the nominal and typical software development life cycle (SDLC), the process involved planning, analysis, design, implementation, testing and integration followed by maintenance after the product deployment (Santander, 2018).

However, with new emerging technologies such as autonomous vehicles, where it is driven by technological innovation – for the productization, it is important to include feedback from the customer as well as related stakeholders in the ecosystem as soon as possible in the life cycle. This will eventually influence better subsequent stages in the life cycle. Figure 2.5 illustrates the potential of service design in helping the inclusion of customer and stakeholder feedback in each stage of the SDLC.

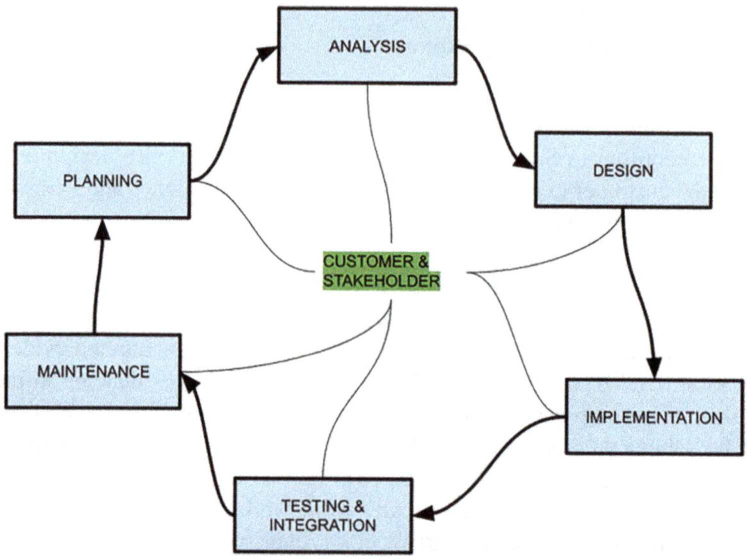


**Fig. 2.5** Integration and fusion of customer and stakeholder feedback in the software product development life cycle can be aided by service design

## *Co-creation for Delivering the Right Experience*

Apple is one of the examples of companies that managed to put customer experience at the centre of product development (Naiman, 2019). Subsequently, it became one of the most significant brands in the world. Oura is another example of a company that has worked with this edge, and they began to have problems when they gained leadership that did not share the same values (Lappalainen, 2022). At Apple, the experience not only involves the software and human interaction but also the purchasing experience on the physical store as well as the website. The reason why Apple managed to be successful is by putting the customer in the back of their head. This same notion is then followed by its subsequent competitors such as Xiaomi (Wong, 2012; Lajoso et al., 2020). For new technologies such as autonomous service robotics, it is important to co-create the technology together with the related customers. This will assure to obtain feedback from customers as soon as possible in the design phase.

## *Transitioning from Innovation to Creating Value for Profitability*

As mentioned in (Korper et al., 2020), service design (SD) focused on human-perspective design can help to guide deep-tech companies especially to put the human touch in their product design. Transitioning from innovation to value creation during the productization, as trivial and straightforward as it may seem, is one of the reasons why start-ups fail to grow and scale their business (Cantamessa et al., 2018; Kalyanasundaram, 2018). And the reason why some start-ups managed to grow is also because they manage to put human perspective in building their product (market and end-user). See the earlier mentioned Uber case in this chapter. Adopting the service design notion in the software product development life cycle for emerging technologies will help to address this.

### *Designing a Structure and Organization for Emerging Technologies' Product Organization*

Transitioning from innovation to value creation also means the necessity to transform the structure of the organization. This is one of the biggest challenges for start-ups and R&D laboratories when discussing productization. During the scaling up and growth period of the company, not only the earthquake and challenges will be experienced from the viewpoint of external (needs to deliver product) but also internally (building the organization). For example, managing and integrating new kind of talents and hiring at a rapid pace is not easy and needs to be handled properly. Designing the needed service for this purpose requires also to maintain communication between these parties.

## Potential Challenges in Adopting Service Design in the Emerging Technologies' Productization

From the previous section, we believe the readers can notice that there are correlations between emerging technologies product development and service design role in improving the process. However, adopting service design into the engineering world is not as easy as it appears. For example, among the challenges that might come include the need for more compassion from the leaders (Muttaqin et al., 2022; Mayer & Wilke, 2022). Changes take time; therefore, the adoption of SD should also be done properly, especially in the time of volatility, uncertainty, complexity and ambiguity (VUCA) (Murugan et al., 2020).

Furthermore, agility is required when dealing with emerging technologies, because it might be that the market and customer are changing their perception towards a certain area due to the changes in world event. Therefore, companies need to address this issue using one of the SD principles mentioned above, that is, betterment oriented. This is also aligned with other philosophies and frameworks in product development such as lean and agile (Murman et al., 2016; Azevedo et al., 2012). Finally, it is important to adopt SD because the failure to do so will lead organizations to the same path as 'collapsing giants' which failed to change during the transformation period (Ab Rahman et al., 2017).

## Summary and Conclusions

This chapter is written to provide the large concept of the book. It is written to guide the readers on the high-level overview of what is emerging technologies and service design, as well as their correlations. The authors try to highlight the emerging technologies' product development challenges, followed by the potential role of service design in improving the process. This chapter is anticipated to provide a sounder interpretation of the knowledge among the readers before proceeding to the next chapters. To summarize, the integration of SD in the product development framework for emerging technologies can yield better products and help organizations to be profitable. The inclusion will not only benefit external customers but also companies and employees as well as internal stakeholders. By improving the product development with the integration of service design knowledge and framework, it will help to expedite the time-to-market for the product, therefore improving the cash conversion cycle rate, and subsequently bringing profitability for the company. In the next chapters, the readers can read in more detail the service design discussions. We hope the readers will enjoy reading this book.

## References

Ab Rahman, A., Hamid, U. Z. A., & Chin, T. A. (2017). Emerging technologies with disruptive effects: A review. *Perintis eJournal, 7*(2), 111–128.

Abdul Hamid, U. Z., Mehndiratta, M., & Adali, E. (2021). *Adopting aviation safety knowledge into the discussions of safe implementation of connected and autonomous road vehicles* (No. 2021-01-0074).

Aust, S. (2022). Vehicle update management in software defined vehicles. In *2022 IEEE 47th conference on Local Computer Networks (LCN)* (pp. 261–263). IEEE.

Azevedo, S. G., Govindan, K., Carvalho, H., & Cruz-Machado, V. (2012). An integrated model to assess the leanness and agility of the automotive industry. *Resources, Conservation and Recycling, 66*, 85–94.

Bough, V., Breuer, R., Maechler, N., & Ungerman, K. (2020). The three building blocks of successful customer-experience transformations. *McKinsey Quarterly Magazine*, 27.

Cantamessa, M., Gatteschi, V., Perboli, G., & Rosano, M. (2018). Startups' roads to failure. *Sustainability, 10*(7), 2346.

Chiesa, V., & Frattini, F. (2011). Commercializing technological innovation: Learning from failures in high-tech markets. *Journal of Product Innovation Management, 28*(4), 437–454.

Daidj, N. (2018). Uberization (or uberification) of the economy. In *Encyclopedia of information science and technology* (4th ed., pp. 2345–2355). IGI Global.

Glushko, R. J., & Tabas, L. (2009). Designing service systems by bridging the "front stage" and "back stage". *Information Systems and e-Business Management, 7*(4), 407–427.

Görgens, M. (2019). *How can Artificial Intelligence use big data to form a better customer experience?* (Bachelor's thesis, University of Twente).
Hamid, U. Z. A. (2022). *Autonomous, connected, electric and shared vehicles: Disrupting the automotive and mobility sectors*. SAE International.
Hopkins, J. C., & Figaro, K. A. (2021). The great resignation: An argument for hybrid leadership. *International Journal of Business and Management Research, 9*(4), 393–400.
Ford, VW-backed Argo AI is shutting down, TechCrunch https://techcrunch.com/2022/10/26/ford-vw-backed-argo-ai-is-shutting-down/
Econlib. (n.d). The Division of Labor. [online] Available at: https://www.econlib.org/library/Enc/DivisionofLabor.html Accessed 17th June 2023.
Econlib. (n.d.). Division of Labor and Specialization. [Online] Available at: https://www.econlib.org/library/topics/highschool/divisionoflaborspecialization.html Accessed 17th June 2023.
Law Insider. (n.d.). Emerging Technologies. [Online] Available at: https://www.lawinsider.com/dictionary/emerging-technologies Accessed 17th June 2023.
U.S. Department of Veterans Affairs. (n.d.). Moral Injury. [Online] Available at: https://www.ptsd.va.gov/professional/treat/cooccurring/moral_injury.asp Accessed 17th June 2023.
SAE International. (n.d.). Next Generation Mobility Productization and Industrialization. [Online] Available at: https://www.sae.org/blog/gp-umar-zakir-abdul-hamid-next-generation-mobility-productization-industrialization Accessed 17th June 2023.
The NextGen Project. (n.d.). [Website] Available at: https://www.thenextgenproject.io/ Accessed 17th June 2023.
Deloitte. (n.d.). Software-Defined Cars: Industrial Revolution on the Arrow. [Online] Available at: https://www2.deloitte.com/cn/en/pages/consumer-business/articles/software-defined-cars-industrial-revolution-on-the-arrow.html Accessed 17th June 2023.
Humble, J., & Kim, G. (2018). *Accelerate: The science of lean software and devops: Building and scaling high performing technology organizations*. IT Revolution.
Kalyanasundaram, G. (2018). Why do startups fail? A case study based empirical analysis in Bangalore. *Asian Journal of Innovation and Policy, 7*(1), 79–102.
Korper, A. K., Patrício, L., Holmlid, S., & Witell, L. (2020). Service design as an innovation approach in technology startups: A longitudinal multiple case study. *Creativity and Innovation Management, 29*(2), 303–323.
Lajoso, J., Sousa, A., Albuquerque, J., Mineiro, R., & Au-Yong-Oliveira, M. (2020). Closed against open innovation: A comparison between Apple and Xiaomi. In *World conference on information systems and technologies* (pp. 436–448). Springer.
Lappalainen, E. (2022, Nov 9). Oura: Ouran Perustajat Jättivät Yhtiön johdon siirryttyä amerikkalaisille – Näin he tienasivat. [Oura: The founders of Oura left the management of the company to the Americans - This is how they made money]. Retrieved 23 Nov 2022, from https://www.hs.fi/visio/art-2000009188037.html
Linzer, M., Griffiths, E. P., & Feldman, M. D. (2022). Responding to the great resignation: Detoxify and rebuild the culture. *Journal of General Internal Medicine, 37*, 1–2.
Mayer, C. H., & Wilke, K. (2022). Leading through VUCA times for a sustainable future of work: Expert views on the global automotive industry. In *Leadership after COVID-19* (pp. 253–268). Springer.
Murman, E., Allen, T., Bozdogan, K., Cutcher-Gershenfeld, J., McManus, H., Nightingale, D., Rebentisch, E., et al. (2016). *Lean enterprise value: Insights from MIT's Lean aerospace initiative*. Springer.
Murugan, S., Rajavel, S., Aggarwal, A. K., & Singh, A. (2020). Volatility, uncertainty, complexity and ambiguity (VUCA) in context of the COVID-19 pandemic: Challenges and way forward. *International Journal of Health Systems and Implementation Research, 4*(2), 10–16.
Muttaqin, G. F., Yunia, D., & Fazri, E. (2022). The impact of empathetic leadership and job innovation in improving operational performance and firm value. *Innovations, 68*, 966.
Naiman, L. (2019). Design thinking as a strategy for innovation. *The European Business Review, 53*, 72–76.

Nuryana, Z., Pangarso, A., & Zain, F. M. (2021). Factor of zoom cloud meetings: Technology adoption in the pandemic of COVID-19. *International Journal of Evaluation and Research in Education, 10*(3), 816–825.

On the road to an automated future. https://www.thatcham.org/what-we-do/automated-driving/

Pelliccione, A. (2020). Younger workforce leads the way for manufacturing. *Control Engineering, 67*(8), 39–48.

Pratama, H., Azman, M. N. A., Kassymova, G. K., & Duisenbayeva, S. S. (2020). The trend in using online meeting applications for learning during the period of pandemic COVID-19: A literature review. *Journal of Innovation in Educational and Cultural Research, 1*(2), 58–68.

Sangiorgi, D., & Prendiville, A. (2014). A theoretical framework for studying service design practices: First steps to a mature field. *Design Management Journal, 9*, 61–73. https://doi.org/10.1111/dmj.12014

Santander, J. R. B. (2018). iNotify&Monitor system with data analytics using RFID and mobile broadband technology. *International Journals of Advanced Research in Computer Science and Software Engineering, 8*, 67.

Slater, S. F., & Mohr, J. J. (2006). Successful development and commercialization of technological innovation: Insights based on strategy type. *Journal of Product Innovation Management, 23*(1), 26–33.

Smith, A. (1937). *The wealth of nations [1776]* (Vol. 11937. na).

Stickdorn, M., Schneider, J., Andrews, K., & Lawrence, A. (2011). *This is service design thinking: Basics, tools, cases* (Vol. 1). Wiley.

Suoheimo, M., Korva, S., Turunen, T., & Miettinen, S. (2022). The first diamond is service design and the second is UX/interaction design: The double diamond model and team roles in making a mobile service application using cross-disciplinary teamwork. In *DMI: Academic design management conference proceedings; DMI: ADMC* (pp. 874–898). Design Management Institute.

Susilawati, S., & Supriyatno, T. (2020). Online learning through WhatsApp group in improving learning motivation in the era and post pandemic COVID-19. *Jurnal Pendidikan: Teori, Penelitian, dan Pengembangan, 5*(6), 852–859.

The Fourth Industrial Revolution: what it means and how to respond | World Economic Forum (weforum.org).

Veenema, T. G., Meyer, D., Rushton, C. H., Bruns, R., Watson, M., Schneider-Firestone, S., & Wiseman, R. (2022). The COVID-19 nursing workforce crisis: Implications for National Health Security. *Health Security, 20*(3), 264–269.

Walthall, R., & Dixit, S. (2022). *Impact of quantum computing in aerospace*. No. EPR2022014.

Wong, S. L. (2012). Challenging apple by imitation. *The New York Times*.

Zhang, Q., Vonderembse, M. A., & Cao, M. (2009). Product concept and prototype flexibility in manufacturing: Implications for customer satisfaction. *European Journal of Operational Research, 194*(1), 143–154.

# Part II
# Frameworks and Research Lenses for Service Design in Emerging Technologies Product Development

# Chapter 3
# Historical Perspectives on Service Design and Technology

**Mari Suoheimo, Paola Trapani, and Satu Miettinen**

## Introduction

Information behaviour and practice (IBP) research can inform technology and service development in different ways even if this relationship is complex (Huvila et al., 2019). In the same way technology can open up opportunities for new service offerings (Bitner, 2001), and service design can also inform the development of different technologies (Kankainen et al., 2012). Adoption of technologies impacts directly on service models such as self-service technology adoption (Curran & Meuter, 2005; Considine & Cormican, 2016) impacted on the delivery of financial services. Take the example of banking accounting; it was a couple decades ago that people would have little notebooks where the withdrawals were registered by hand, and later the same work was done with an onsite printer. Then, the service developed into a website, later an ATM and lately into mobile accounting services. Regardless of what is the driving force behind these developments, a thoughtful service design approach is needed to support the desired service experience. Further, it is obvious that changes in the technologies have an impact on business environment and thus creation and development of new business models (Palo & Tähtinen, 2011). When discussing the historical perspectives of service design and technological developments, it's good to notice that (1) adoption of technologies' impacts on service models and thus, for example, the delivery models of services and (2) creation of business models.

---

M. Suoheimo (✉) · P. Trapani
Service Design, Oslo School of Architecture and Design, Oslo, Norway
e-mail: mari.suoheimo@aho.no; paola.trapani@aho.no

S. Miettinen
Service Design, University of Lapland, Rovaniemi, Finland
e-mail: satu.miettinen@ulapland.fi

 
U. Z. A. Hamid, M. Suoheimo (eds.), *Service Design for Emerging Technologies Product Development*, Springer Series in Design and Innovation 29,
https://doi.org/10.1007/978-3-031-29306-1_3

Different technologies or technological solutions have a strong role in the practice of service design. These can be established and available such as an app store that enabled the third-party developer to deliver a plenitude of services (Basole & Karla, 2012) or emerging technologies such as AI and machine learning (Toreini et al., 2020). Emerging technologies are characterised by radical novelty, relatively fast growth, coherence, prominent impact and uncertainty and ambiguity (Palo & Tähtinen, 2011). Technological solutions play an operational role in the designing of service journeys or moments, for example, with choice of service channel. Different technological solutions would enable this to be online, mobile, organised through a robot or automated human-computer interaction. New innovative or emerging technologies could have a more strategic role as these would directly impact on business models.

Service design is a maturing field within academia and there are many views on what it is. For practitioners and researchers, this can create confusion regarding what view of service design to employ. In response, the aim of this chapter is to present four perspectives on service design: 'service design', 'product-service system design (PSSD)', 'design for service' and 'service ecosystem design'.

*Service design* has sprung with the need of designing user experiences in a holistic and co-creative manner with the stakeholders in question (Stickdorn et al., 2011), and it has its birth from cognitive psychology and interaction design (Rytilahti et al., 2015). *Design for service* involves creating the conditions for value-in-use through codesign with staff and service users (Meroni & Sangiorgi, 2011; Wetter-Edman et al., 2014). *Service ecosystem design* involves actors intentionally shaping social structures to create the emergence of desired value propositions (Vink et al., 2021a, b). *Product-service system design* is the design of 'an innovation strategy, shifting the business focus from designing and selling physical products only, to selling a system of products and services which are jointly capable of fulfilling specific client demands' (Vezzoli & Manzini, 2008). Many of the perspectives presented here have sprung from the need to situate the approach into a certain context, and others have emerged from an evolving theoretical understanding. These different service design perspectives can be helpful to consider when doing service design with different technological solutions.

In this chapter, we will not do a deep dive into how these perspectives can be used in the context of making new products and services when applying different and sometimes even emerging technologies, but rather paint the landscape of how each of these perspectives can be applied in such contexts. Other chapters in this book give examples of how the perspectives are being currently applied in academia and in practice. We introduce the perspectives here so that the design, marketing and engineering communities can make use of them when doing service design in relation to different and sometimes even emerging technologies. The presentation of each perspective is succinct, and, thus, we suggest that the readers dig deeper in the literature cited for further details.

## Historical Perspectives

In this section, we will present the five different perspectives and how they have emerged in the history of service design. As Fig. 3.1 illustrates, each perspective has emerged at different times attending to some needs or paradigms that have been seen as useful in service design. The history of designing services is fuzzy, and we do not wish to make flag posts on who initiated first, what and where as things are blurrier, and it is not the aim of this chapter.

The first idea of service blueprints, a tool that is widely used by service designers today, was introduced in the 1980s by Shostack (1982, 1984). Current definitions of service design are closely linked to the developments in conceptualisations of service. During the 1980s, services were defined based on how they differed from goods (i.e. tangible products) through intangibility, heterogeneity, inseparability and perishability (IHIP model, Zeithaml et al., 1985). However, the twenty-first century marked a move from a goods-dominant logic to a service-dominant logic (Vargo & Lusch, 2004) which views service as the fundamental basis of exchange (further elaborated later in this chapter). Value is no longer considered to be embedded in goods, but rather co-created by multiple actors through resource integration and is always uniquely and phenomenologically determined by the beneficiary (Vargo & Lusch, 2004, 2016). This novel perspective on service affected the views on what could be designed. Materials of service design have thus evolved from servicescapes (Bitner, 1992) and touchpoints (Clatworthy, 2011) where the focus was mainly on the physical dimension and other tangible elements of a service to sociomaterial configurations (Kimbell, 2011; Vink & Koskela-Huotari, 2021, 2022; Vink et al., 2021a, b) which transcend the merely tangible dimension (Blomkvist et al., 2016).

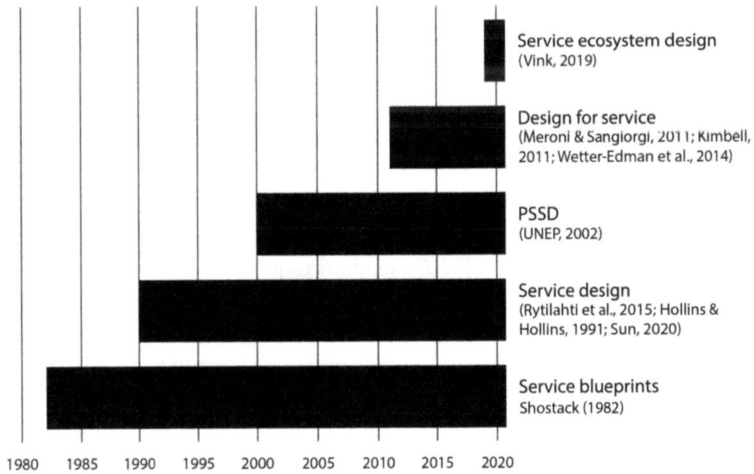


**Fig. 3.1** Different service design perspectives and the beginning of the use of service blueprints in the history

The basic service design concepts – service blueprints, service journeys and servicescapes – have prompt technological solutions for service design. For example, Stickdorn with his team was developing a mobile ethnography application for documenting a service experience called myServiceFellow/Experience fellow between 2007 and 2020[1] (Stickdorn et al., 2014). After this the team focused on developing service journey mapping tool called Smaply[2] (Bosio et al., 2021). Concretising service journeys, service moments and touchpoint as well as a more strategic view on decision-making has also inspired the SINCO lab,[3] service prototyping lab at the University of Lapland. This lab is using agile technologies that enable prototyping, testing and evaluating service proposals (Miettinen et al., 2012; Kuure & Miettinen, 2013). The SINCO lab was initiated through two European regional development projects during the years 2009–2013. The SINCO lab has contributed to the digitalisation of design service in large-scale corporations. The lab has been iterated into a totally digitised service design process at 10X Lab in collaboration with Volkswagen (Schaaf, 2021). The digitalisation and the development of technologies do not only impact on service and business models but also into ways in which the services are designed and prototyped.

Many recognise that *service design* sprung from the interaction community and cognitive psychology (Kimbell, 2011; Koskinen et al., 2011; Rytilahti et al., 2015), and others state how service design has accepted theories and practices of interaction (Junginger & Sangiorgi, 2011). Service design started to be viewed as a separate research domain in design academia around the same time in the 1990s when KISD introduced service design courses (e.g. Hollins & Hollins, 1991; Sun, 2020). From then, service design has grown in importance in art and design schools across Europe and in the world. Some examples are the introduction of service design courses at the faculty of Arts and Design at the University of Lapland in 2008 (Tuomaala, 2008) and first academic thesis being published in different universities, for example, at Aalto University (Koivisto, 2007; Miettinen, 2007). We can argue that the history of design itself has a background in arts (Norman, 2013) and has had influence in the way service design has taken form by being practised in many faculties of art and design. There are still institutes or universities where service design is on the contrary encountered in the departments of technology, engineering or marketing.

The core of this perspective to design of services is the user- or human-centredness and placing their needs in the centre of service development (Stickdorn et al., 2011). The ISO 9241-210:2010[4] standard requires and recommends human-centred design principles and activities throughout the life cycle of computer-based interactive systems. This has created a strong link between human- or user-centredness and service design. For example, the approaches introduced by

[1] https://www.smaply.com/experiencefellow

[2] https://www.smaply.com/

[3] http://sinco.fi/

[4] https://www.iso.org/standard/52075.html

Stickdorn et al. (2011) takes the user experience as an important aspect of making services but also the need of involving stakeholders and codesigning with them. One can also question if we should present it as a separate perspective or not, but we see that the way it is practised from the three other perspectives of this chapter is different, thus needs an opening to readers that are new with this service design practice.

Kimbell makes a good distinction on how the service design practised in the design fields is different and more user-centred compared to, for example, service engineering (Kimbell, 2011). We can also argue that all the other two service design perspectives presented in this chapter, that is, 'design for service' and 'service ecosystem design', have sprung or gained influence from the design field. The user-centredness is one aspect that has been also questioned lately as we have more planetary issues to consider also through service design (Sustar, 2022) in the midst of global warming and if non-humans should be considered as stakeholders (Carvalho & Riquito, 2022).

The emergence of *PSSD* can be traced to the early 2000s (UNEP, 2002). The early service design community has been investigating the 'product-service system', where companies offer a mix of both products and services. Halen et al. (2005, p. 21) have summarised this understanding as 'a marketable set of products and services capable of jointly fulfilling a user's needs'.

PSSD originated in the sector of environmental management that, over the decades, has evolved from applying strategies of damage *repair* to damage *prevention*. Initially, attempts were made to limit and remediate pollution caused by production processes, for example, toxic chemicals poured into river waters. In the essence of PSSD, it is no longer sufficient to continue dematerialising products to replace them with services. It is necessary to critically redefine what sustainable consumer *needs* should look like, along with the concepts of *wellbeing* and *satisfaction*, and how these issues can be met. Lastly, it is mandatory to reconsider the role of the providers who jointly produce the system that satisfies these needs. PSSD seems to provide the right strategy to make the leap, rather than the slow transition, towards sustainable lifestyles (Ceschin, 2014).To give an example, in the traditional scenario of selling products, a school buys lights to illuminate classrooms. In innovative PSSD scenarios, the school rents the lights or even pays a fee to light the classrooms to a third-party company that is free to determine the most appropriate products and methods to meet the school's needs.

In contrast to the historical understanding of service design as a phase in the development new services, *design for service* was introduced as a view during the 2010s when Meroni and Sangiorgi (2011), Kimbell (2011) and Wetter-Edman et al. (2014) made some new propositions to a framework, which is today known as 'design for service'. This perspective on service design is informed by the service-dominant logic. Wetter-Edman et al. (2014) makes a distinction on how 'design for service' is different from 'design of service', or 'service design'. 'Design for service' shifted the angle from viewing service design as a stage in the new service development (e.g. Edvardsson, 1997) to an iterative process, thus capturing an ongoing dialogue and co-creation with various actors involved (Blomkvist et al.,

2010; Meroni & Sangiorgi, 2011). This was further highlighted in the 'designing for service' perspective (e.g. Kimbell, 2011; Sangiorgi & Prendiville, 2017) by replacing 'design' with a more active and continuous verb form 'designing'. We can see that the field on 'service design' is not homogenous, but heterogenous with different perspectives that are being applied to design services.

Already in the late 1990s, Pacenti (1998) brought up how services could be viewed as complex organisations. Furthermore, a new understanding of service emerged in the early 2000s that shifted the understanding from focusing on 'services' as an intangible market offering to 'service' as the exchange of knowledge and skills (Vargo & Lusch, 2004, 2008a, 2008b). As a result in the years that followed, there has been an ongoing trend in the service design literature of integrating increasing complexity, including through the introduction of 'design for service' in the 2010s (Kimbell, 2011; Meroni & Sangiorgi, 2011; Wetter-Edman, 2014) as well as 'service ecosystem design' in the 2020s (Vink et al., 2021a, b). One is able to read more publications on service-dominant logic in marketing research (Vargo & Lusch, 2016), but also in recent service design literature (e.g. Vink et al., 2021a, b; Pyykkö et al., 2021; Wetter-Edman, 2014). *Service ecosystem design* as a perspective got its birth around 2020 when Vink (2019) introduced the concept as part of her Ph.D. thesis. The perspective is informed by service-dominant logic, and it focuses on the design of social structures that are recognised as service design materials. In 'service ecosystem design', the service designers create value through multiple actors and provide a systemic understanding about how service design can support long-term change (Vink, 2019).

As an emerging and multidisciplinary area of research and practice, service design tends to be challenging to define. Meroni and Sangiorgi (2011) refer to it as a human-centred, creative and iterative approach to service innovation (see also Wetter-Edman et al., 2014). Others emphasise its participatory and transformative nature (e.g. Karpen et al., 2017; Trischler et al., 2018; Patrício et al., 2020). Ostrom et al. (2010, p. 17) suggest that 'service design involves the orchestration of clues, places, processes, and interactions that together create holistic service experiences for customers, clients, employees, business partners, or citizens'. Service design can thus be seen as a process of transforming the materials of service (Vink & Koskela-Huotari, 2021, Vink & Koskela-Huotari, 2022) including tangible and intangible, social and material.

Junginger and Sangiorgi (2011) recognise how service design works in many of the orders of design introduced by Buchanan (1992) especially with the second, third and fourth when they write in the context of public services. The first order is often understood as the visual or graphic design and the second as industrial design or making physical objects. The third order designs interactions and service more predominantly, and the fourth is more related to the systems and complex environments. Despite a progression to integrate increasing complexity, we still see that service design continues to work in the first order as well, for example, by working with the service touchpoints. Service touchpoints are the physical contact points for users, such a check-in desk at a hotel reception. It is not in vain how Polaine et al. (2013, p. 32) use the musical metaphor how 'each musician must play to the best of his or her abilities, yet at the same time play in harmony and keep time with the others. Things can quickly go awry if each musician simultaneously tries to play as a soloist'. This way we could argue that the service design takes the role as a facilitator to bring harmony among the different players.

## Key Perspectives on Service Design

This next section will introduce and discuss four perspectives on service design. These are 'service design', 'product-service system design (PSSD)', 'design for service' and 'service ecosystem design'. Before introducing the perspectives, we will also elaborate briefly on service-dominant logic, which led to a fundamental shift in terms of the focus and approaches to service design. It is presented first since many of the perspectives are influenced by its evolution, but it itself is not a separate perspective.

### *Service-Dominant Logic Informing Service Design*

In service research, service-dominant logic (S-D logic) had a significant influence on the development and approach to service design (Kimbell, 2011). This is because S-D logic introduced a new perspective on service and how value is created by moving away from *services* as a specific product category towards *service* as a perspective on value creation (Edvardsson et al., 2005). Specifically, S-D logic uses the term service to describe the process of doing something beneficial for and in conjunction with some entity rather than units of output as implied in the plural 'services' (Vargo & Lusch, 2008a, 2008b). During this process, value is not seen as something that is produced into products and services and then exchanged/delivered on the marketplace, but as something that is created during use and thus 'uniquely and phenomenologically determined by the beneficiary' (Vargo & Lusch, 2016, p. 8). This shift also means that the focus of service design is to understand possible future-use situations through insights into customers' dreams, drivers and needs (Trischler et al., 2018), while the outcome of the service design process cannot actually be a finished service product or predefined solution, but rather is a supporting configuration of resources that customers can transform into value through use (Patricio et al., 2011). With that in mind, different scholars have suggested *design for service* as the more suitable term describing what service design is and does (Kimbell, 2011; Meroni & Sangiorgi, 2011; Wetter-Edman et al., 2014).

With the focus on exploring and designing for supporting users to create value in future-use situations, much of service design research took a human-centred, or more narrowly user-centred, approach. For example, participatory design approaches such as customer co-creation (Witell et al., 2011) or codesign (Trischler et al., 2018) became a hot topic in service research because it allowed service providers to tap into the customers' unique knowledge on service usage and needs. However, at the same, time S-D logic evolved further and increasingly took on a systemic lens (Vargo & Lusch, 2016). This is because S-D logic asserts that 'value is not completely individually, or even dyadically, created but, rather it is created through the integration of resources, provided by many sources, including a full range of market-, private- and public-facing resources and actors' (Vargo & Lusch, 2016, p. 9). In fact, and especially in the age of digitalisation, a customer or user more often than not co-creates value by integrating sources from many different sources, rather than relying on a single or multiple service providers (see, e.g. McColl-Kennedy et al., 2012; Trischler & Westman Trischler, 2021).

In addition to the multi-actor stance, Vargo and Lusch (2016, p. 8) emphasise that '[v]alue co-creation is coordinated through actor-generated institutions and institutional arrangements', thus pointing towards the importance of institutional mechanisms. These developments are reflected in the service ecosystem concept which was introduced as the unit of analysis for value co-creation among multiple actors, and is defined as 'a relatively self-contained, self-adjusting system of resource-integrating actors connected by shared institutional arrangements and mutual value creation through service exchange' (Vargo & Lusch, 2016, p. 10–11). As we show further below, these developments driven by S-D logic not only informed the design for service perspective (Wetter-Edman et al., 2014) but more recently also service ecosystem design (Vink et al., 2021a, b).

## Service Design

Service design, how it is taught in many faculties of art and design, can be little different, for example, from how it is introduced in marketing, engineering or PSSD. The five common principles applied in this service design perspective are user-centredness, co-creation, sequencing, evidencing and holism (Stickdorn et al., 2011). The essence is to place the *users'* or the *humans'* needs in the centre of the development process (Miettinen, 2016). The users take part of the process through co-creational and participatory approaches.

The work often begins by making a stakeholder map to understand what actors should be involved in the development process. Ecosystem mapping is another recommended tool to map the actors and their involvement in a larger concept. As the process involves different stakeholders, the essence is always holistic. For example, in designing a service for carpooling, one might need to involve the users, the car company in question, municipality, engineers, software developers and so on. The service designer takes a bottom-up role to facilitate the co-creational spaces (Suoheimo, 2020) often through workshops or applying empathy building tools as user interviews or observation (Stickdorn et al., 2011).

The sequencing is applied to understand a system, for example, through service journeys – how a service begins and ends. At the same time, there is evidencing as the intangible issues are made tangible often through different visualisations. Double diamond or design thinking are generally applied methodologies in the process (Suoheimo et al., 2022; Pyykkö et al., 2021). Mager (2004) has advocated that services should be designed with similar attention to processes in product development. She has also been bridging service design with design management.

Service design can be applied in both private or public spaces. Often the service design especially in the public services or designing with communities can overlap with social design (Young, 2012). Kuure and Miettinen (2017) have written how service design can create understanding for challenging life situations. Also, service design can work in the context of organisational development (Miettinen & Vuontisjärvi, 2016; Suoheimo, 2019).

## *Product-Service System*

Previous studies mostly defined product-service system design as an approach to design that shifts the focus from designing just material things (e.g. products) or immaterial ones (e.g. services) to creating an integrated system of both to fulfil a specific customer need (UNEP, 2002). This definition takes into account the different components of a product-service system, namely, a bundle of *products*, which are the tangible artefacts of the system; *services* that make the products available (e.g. sales services, renting and sharing); already existing *infrastructures* such as roads, communications backbones and distribution and collection systems); and *actors' networks*, that is, all the socio-economic key players involved in the PSS' production and delivery who are often organised in partnerships and collaborations (Mont, 2002).

Product-service system design can be seen as a sort of life cycle design taken to its extreme. Common to both is the core principle of the *functional economy*, according to which it is mandatory to create the maximum use value for the longest possible time, while consuming as little energy and material as possible (Stahel, 1997).

However, it is worth noting that not all product-service systems are sustainable – only those specifically designed, developed and delivered to generate fewer material flows and emissions. It may sound like a paradoxical statement, but the most sustainable products are those that are not even produced, distributed, consumed and recycled. In this respect, sustainable product-service systems (sPSS) prioritise designing access to the function of a shared product rather than selling it as an exclusive-use item. Compared to the current consumer-economy based on the production and consumption of goods, the functional economy populated by product-service systems optimises the use of goods and services and thus emphasises the management of existing wealth (Stahel, 1997). PSSD involves a paradigm shift from *consumerism* to the so-called satisfaction economy whose keystone is wellbeing rather than the ownership of products. This scenario can theoretically be extended to include many products of daily use: for example, the possibility of access to drinking water can reduce the use of the disposable plastic bottle; the possibility of having clean linen can avoid the necessity for a personal washing machine; it is not necessary to own a private bike if there is a bike-sharing. We must move from measuring wellbeing through producing and consuming goods to living better by consuming less (Vezzoli & Manzini, 2008).

According to the business model around which they are organised, it is possible to distinguish three main types of PSS:

1. *Product-oriented PSSs*. Examples include product maintenance, repair, upgrading, substitution and take-back (UNEP, 2002). At the core of the value offering, there is still a product sold to the customer. However, the company provides additional services to guarantee its life cycle performance.
2. *Use-oriented PSSs*. Examples include rental, sharing, pooling and pay-per-use. In this case, the value proposition core is shifted from the sale of products to the access to their function and performances, enabling customers to get the wanted results. The access requires payments only for the time the product is used, whereas ownership is unnecessary.
3. *Result-oriented PSSs*. Examples include activity management or outsourcing and pay-per-service unit. The value offer requires the company to provide a customised mix of services to provide a specific final result, that is, an integrated solution to meet the customer's satisfaction.

The mix of services does not require the client to acquire the means necessary to get the result. The producer retains the ownership of the products and is paid by the client only for providing the agreed results. The customer benefits by being freed from the problems and costs involved in the acquisition, use and maintenance of equipment and products (UNEP, 2002).

At this point, one might wonder what the benefits of PSS are. Modern, contemporary societies need to quickly find sustainable production and consumption patterns. Life cycle design has retaught the entire life cycle of products to increase their eco-efficiency and sustainability. However, paradoxically, the benefit of an eco-efficient washing machines is largely offset by the fact that more washing machines are sold in the world as the world's population becomes more affluent. The crux of the matter is a paradox whereby one must reconcile reducing product circulations with maintaining, if not improving, the standard of living.

At the core of PSS, there is a transfer of responsibility from the user to the producer. If the product is a property of the PSS provider, who also has to deal with the disposal, the company is motivated to design for durability and repair. The relationship with customers is continuous and caters for market differentiation: revenues come from service subscriptions which favours the increase of customers' loyalty.

Despite all the theory and knowledge, research and case studies, there are relatively few actual cases of successful sustainable product-service systems. The reason they don't easily become mainstream is their objective difficulty, due to the necessary collaboration of very different actors. sPSS are as ambitious as they are risky. Their appearance manifests as a radical innovation which impacts the market as an earthquake. Successful breakthroughs infrequently occur within any area, perhaps once every 5–10 years (Norman & Verganti, 2014).

## *Design for Service*

The 'design for service' emerged in the service design literature in the early 2010 as noted above. Kimbell (2011) illustrates how the shift from services to service changes the approach and focus of service design. When considering that value cannot be designed into services but is created during use, one cannot have a finished service product as an outcome of the service design process. Instead, and as highlighted by the design for service perspective, the focus should be on designing for service, that is, for a possible future-use situation (e.g. Kimbell, 2011; Wetter-Edman et al., 2014). Meroni and Sangiorgi (2011) highlight that it is about creating the conditions for the desired forms of value co-creation to exist and that this, as previously mentioned an iterative and ongoing process, including in use.

Building on this perspective, Wetter-Edman et al. (2014) propose four ideas that make up 'design for service'. First is to understand service systems from the actors' perspective. The actors should be understood by their value co-creation activities, experiences and assessments as the aim when designing new future service systems. Second, it approaches understanding of actors, users in the service in the ways they experience and how the experience is formed in specific contexts, which all depend on how the resources are integrated or operated. Third, 'design for service' applies or includes market-facing, public and private resources in its different practices that may be, for example, tools and approaches. New service systems will be created and value is co-created in the process. Fourth, service-dominant logic provides a theoretical framework for design for service by analysing and understanding it.

While Wetter-Edman et al. (2014) and others (e.g. Patricio et al., 2011) acknowledge that service design requires a systemic approach, design for service still very much draws on a user-centred design tradition. This is, for example, evident in its link to service logic, which has a tradition of a strong customer focus (Grönroos, 2006). In addition, a central element of design for service has become participatory design, and in particular codesign, which is an approach that 'allows selected customers or users to become part of the design team as experts of their experiences', and in so doing, taps their unique knowledge about usage and latent needs (Trischler et al., 2018, 75). Participatory approaches are further applied in order to ensure that 'the user should feel empowered, having control and ownership of the situation and information' (Wetter-Edman, 2014, p. 104), which is especially important in public service context and the involvement of vulnerable users (Trischler et al., 2019).

## *Service Ecosystem Design*

Extending the design for service view, service ecosystem design is an emerging perspective on service design that is informed by service-dominant logic's service ecosystems perspective. The service ecosystems perspective highlights that value is co-created by multiple actors in exchange systems that are guided by institutionalised social structures (Vargo & Lusch, 2016). Integrating this premise, service ecosystem design provides a systemic, embedded understanding of how service design supports intentional, long-term change. Service ecosystem design involves the intentional shaping of institutionalised social structures and their physical enactments by actor collectives (Vink et al., 2021a, b). In this view, social structures, including shared and entrenched norms, rules, roles and values, and their physical enactments, including symbols, artefacts activities and relations that are carriers of social structures, become the central material of service design (Vink & Koskela-Huotari, 2021).

Most of the time, in service ecosystems, actors are reproducing existing social structures, in that without any awareness of these taken-for-granted structures – they simply enact the routinised, expected behaviours implied by these structures (Greenwood et al., 2008). Thus, service ecosystem design is an embedded process that happens only when actors work to intentionally shape the very structures that guide their behaviour. These structures are shaped through the core processes of reflexivity and reformation to facilitate the emergence of desired value co-creation forms (Vink et al., 2021a, b).

The process of reflexivity involves building an awareness of existing social structures inhabited by oneself and others (Suddaby et al., 2016). This means making the invisible, taken-for-granted social structures more visible so that they can be shaped. This process of building reflexivity can be aided through the use of various service design methods to tap into different modes of reflexivity including temporal, material, corporeal, relational, cultural and cognitive (Vink & Koskela-Huotari, 2022). For example, using role playing can aid actors in perceiving social structures of a particular situation using their bodily experiences, tapping into the corporeal mode of reflexivity. Once there is greater awareness of the existing social structures, reformation becomes possible.

Reformation involves intentionally reshaping social structures towards desired forms of value co-creation (Vargo & Archpru Akaka, 2012). This process involves actors intentionally creating, disrupting and maintaining social structures (Lawrence & Suddaby, 2006). To do so, actors can craft the physical enactments of social structures to purposefully alter the underlying social structures. For example, actors might shift the symbols and artefacts in a situation to aid in supporting a new norm or role. A number of designerly approaches are being experimented with to more explicitly support the process of reformation, such as tiny tests where actors are encouraged to challenge social structures in their everyday lives and reflect on the ripple effects of their actions (Vink et al., 2021a). However, it is important to recognise that the outcomes of such a process are emergent and cannot be fully controlled. They are realised through the interplay of actors' focal design process with both conflicting and aligning reproduction (non-design) and design processes happening within the system.

In this way, service ecosystem design provides an alternative process to the more linear 'double-diamond' model that is prominent in service design. Instead, it recognises that design is an embedded feedback loop of reflexivity and reformation that happens within the ongoing processes of reproduction in the service ecosystem. By recognising this embeddedness, this view of service design better accounts for some of the struggles and resistance that actors face when doing service design. Furthermore, this understanding presents service design as an ongoing, collective process that all actors are involved in within service ecosystems, not just when they are involved in a service design project but any time they are working to intentionally shape the structures around them towards desired futures. This opens up the potential for service design to better mobilise collective efforts towards service ecosystem transformation and more strategically cultivate long-term change by focusing on shaping the underlying social structures that guide actors' behaviour in service ecosystems.

## Discussion

In this chapter, we have presented what service design can be from various different perspectives. It is hard to make general descriptions of each perspective, and the way we present them might be stereotyping them since depending on the different universities or design agencies practising these perspectives, they certainly will bring some specific orientations to them. All the perspectives have different fields that they were born in, which makes it extremely difficult to make generalisations or comparisons between them. Each has different philosophical standpoints and ways of practising service design.

We see that making further studies on how to understand their similarities or differences would be valid, since the space of this chapter would not be enough to make such a reflection. It would require a larger work as a Ph.D. thesis where a researcher immerses into each service design perspective to fully understand their differences. Table 3.1 illustrates how the four different perspectives all have

**Table 3.1** The perspectives on service design and the field they emerged

| Perspective on service design | Service design | PSSD | Design for service | Service ecosystem design |
|---|---|---|---|---|
| Field that it emerged | Cognitive psychology and interaction design (Rytilahti et al., 2015) | Environmental management (UNEP, 2002) | Intersection of design and service marketing (Meroni & Sangiorgi, 2011; Kimbell, 2011; Wetter-Edman, 2014) | Intersection of design, service marketing and systems thinking (Vink, 2019) |

different 'homes' where they sprung. For example, the PPSD is from environmental management, and the design for service is in the intersection of design and service marketing. These issues, 'the places of birth' can in one way explain why it is hard to make generalisations of them or try to find similarities and differences. They are also points that can create tensions when trying to understand and make sense on how each perspective is similar or different from each other.

To better understand how each of the different service design perspectives work in the case of designing services with emerging technologies, we created Table 3.2. In the first column, there is a service design perspective in chronological order of

**Table 3.2** Service design perspectives and their possible application on designing services with emerging technologies

| Service design perspective | Context in applying it with emerging technologies |
|---|---|
| Service design | When using service design perspective in autonomous driving, the service designers can begin the process by mapping first the complexity involved through complexity mapping or mess mapping (Suoheimo & Miettinen, 2018). It is vital to understand the stakeholders and ecosystem involved in the design process. Taking the user that can be multiple is important. How they all as service providers or the end-users experience the service are key issues to uncover. This can be made through service journeys to create empathy toward all the actors in the process. Often personas are created to better attend the groups involved. The level of complexity can be from simple, complex or wicked depending on the complexity of the problem in question. The perspective and the tools could provide new ways of understanding how autonomous cars can be used for carpooling, thus making the society transfer from car owning to car sharing |
| PSSD | One example discussed in another chapter of this book is the product and food service system on board airliners. In the traditional way of doing business, this PSS results from the synchronised collaboration of the airline, catering company, raw material suppliers, distributors, etc. Standardisation sacrifices the passenger's culinary experience to food safety and hygiene.<br>Emerging technologies, like AI and robotics, can introduce a new level of innovation in the way the PSS itself is designed. The data stream on passenger purchases continuously updates the knowledge of tastes and eating habits to predict future purchasing preferences. Meal design no longer requires months of meetings and human resource-intensive work and does not limit the choice between the two or three options |
| Design for service | Using a design for service perspective to improve the mental health of youth might involve codesigning a virtual reality (VR) world with youth and counsellors. Recognising that mental health cannot be fully controlled, the intention could be to codesign a world that supports mindfulness and calmness in a way that resonates for youth, informed by the knowledge of therapists, and potentially becomes a space where youth and counsellors could meet virtually. This approach would enable a platform for youth to relax, release and connect with counsellors, but not necessarily overdetermine what would happen within such a service exchange |

(continued)

**Table 3.2** (continued)

| Service design perspective | Context in applying it with emerging technologies |
|---|---|
| Service ecosystem design | If service ecosystem design were being applied to a case where service robots were proposed to be used in seniors' housing, the focus would be first on building reflexivity of the actors affecting and affected by such a change. This involves building collective awareness of the existing social structures, including shared norms, rules, roles, values and beliefs, connected with seniors' care, and then critically examining together how the involvement of robots might influence those social structures. If considered appropriate and in support of the collective desired forms of value co-creation, there would be efforts made to intentionally reshaping the social structures within the seniors' housing in a way that enabled the desired forms of care with the robots. This approach recognises that a vast multiplicity of actors are already influencing the related social structures and there may be conflict in their desired forms of value co-creation that would need to be considered and likely worked through. This approach would also necessitate a careful, ongoing analysis of unintended consequences of the emergent properties within the service ecosystem and adapting the focal design process to better facilitate the desired emergent outcomes |

their emergence, and then on the right side is how the perspective would work in the case of designing with emerging technologies by providing a practical example although some may be more theoretical approaches, but still the theory will be shaping the practice.

All the service design perspectives presented here can play a role when designing with emerging technologies in the creation of new services, experiences, interactions or tangible products. As shown in Table 3.2, each one of them takes a unique perspective. One aspect that is common for many of these perspectives as 'service design', 'PSSD', 'design for service' and 'service ecosystem design 'for service is the co-creation of value with the users, and it can be seen also in the table with the examples provided. Another is user-centredness that is an essential value when designing in the context of emerging technologies. The needs of the user are in the centre rather than the technology – as the emerging technology comes to satisfy those needs in a whole new level by creating innovations.

Technology may play a part in a different perspective in each of the service design perspectives presented here. Although we presented the different service design perspectives as forms of making service design, it is often that the designers make a bricolage of them (Matthews, 2021), taking what is essential for them in each specific case.

## Conclusions

We have shared alternative perspectives on service design that can inform the process of designing in relation to emerging technologies' product and service development. We still think that depending on the case or the scope of the process, there may be different perspectives that are better suited to inform the practical service design process. This has not been an exhaustive presentation of all possible perspectives, but this chapter covers some of the most commonly used today in the service design practice and research.

The perspectives show how the service design history has developed and still evolving into new perspectives. We recommend future research to investigate the similarities and differences of the perspectives presented in this chapter which are 'service design', 'PSSD', design for service and service ecosystem design. We can conclude that 'service design', 'PSSD', 'design for service' and 'service ecosystem design' play an important role in defining the user needs when designing in the context of emerging technologies. By combining the needs and the technology, new innovations are possible. We also argue that applying the user's perspective whether they are one or a whole ecosystem or actors right from the start will better enable the implementation of the products and services as they have been designed to attend those needs.

**Acknowledgements** We would like to acknowledge the following research colleagues. We wish to thank associate professor Josina Vink for informing about service ecosystem design, associate professor Jakob Trischler for service-dominant logic and assistant professor Martina Čaić for the history of service design. Your insights on these topics as the views on this chapter in general have been precious in bringing more solid perspectives.

## References

Basole, R. C., & Karla, J. (2012). Value transformation in the mobile service ecosystem: A study of app store emergence and growth. *Service Science, 4*(1), 24–41.

Bitner, M. J. (2001). Service and technology: Opportunities and paradoxes. *Managing Service Quality: An International Journal, 11*, 375.

Bitner, M. J. (1992). Servicescapes: The impact of physical surroundings on customers and employees. *Journal of Marketing, 56*(2), 57–71.

Blomkvist, J., Clatworthy, S., & Holmlid, S. (2016). Ways of seeing the design material of service. In *Service design geographies. The ServDes. 2016 conference, Copenhagen 24–26 May 2016* (Vol. 125, pp. 1–13). Linköping University Electronic Press.

Blomkvist, J., Holmlid, S., & Segelström, F. (2010). Service design research: Yesterday, today and tomorrow. In M. Stickdorn & J. Schneider (Eds.), *This is service design thinking* (pp. 308–315). BIS Publishers.

Bosio, B., Rainer, K., & Stickdorn, M. (2021). Mobile ethnography: A customer experience research method for innovation. In *Handbook of research methods for marketing management*. Edward Elgar Publishing.

Buchanan, R. (1992). Wicked problems in design thinking. *Design Issues, 8*(2), 5–21.

Carvalho, A., & Riquito, M. (2022). Listening-with the subaltern: Anthropocene, Pluriverse and more-than-human agency. *Nordia Geographical Publications, 51*(2), 37–56.
Ceschin, F. (2014). How the design of socio-technical experiments can enable radical changes for sustainability. *International Journal of Design, 8*(3), 1–21.
Clatworthy, S. (2011). Service innovation through touch-points: Development of an innovation toolkit for the first stages of new service development. *International Journal of Design, 5*(2), 15–28.
Considine, E., & Cormican, K. (2016). Self-service technology adoption: An analysis of customer to technology interactions. *Procedia Computer Science, 100*, 103–109.
Curran, J. M., & Meuter, M. L. (2005). Self-service technology adoption: Comparing three technologies. *Journal of Services Marketing, 19*, 103.
Edvardsson, B., Gustafsson, A., & Roos, I. (2005). Service portraits in service research: A critical review. *International Journal of Service Industry Management, 16*(1), 107–121.
Edvardsson, B. (1997). Quality in new service development: Key concepts and a frame of reference. *International Journal of Production Economics, 52*(1–2), 31–46.
Greenwood, R., Oliver, C., Sahlin, K., & Suddaby, R. (2008). Introduction. In R. Greenwood, C. Oliver, K. Sahlin, & R. Suddaby (Eds.), *SAGE handbook of organizational institutionalism* (pp. 1–46). SAGE.
Grönroos, C. (2006). Adopting a service logic for marketing. *Marketing Theory, 6*(3), 317–333.
Halen, C. V., Vezzoli, C., & Wimmer, R. (2005). Methodology for product service system innovation. *Koninklijke Van Gorcum BV, Assen.*
Hollins, G., & Hollins, B. (1991). *Total design: Managing the design process in the service sector*. Pitman.
Huvila, I., Enwald, H., Eriksson-Backa, K., Liu, Y. H., & Hirvonen, N. (2019). Information behaviour and practises research informing technology and service design. *Proceedings of the Association for Information Science and Technology, 56*(1), 541–545.
Junginger, S., & Sangiorgi, D. (2011). Public policy and public management: contextualizing service design in the public sector.
Kankainen, A., Vaajakallio, K., Kantola, V., & Mattelmäki, T. (2012). Storytelling group–a co-design method for service design. *Behaviour & Information Technology, 31*(3), 221–230.
Karpen, I. O., Gemser, G., & Calabretta, G. (2017). A multilevel consideration of service design conditions: Towards a portfolio of organisational capabilities, interactive practices and individual abilities. *Journal of Service Theory and Practice, 27*(2), 384–407.
Kimbell, L. (2011). Designing for service as one way of designing services. *International Journal of Design, 5*(2), 41–52.
Koivisto, M. (2007). *Mitä on palvelumuotoilu? Muotoilun hyödyntäminen palvelujen suunnittelussa. Taiteen maisterin lopputyö*. Taideteollinen korkeakoulu.
Koskinen, I., Zimmerman, J., Binder, T., Redstrom, J., & Wensveen, S. (2011). *Design research through practice: From the lab, field, and showroom*. Elsevier.
Kuure, E., & Miettinen, S. (2017). Social design for service. Building a framework for designers working in the development context. *The Design Journal, 20*(sup. 1), S3464–S3474. https://doi.org/10.1080/14606925.2017.1352850
Kuure, E., & Miettinen, S. (2013, October). Learning through action: Introducing the innovative simulation and learning environment service innovation corner (SINCO). In *E-learn: World conference on E-learning in corporate, government, healthcare, and higher education* (pp. 1536–1545). Association for the Advancement of Computing in Education (AACE).
Lawrence, T. B., & Suddaby, R. (2006). Institutions and institutional work. In S. Clegg, C. Hardy, T. B. Lawrence, & W. R. Nord (Eds.), *The SAGE handbook of organization studies* (2nd ed., pp. 215–254). SAGE.
Mager, B. (2004). *Service design: A review*. Köln International School of Design.
Matthews, T. (2021). *Exploring sacred service design*. [Doctoral dissertation, Oslo School of Architecture and Design]. Adora. https://hdl.handle.net/11250/2758408

McColl-Kennedy, J. R., Vargo, S. L., Dagger, T. S., Sweeney, J. C., & van Kasteren, Y. (2012). Health care customer value cocreation practice styles. *Journal of Service Research, 15*(4), 370–389.

Meroni, A., & Sangiorgi, D. (2011). *Design for services*. Gower Publishing.

Miettinen, S. (2007). *Designing the creative tourism experience*. University of Art and Design Helsinki.

Miettinen, S. (2016). Introduction to industrial service design. In *An introduction to industrial service design* (pp. 21–32). Routledge.

Miettinen, S., Rontti, S., Kuure, E., & Lindström, A. (2012). *Realizing design thinking through a service design process and an innovative prototyping laboratory: Introducing Service Innovation Corner (SINCO).*

Miettinen, S. A., & Vuontisjärvi, H. (2016). Service design for participatory development. In P. M. Rytilahti & S. A. Miettinen (Eds.), *For profit, for good: Developing organizations through service design* (pp. 68–73). University of Lapland.

Norman, D. (2013). *The design of everyday things: Revised and expanded edition*. Basic books.

Mont, O. K. (2002). Clarifying the concept of product–service system. *Journal of Cleaner Production, 10*(3), 237–245. https://doi.org/10.1016/S0959-6526(01)00039-7

Norman, D., & Verganti, R. (2014). Incremental and radical innovation: Design research vs. technology and meaning change. *Design Issues, 30*, 78–96. https://doi.org/10.1162/DESI_a_00250

Ostrom, A. L., Bitner, M. J., Brown, S. W., Burkhard, K. A., Goul, M., Smith-Daniels, V., et al. (2010). Moving forward and making a difference: Research priorities for the science of service. *Journal of Service Research, 13*(1), 4–36.

Pacenti, E. (1998). *Il progetto dell'interazione nei servizi. Un contributo al tema della progettazione dei servizi.* [Doctoral dissertation, Politecnico di Milano].

Palo, T., & Tähtinen, J. (2011). A network perspective on business models for emerging technology-based services. *Journal of Business & Industrial Marketing, 26*(5), 377–388. https://doi.org/10.1108/08858621111144433

Patricio, L., Fisk, R. P., Falcao e Cunha, J., & Constantine, L. (2011). Multilevel service design: From customer value constellation to service experience blueprinting. *Journal of Service Research, 14*(2), 180–200. https://doi.org/10.1177/1094670511401901

Patrício, L., Sangiorgi, D., Mahr, D., Čaić, M., Kalantari, S., & Sundar, S. (2020). Leveraging service design for healthcare transformation. *Journal of Service Management, 31*(5), 889–909.

Polaine, A., Løvlie, L., & Reason, B. (2013). Service design: From insight to implementation. .

Pyykkö, H., Suoheimo, M., & Walter, S. (2021). Approaching sustainability transition in supply chains as a wicked problem: Systematic literature review in light of the evolved double diamond design process model. *PRO, 9*(12), 2135.

Rytilahti, P., Miettinen, S., & Vuontisjärvi, H. R. (2015). The theoretical landscape of service design. In *Design, user experience, and usability: Design discourse* (pp. 86–97). Springer.

Sangiorgi, D., & Prendiville, A. (2017). *Designing for service: Key issues and new directions*. Bloomsbury.

Schaaf, L. (2021). *DIGITAL CO-CREATION digitalization within service design: Transformation from analog thinking towards digital doing*. University of Lapland.

Shostack, G. (1982). How to design a service. *European Journal of Marketing, 16*, 49–63.

Shostack, G. L. (1984). Designing services that deliver. *Harvard Business Review, 62*, 133–139. https://doi.org/10.1225/84115

Stahel, W. R. (1997). *The functional economy: Cultural and organizational change*. National Academies Press.

Stickdorn, M., Schneider, J., Andrews, K., & Lawrence, A. (2011). *This is service design thinking: Basics, tools, cases* (Vol. 1). Wiley.

Stickdorn, M., Frischhut, B., & Schmid, J. S. (2014). Mobile ethnography: A pioneering research approach for customer-centered destination management. *Tourism Analysis, 19*(4), 491–503.

Suddaby, R., Viale, T., & Gendron, Y. (2016). Reflexivity: The role of embedded social position and entrepreneurial social skill in processes of field level change. *Research in Organizational Behavior, 36*, 225–245.

Sun, Q. (2020). Towards a new agenda for service design research. *The Design Journal, 23*(1), 49–70.

Suoheimo, M. E., & Miettinen, S. A. (2018). Complexity mapping and mess mapping tools for decision-making in transportation and Maas development. In *proceedings of the. DMI: Academic design management conference* (pp. 1176–1188). Design Management Institute.

Suoheimo, M. (2020). *Approaching wicked problems in service design.* [Doctoral dissertation, University of Lapland]. Acta electronica Universitatis Lapponiensis. https://lauda.ulapland.fi/handle/10024/64421

Suoheimo, M. (2019). Strategies and visual tools to resolve wicked problems. *The International Journal of Design Management and Professional Practice, 13*(2), 25–41. https://doi.org/10.18848/2325-162X/CGP/v13i02/25-41

Suoheimo, M., Korva, S., Turunen, T., & Miettinen, S. (2022). The first diamond is service design and the second is UX/interaction design: The double diamond model and team roles in making a mobile service application using cross-disciplinary teamwork. In *DMI: Academic design management conference proceedings; DMI: ADMC* (pp. 874–898). Design Management Institute.

Suoheimo, M., Vasques, R., & Rytilahti, P. (2020). Deep diving into service design problems: Visualizing the iceberg model of design problems through a literature review on the relation and role of service design with wicked problems. *The Design Journal, 24*(2), 231–251.

Sustar, H. (2022). Service design as a sustainability assessment approach for circular business models. In *Proceedings of the NBM @ ROME 2022* (pp. 685–705).

Toreini, E., Aitken, M., Coopamootoo, K., Elliott, K., Zelaya, C. G., & Van Moorsel, A. (2020, January). The relationship between trust in AI and trustworthy machine learning technologies. In *Proceedings of the 2020 conference on fairness, accountability, and transparency* (pp. 272–283).

Trischler, J., Pervan, S. J., Kelly, S. J., & Scott, D. R. (2018). The value of codesign: The effect of customer involvement in service design teams. *Journal of Service Research, 21*(1), 75–100.

Trischler, J., Dietrich, T., & Rundle-Thiele, S. (2019). Co-design: From expert-to user-driven ideas in public service design. *Public Management Review, 21*(11), 1595–1619.

Trischler, J., & Westman Trischler, J. (2021). Design for experience–a public service design approach in the age of digitalization. *Public Management Review, 24*, 1–20.

Tuomaala, M. -S. (2008). Opinto-opas 2008–2010. Taiteiden tiedekunta. Lapin yliopisto. Downloaded 12.12.2022 http://www.medialappi.net/oppaat/TTKopas.pdf

UNEP. (2002). *Product-service systems and sustainability.* UNEP. https://wedocs.unep.org/20.500.11822/8123

Vargo, S. L., & Archpru Akaka, M. (2012). Value cocreation and service systems (re)formation: A service ecosystems view. *Service Science, 4*(3), 207–217.

Vargo, S. L., & Lusch, R. F. (2004). Evolving to a new dominant logic for marketing. *Journal of Marketing, 68*(1), 1–17.

Vargo, S., & Lusch, R. (2008a). Why "service"? *Journal of the Academy of Marketing Science, 36*(1), 25–38.

Vargo, S. L., & Lusch, R. F. (2008b). Service-dominant logic: Continuing the evolution. *Journal of the Academy of Marketing Science, 36*(1), 1–10.

Vargo, S. L., & Lusch, R. F. (2016). Institutions and axioms: An extension and update of service-dominant logic. *Journal of the Academy of Marketing Science, 44*(1), 5–23.

Vezzoli, C., & Manzini, E. (2008). Review: Design for sustainable consumption and production systems. In *System innovation for sustainability 1*. Routledge.

Vink, J. (2019). *In/visible-conceptualizing service ecosystem design* (Doctoral dissertation, Karlstads universitet).

Vink, J., & Koskela-Huotari, K. (2021). Social structures as service design materials. *International Journal of Design, 15*(3), 29–43.

Vink, J., & Koskela-Huotari, K. (2022). Building reflexivity using service design methods. *Journal of Service Research*. Online First. https://doi.org/10.1177/10946705211035004

Vink, J., Wetter-Edman, K., & Koskela-Huotari, K. (2021a). Designerly approaches for catalyzing change in social systems: A social structures approach. *She Ji: The Journal of Design, Economics, and Innovation, 7*(2), 242–261.

Vink, J., Koskela-Huotari, K., Tronvoll, B., Edvardsson, B., & Wetter-Edman, K. (2021b). Service ecosystem design: Propositions, process model, and future research agenda. *Journal of Service Research, 24*(2), 168–186.

Wetter-Edman, K. (2014). *Design for service: A framework for articulating designers' contribution as interpreter of users' experience.* [Doctoral dissertation, University of Gothenburg]. Litorapid Media AB. ISBN: 978-91-979993-9-7.

Wetter-Edman, K., Sangiorgi, D., Edvardsson, B., Holmlid, S., Grönroos, C., & Mattelmäki, T. (2014). Design for value co-creation: Exploring synergies between design for service and service logic. *Service Science, 6*(2), 106–121.

Witell, L., Kristensson, P., Gustafsson, A., & Löfgren, M. (2011). Idea generation: Customer co-creation versus traditional market research techniques. *Journal of Service Management, 22*(2), 140–159.

Young, R. A. (2012). Refocusing the practice of service design in socially responsible contexts. In S. Miettinen & A. Valtonen (Eds.), *Service design with theory. Discussions on change, value and methods* (pp. 81–92). Lapland University Press.

Zeithaml, V. A., Parasuraman, A., & Berry, L. L. (1985). Problems and strategies in services marketing. *Journal of Marketing, 49*(2), 33–46.

# Chapter 4
# Macro-Trend Study Under Service System: Preliminary Research in Service Innovation and Emerging Technology

**Sheng-Hung Lee, Maria C. Yang, Olivier L. de Weck, Chaiwoo Lee, Joseph F. Coughlin, and Eric Klopfer**

## Introduction

In our daily life and work, we are immersed in many types of services, most of which are invisible experiences around us (Vink & Koskela-Huotari, 2021). Services can be viewed as an accumulation of a series of interactions between people and objects under predetermined sequences and frequencies that occur over time (Penin, 2017).

S.-H. Lee (✉)
Department of Mechanical Engineering, Massachusetts Institute of Technology, Cambridge, MA, USA

AgeLab, Massachusetts Institute of Technology, Cambridge, MA, USA
e-mail: shdesign@mit.edu

M. C. Yang
Department of Mechanical Engineering, Massachusetts Institute of Technology, Cambridge, MA, USA

Ideation Lab, Massachusetts Institute of Technology, Cambridge, MA, USA

O. L. de Weck
Department of Aeronautics and Astronautics, Massachusetts Institute of Technology, Cambridge, MA, USA

Engineering Systems Laboratory (ESL), Massachusetts Institute of Technology, Cambridge, MA, USA

C. Lee · J. F. Coughlin
AgeLab, Massachusetts Institute of Technology, Cambridge, MA, USA

E. Klopfer
Scheller Teacher Education Program and The Education Arcade, Massachusetts Institute of Technology, Cambridge, MA, USA



U. Z. A. Hamid, M. Suoheimo (eds.), *Service Design for Emerging Technologies Product Development*, Springer Series in Design and Innovation 29,
https://doi.org/10.1007/978-3-031-29306-1_4

For example, we might take public transportation to work to utilize mobility service; we might go to restaurants to enjoy the dining service with friends and family; we might study at school to use education service (Wolfe, 2020) or government service; we might want to go to movies, theater, concerts, and sports events to utilize entertainment service. These life-relevant services not only continue to evolve but also shape our lives, describe and influence our perceptions, and adapt to and potentially enact culture (Duan et al., 2021).

Most of the time, we take it for granted. When things do not go well or the way we expect them to be, it can turn out to be a bad or at least unsatisfactory user experience. And in recalling our memory of the experience, we might want to articulate our pain points and make suggestions to service providers. Ideally, users hope the services they purchase/receive can be improved or even innovated with their feedback.

Prior to discussing the concept of service design, we understand that the term design simply is about establishing preferable ways of being in the world by materializing stages that are associated with change (Wizinsky, 2022). Design is synonymous with the transformation and adaptation processes.

Service design seamlessly penetrates people's life and work in multiple ways and contexts with different levels of influence, for example, hospitality services, hospital services, financial planning services, autonomous delivery services, tour guide services, house rental services, airport self-check-in services, museum audio guide system service, and even organizational services, namely, leadership and culture building (Almossawi, 2022; Gheerawo, 2022; Bethune, 2022). Typically, when we define the concept of service design, we understand it from an outside-in perspective during the process of service design and development (Mager, 2004).

Inevitably, what we purchase is not only physical products but also invisible services around physical products that generate service value for us (Telalbasic, 2021). Consumers essentially purchase beyond owning a brand-new iPhone from Apple. Instead, they use the physical product as a tangible "medium" to allow them to subscribe to Apple cloud services, for example, iMessage, FaceTime, iTunes, and Apple TV. Therefore, in order to improve people's quality of life through service innovation, it is critical to understand the history of service design, its comprehensive definition, and creative application.

Thus the first part of the research is to offer an overview of the history and concept of service design and its application by selecting 30 leading academic research papers and articles as a main resource to study, analyze, and reframe the connection between service design, service innovation, and social-technological challenges. The second part of the research is to discuss the benefits and weaknesses of the selected service design framework applied to three macro-trend models: (1) the people-centered model (PM), (2) the technology-centered model (TM), and (3) the hybrid model (HM), a mix of PM and HM, to help us better understand, reframe, and analyze social-technological challenges through a service design lens (Amatullo et al., 2022).

Some well-known service innovation examples, like Airbnb and Uber, inspire us to push the boundary of traditional industries by conducting design experiments from the angles of product development, service innovation, and experience design covering both tangible and intangible service design touchpoints. In addition, we mention the topic of service design in terms of measuring its quality of outcome, the adaptability to scale, and the effectiveness of its social impact to complete this experimental research.

## Literature Review

We studied this research topic using the literature on (1) the history of service design; (2) the definition of service design; (3) the measurement and effectiveness of service design quality, (4) macro-trend models: PM, TM, and HM; and (5) service systems: product, service, and experience design.

### *The History of Service Design*

Lynn Shostack coined the term "service design" in 1982 in his article "How to Design a Service" published in the *European Journal of Marketing* Vol. 16. Instead of applying service design in the creative and design field, Shostack used the term to discuss the task description and market planning (Catalanotto, 2018).

The concept of service design began in the 1990s, which was almost the same time as the development of interaction design (Holmlid & Evenson, 2008). Service design was defined as a discipline, which was positioned close to industrial design in the very beginning (Mager, 2008). At the beginning of the 1990s, service design was still a relatively new concept and methodology for academia and industry. Especially when first introduced to the University of Applied Science in Cologne as academic design research, people misunderstood and confused the concept of service design with other design disciplines (Mager, 2009).

At that moment, people's impression of design/industrial design focused on making products aesthetically appealing and ergonomic to increase the sales price of the business. Service design thus has a different emphasis in terms of its purpose, values, thinking process, and research approaches.

Service design, like other design disciplines in general, is an applied science. It is the application of design thinking and design methodologies (Mager, 2013). Service blueprint (Kuang & Chou, 2017; Chuang, 2007; Shostack, 1984) and service concept (Goldstein et al., 2002) can be useful tools and frameworks when service designers or service providers want to design and plan new services or improve the current service models.

Part of service design, in essence, is a human-centered design process with an amplifying participatory section, which naturally infuses new design methods and perspectives to innovate the field of service design (Holmlid & Evenson, 2008). We can recognize that service design starts with the desirability of service recipients (Kuang & Chou, 2017).

Interestingly, service design, as a design profession, lives and learns through design applications, which give more weight to using a human-centered design process (Brown, 2008), co-creation approaches, and participatory methodologies to have a more holistic perspective in service design projects (FJORD, 2017). Meanwhile, the service design domain also has significantly expanded its scope of studies and evolved by integrating other fields of knowledge and applications, for example, big data (Sun & Park, 2017), machine learning, power distribution (Goodwill & Bendor, 2021), public service (Trischler et al., 2019), psychology (Kim, 2021; Kim et al., 2019), and other emerging technologies (Lee, 2022a).

## *The Definition of Service Design*

Everyone has experienced or consumed the outcome of service and might also offer service to others. What is service design by definition from academic and industry perspectives? Services, in general, are purchased and consumed simultaneously, and normally require people, including service recipients (e.g., users), service providers, and employees to connect with each key stakeholder during services (Bitner, 1992).

"Service design gives shape to experiences that have no form," said Jamin Hegeman, VP, Experience Strategy at Capital One (Hegeman, 2017). According to *The Economist*, a British weekly newspaper, it wrote "service design can be understood as the design of products of economic activity that you can't drop on your foot, ranging from hairdressing to websites."

Tomiyama et al. gave a more formal academic definition of service: "a service is an activity that a service provider offers to a service receiver in a service environment and generates values for the service receiver," which can better inform us about the concept of service design (Tomiyama et al., 2004).

The people-oriented design concept is the key characteristic of service design (Sung, 2014). Holmlid and Evenson concluded in their research that by nature service design is human-centered and participatory that can provide meaningful perspectives to service information (Holmlid & Evenson, 2008). Therefore, service design typically starts from the needs of users, and service recipients (Kuang & Chou, 2017), and can be applied to address the form and functionality of services through the lens of service recipients. To service recipients, the purpose of service design is to ensure the service interfaces and interactions are useful, usable, and desirable, whereas for service providers, service offerings are effective, efficient, and distinctive (Mager, 2008).

To improve and innovate services, we also need to consider not only service recipients but also other key stakeholders, and service providers as well as their profit, because they want their service to be unique to sell on the market to form their special selling proposition while competing with other types of service offerings (Mager, 2013).

Anderson et al. argued that service design is an essential innovation tool and can also work as a mindset, solving transformative service design challenges, for example, redesigning healthcare service systems (Anderson et al., 2018). Wolfe also proposed that service designs are critical and impactful approaches with the purpose to enable cutting-edge social-technological innovations by serving both public services and private enterprises (Wolfe, 2020).

We also need to think about what is good service design in general. To put it simply, good service can maintain consistent service quality, and its processes and results are more precise and predictable from service providers' perspectives, whereas bad service is relatively inconsistent and contradictory across the entire customer experience, which is difficult to predict, scale, and even replicate its service (Penin, 2017).

Good service design can be decomposed into design aesthetics, the meaning of the design process for potential users, and people's emotional attachment on their journey (Simonse et al., 2019). Good service design is also used with an integrated service design model to be more inclusive to support various perspectives of service providers and service recipients across critical service touchpoints in their journeys to enable more innovations and make improvements.

## *The Measurement and Effectiveness of Service Design Quality*

Understanding the history of service design and its definitions, we are curious about exploring how to measure the effectiveness and quality of the service design process and outcome (Sun, 2020). Do we have universal standards to follow or evaluation criteria for the measurement in the service design domain? What is an evidence-based framework for evaluating a service design? What is a scientific framework for judging whether a service design is innovative and a good solution? (Furrer et al., 2016)

One example in Table 4.1 is the framework of understanding environment-user relationships in service organizations that show the structure and flow of how to measure people's responses as a guide to measuring cognitive, emotional, and physiological responses to environments (Bitner, 1992).

In 2004, Hevner et al. proposed seven useful principles (design as an artifact, problem relevance, design evaluation, research contributions, research rigor, design as a search process, communications of research) to conduct service design research focusing on information systems and academia. Furrer et al. discussed the service measurement from a marketing and business perspective by proposing four angles (activities, marketing role, customer role, and design science concepts) from the innovative service design framework in 2016. Simonse et al. proposed four key criteria (analyze, experience, co-design, and evaluate) for their integrated service design framework in 2019 to experiment in care service and patient journey.

While designing new types of services, we recognize that service is a process and do not only consider one touchpoint, single task, or situation across the process

**Table 4.1** The service design frameworks to measure the service quality and effectiveness

| Framework | Measure criteria or evaluation approach | Research context | Material source |
|---|---|---|---|
| Environment-user relationship service framework | *Cognitive* response to environments<br>*Emotional* response to environments<br>*Physiological* response to environments | Marketing<br>Organization structure | Bitner (1992) |
| Integrated service design | *Analyze* the care service system<br>*Experience* the journey yourself, observe, and sketch it<br>*Co-design* the patient journey: craft the journey toolkit, interview, and synthesize<br>*Evaluate* the patient journey for integrated service design | Patient journey<br>Care service | Simonse et al. (2019) |
| Seven guidelines for conducting research on service design | Guideline 1: design as an artifact<br>Guideline 2: problem relevance<br>Guideline 3: design evaluation<br>Guideline 4: research contributions<br>Guideline 5: research rigor<br>Guideline 6: design as a search process<br>Guideline 7: communications of research | Information systems research<br>Academic research on service design | Hevner et al. (2004) |
| Innovative service design | *Activities* (e.g., problem surfacing, problem structuring, solution imagining, innovation creating, innovation creating, innovation optimizing, value proposition developing, and value delivering)<br>*Marketing role* (e.g., coach, analyst, experimenter, role play customer, customer engineer, value optimizer, deliver point provider)<br>*Customer role* (e.g., client, usage subject matter expert, sounding board, role player, co-designer/validator, value validator, value co-creator, and benefiter)<br>*Design science concepts* (e.g., psychometric measurement and analysis, invention axiom, information or comprehensiveness axiom, system optimization, feedback) | Marketing and business | Furrer et al. (2016) |
| SERVQUAL | Reliability<br>Assurance<br>Tangibles<br>Empathy<br>Responsiveness | Customer satisfaction measurement | Parasuraman et al. (1988, 1991) |
| Voice of customer (VOC) | Identify customer needs<br>Structure customer needs<br>Provide priorities for customer needs | Marketing research | Griffin and Hauser (1993) |

(continued)

**Table 4.1** (continued)

| Framework | Measure criteria or evaluation approach | Research context | Material source |
|---|---|---|---|
| Lead user analysis (LUA) | Specify lead user indicators<br>Identify lead user group<br>Generate concept (product) with lead users<br>Test lead user concept (product) | Understand the role of users in innovations | Urban and von Hippel (1988), von Hippel (1986) |
| Service-dominant logic | Service is the fundamental basis of exchange<br>Indirect exchange masks the fundamental basis of exchange<br>Goods are a distribution mechanism for service provision<br>Operant resources are the fundamental source of strategic benefit<br>All economies are service economies<br>Value is co-created by multiple actors, always including the beneficiary<br>Actors cannot deliver value but can participate in the creation and offering of value propositions<br>A service-centered view is inherently customer-oriented and relational<br>All social and economic actors are resource integrators<br>Value is always uniquely and phenomenologically determined by the beneficiary<br>Value co-creation is coordinated through actor-generated institutions and institutional arrangements | Help understand the new definition and concept of service to be measured and perceived in the domain of economics<br>New paradigms shift for production and consumption under value systems | Vargo and Lusch (2004, 2008, 2016), Lusch and Vargo (2014) |

(Vargo & Lusch, 2008). It is actually a multifaceted measuring process to review the effectiveness and quality of the service design outcome. Thus, it indicates that the critical role and responsibility of the service designers is to illustrate a bird's-eye view of the entire service experience and its relevant elements merging with the details of each component within the service systems (Kankainen et al., 2012).

Since service design is a complicated process interconnecting service providers, service recipients, and other key stakeholders within service systems, service design can involve many stakeholders with different needs and incentives (van der Bijl-Brouwer, 2022; Seravalli & Witmer, 2021). Therefore, it is critical to understand the key stakeholders' expectations and goals upfront in the user journey to help us validate the outcome and value of service innovation (Simonse et al., 2019).

## *Macro-Trend Models: People-Centered, Technology-Centered, and Hybrid*

Complicated social-technological and systemic challenges have transformed the service design conditions under the different macro-trends models (Amatullo et al., 2022; Vink & Koskela-Huotari, 2021). Advanced technological developments and global social issues have made people sophisticated in various aspects: desirability, purchasing behavior, lifestyle, education, and expectations of work and family, which enable the emergence of new types of services and business models beyond dyadic interactions between service providers and service recipients (Patrício et al., 2018a).

For example, three different typologies of servicescapes have been proposed from the academic research (Bitner, 1992): self-service (requires customers only), interpersonal service (requires both customers and employees), and remote service (requires employees only).

While Bitner mentioned the above three types of services designed for service recipients (e.g., users and customers), in this study, we reinterpret and categorize them into three models to discuss the macro-trends of service design: (1) People-centered model: Service providers use labor or people services to have customers served by people, for example, waiters, guides, and nurses, and make them feel a sense of privilege naturally (Patrício et al., 2018b). (2) Technology-centered model: Service providers use technologies to replace the majority of labor costs and services, for example, autonomous vehicle or autonomous delivery services. (3) Hybrid model: Service providers use self-service to empower the customers to make their own decisions. Some of the service design touchpoints might have staff or machines to help customers finish their tasks, for example, airport self-check-in kiosks supported by ground staff.

We mention the three macro-trend models under the influence of technological developments and global social impacts to give a clear explanation of the ideas of service design in the social-technological context (see sections "People-Centered Model", "Technology-Centered Model", and "Hybrid Model"). In Table 4.4, we provide a brief explanation of each model and relevant examples and compare their advantages and weakness through the lens of service.

We also introduce the term "service currency" in this section, which stands for the cost of the objects (e.g., physical and informational) or resources (e.g., labor and financial) exchanged between service providers and service recipients (e.g., users and consumers).

In addition, we define two key types of stakeholders in the study:

1. Service providers: companies/enterprises which offer service and experience to customers across user journey touchpoints through visible and invisible artifacts. Some researchers have suggested that service providers might act as creative secretaries to facilitate the conversation between customers and companies (Kankainen et al., 2012).

2. Service recipients: people in general who consume or experience the services provided by service providers. We often called them users or customers in the service system.

## *Service Systems: Product, Service, and Experience Design*

Design and the design process can be considered a social systemic practice with impacts (Jones & Van Ael, 2022), whereas services can be viewed as the soft infrastructure of society which is composed of many different layers of complex systems and subsystems (Penin, 2017). In this study, we can think about a system as a set of technical artifacts under well-established behaviors and constraints with the purpose to interact with other elements/artifacts (De Weck et al., 2012).

If we zoom out a bit to look at a concept of system in a relatively bigger context, we can consider a system as a combination of various types of elements/artifacts that can generate greater value and outcomes which cannot be produced by a single element/artifact (Engel, 2018; Rechtin, 1991).

Other features of systems can also help us decompose service systems, including system boundary/environment, system structure, the concept of system of system (SoS), system hierarchy, and many other different system types (Haberfellner et al., 2019; Crawley et al., 2016).

While we reconsider service design at a system level (e.g., social innovation service systems, technological ecosystem systems, healthcare service systems, financial planning service systems, and public transportation service systems), we will have a more holistic view to explore and renavigate the relationship between service systems with three macro-trend models—PM, TM, and HM—which purposefully enables us to reconnect and reframe the social-technological types of service design challenges (Rodrigues et al., 2021).

Anderson et al. mentioned the above concept as a product-service system, which can also be viewed as part of the transformative service research, TSR (Anderson et al., 2018). Others talked about service systems, which are defined from an organizational angle by considering people, communication, context, and technology and linked to other service systems by organizational or social value propositions (Maglio et al., 2009; Spohrer et al., 2007).

However, in the study, we separated the concept of service systems into three categories: (1) product design, (2) service design, and (3) experience design (Table 4.2). Product design can broadly represent the visible part of the service systems (Bitner, 1992), for example, autonomous vehicles, airport self-check-in kiosks, or architectural spaces like libraries and hospitals, whereas experience design is the non-visible part of the service systems, including people's feelings, perceptions, vibe of the environment, or activities hosted in the public space.

Service design is a series of predefined interactions between service providers and service recipients (e.g., users and customers). In addition, service design can act as a medium to connect, activate, and communicate between the other two service

**Table 4.2** The three service systems: product, service, and experience

| Service system | Product design | Service design | Experience design |
|---|---|---|---|
| Explanation | Product design can broadly represent the visible assets within the service systems, and we hypothesize that most visible assets can give people a tangible experience | Service design is a series of predefined interactions between service providers and service recipients. Service design can act as a medium to connect, activate, and communicate between the other two service systems to generate the benefit and create value out of the services | Experience design is to describe the non-visible part of the service systems. We can leverage the multiple physical and digital service touchpoints to curate the experience |
| Examples | Autonomous vehicles, airport self-check-in kiosks, or architectural space like libraries and hospitals | The self-check-in service, interface, apps, websites, and the airline staff can be all accounted for as part of the service design | People's feelings, feedback, ideas, perceptions, vibe of the environment, or activities hosted in the public space |

systems, product design and experience design, to generate the benefit and create value out of the services or to cause negative effects.

For example, people head to the airport to finish the self-check-in through the kiosk or with the help of ground staff from the airline. The self-check-in kiosk belongs to the category of product design; the overall self-check-in user journey is part of the experience design, including other key moments: prepare the travel plan, get to the airport, arrive at the airport, board, in-flight, arrive at the destination (Lai et al., 2022), and finally, the self-check-in service. Interface, apps, websites, and the airline staff can be all accounted for as part of the service design category.

In addition to discussing three types of service systems, we also want to understand how to improve the overall service experience within its systems through service design approaches. Service co-creation can be one great option, since there are many different types of service design methodologies. For example, service design for value networks (SD4VN) can be applied to the target subjects by enabling other interactive elements within networks to create more beneficial value through service system co-creation (Patrício et al., 2018a). Part of the SD4VN approach also follows a three-step design thinking process, including inspiration, ideation, and implementation (Brown, 2008).

However, when improving the service experience both for service providers and service recipients or designing new service offerings, we should consider widely about the service system. How can we reframe, solve, and refine the service design challenges through the lens of ecosystems of services, since service can be viewed as the projection of part of networks and a series of experience touchpoints from people's behaviors, which might inform and inspire to have better human-centered service design solutions (Kankainen et al., 2012)?

**Fig. 4.1** The map of macro-trend models with service systems

**Table 4.3** Service design opportunity areas according to service systems and macro-trends

| Macro-trend | People-centered model | Technology-centered model | Hybrid model |
|---|---|---|---|
| Product design | Opportunity area 1 | Opportunity area 2 | Opportunity area 3 |
| Service design | Opportunity area 4 | Opportunity area 5 | Opportunity area 6 |
| Experience design | Opportunity area 7 | Opportunity area 8 | Opportunity area 9 |

We applied two categories: (1) service systems and (2) macro-trends as the hypothetical axes to experimentally build a 2X2 diagram shown in Fig. 4.1 and to map out nine service design opportunity areas in Table 4.3 to discuss different emerging service economics in Fig. 4.3.

Different types of service economics have emerged at various paces based on three service systems under three macro-trends. Service economics also influences most aspects of people's lives including our working and living environment, complicated social structures, education systems, governmental power, and culture. The intention of creating the map of macro-trend models with service systems is to better help us explore, discuss, and even shape these invisible socioeconomic structures with service design components as well as considering visible service touchpoints.

In Fig. 4.1, we do not put much emphasis on the dimension of time connecting to service models and the design process. We recognize that normally the service will gradually evolve over time because of the changes in environment, the growth of people, the culture of organizations, and many other factors. For example, a customer going to a bar for his/her first time will have a different experience compared with people going to a bar multiple times. An experienced customer might know a bartender, and the preferred food and drink he or she wants to order. The dimension of time will impact not only the service design and its process but also the service quality and customers' expectations.

For further research inspired by Fig. 4.1, we can also consider when we apply service design for the public sector versus the private sector, what are the different problems that we need to think of and address? Do we need to create extra service design criteria to complete Fig. 4.1?

## Research Methods

We conducted the research flow through a rigorous and direct systemic approach in four steps: (1) relevant material selection, (2) framework analysis, (3) content synthesis, and (4) insight generation. Unlike linear study processes, the above four steps are dynamic interconnected stages that help us understand the service design research topic in a more comprehensive and flexible way, including analyzing the data and material, synthesizing the content from multiple angles, concluding the research result with evidence-based perspectives, reflecting on the learnings, and suggesting future research areas. Figure 4.2 shows the overview of the research flow.

Even though each step is connected through four arrows in Fig. 4.2, that does not mean it is a linear approach. In fact, the overall research flow and study process are interconnected, organic, interactive, and naturally blended with four actions: inform, distill, conclude, and reflect to complete this study.

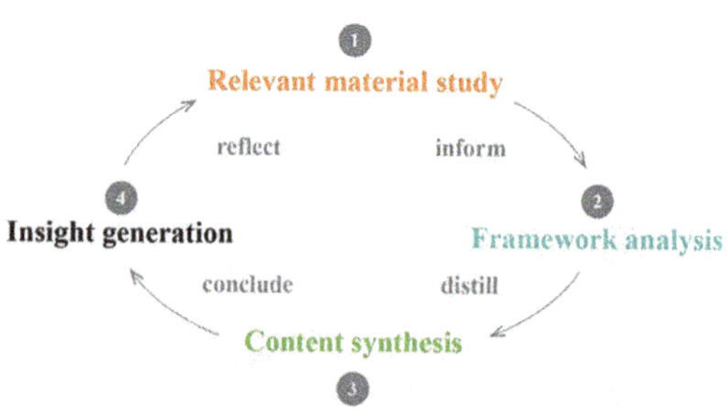


**Fig. 4.2** Research method overview

## *Step 1: Relevant Material Study*

We searched for relevant materials in response to our two research questions: (1) What is service design? (2) How do we use the service design lens to analyze three macro-trend models—PM, TM, and HM—to help us better understand the social-technological challenges? We applied the three selection criteria: (1) abstract, (2) keywords, and (3) conclusion from books, journal papers, and conference papers to filter out the critical sources for this study. Due to the three-month scope of this study, we ended up with ten journal papers, 20 conference papers, and eight books as our reference study materials.

## *Step 2: The Framework Analysis*

Framework analysis includes different types of service design frameworks (Sangiorgi, 2009), methodologies, and evaluation criteria of service design quality, for example, service blueprint, service concept, integrated service design, SERVQUAL, voice of customer (VOC), lead user analysis (LUA), and service-dominant logic. Thus, in this step, we emphasize the service designs by analyzing and comparing their design intention, process, and modeling to help us better explore the potential integrated service design framework suitable for reframing, ideating, or even solving some of the social-technological challenges in the near future.

## *Step 3: Content Synthesis*

In addition to exploring the service design frameworks, methodologies, and evaluation criteria analysis, we also care about the content and context of each selected service design approach and theory, including authors' perspectives on service design, their design intention, purpose, and motivation for creating or refining the service design methodologies and their decision-making process under some critical social-technological factors which might influence them and the service design result.

## *Step 4: Insight Generation*

After gathering relevant resources, analyzing materials, and synthesizing the team discussion and data, we distilled four key insights inspired and generated by mapping three proposed macro-trends: PM, TM, and HM with three service design systems: product, service, and experience design (Figs. 4.1, 4.3, and 4.4). The

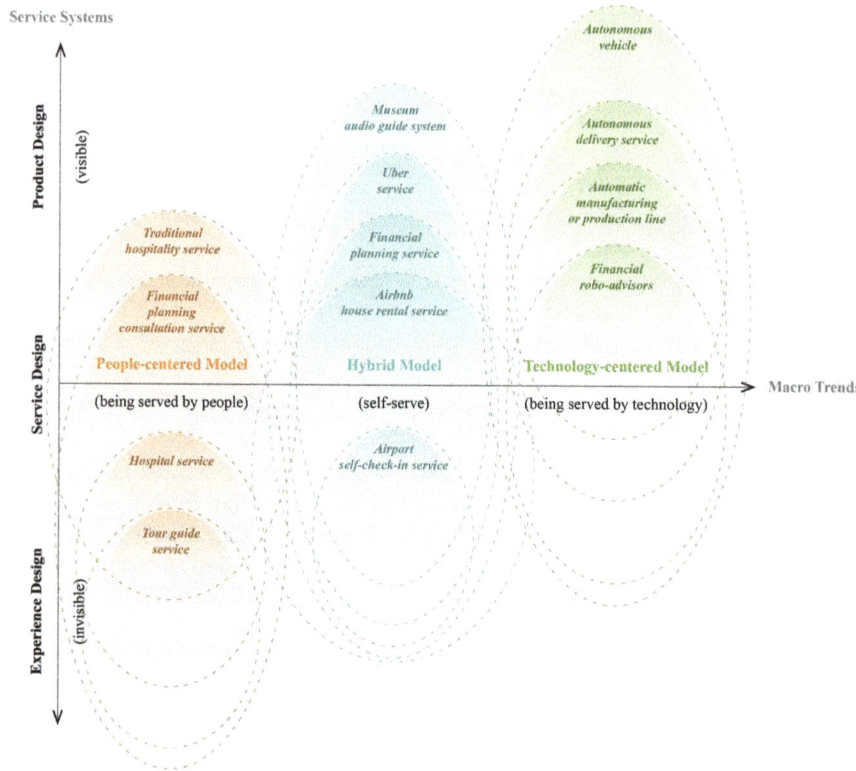


**Fig. 4.3** The example of macro-trend models in service systems

suggested further research areas and research discussion are also associated with the four key insights covered in section "Conclusion".

## Research Result

After synthesizing the learnings from literature reviews and studies, we proposed the concepts by illustrating the diagram of the relationship between three service systems: product, service, and experience design and three trend models: PM, TM, and HM and ten relevant examples (Fig. 4.3).

We used financial service as one example to discuss across three macro-trends with three service systems. Under PM, let us imagine a typical financial planning service that is designed and provided by professional financial advisors. People possibly need to go to their office building located in the city center to book the consulting time for financial advice. The consultation services possibly happen through a series of intimate and in-person conversations with then financial advisors.

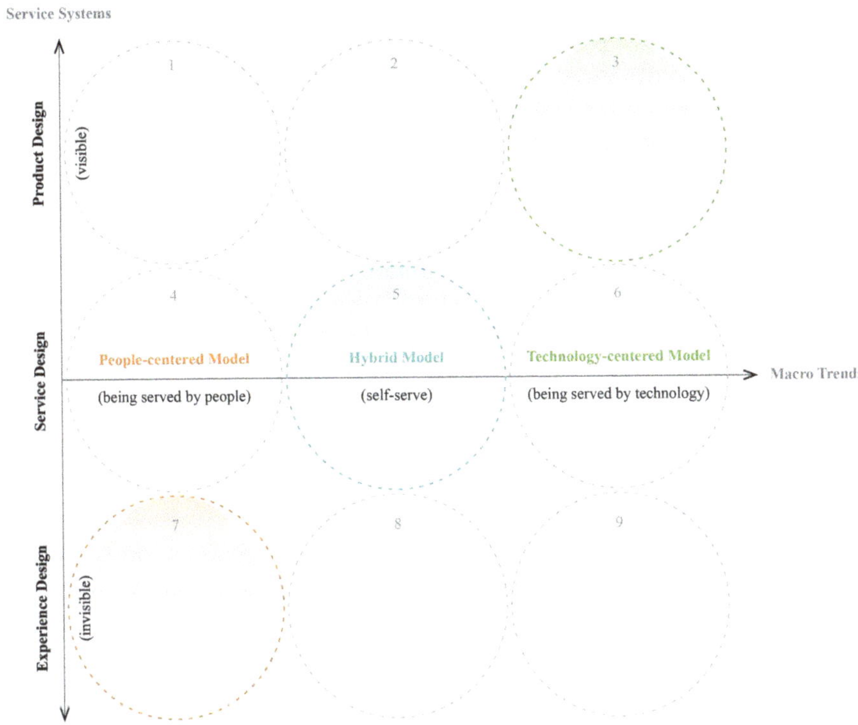


**Fig. 4.4** The opportunity areas of macro-trends in service systems

People's impression of the overall financial service experience is more based on their financial advisors, and it is a highly tailored customer service offered by a group of people with professional domain knowledge. Financial advisors might sell customers' financial product potential, which is mainly service-based experience design and less physical product design solutions to advise and manage their investment and financial conditions.

In TM, with the emergent technologies, digital transformation, cultural- and social- structure change, and political and economic restructuring, people as consumers are actually empowered and even encouraged to pursue more options to select their financial services through multiple platforms: in-person, digital, or hybrid. Financial robo-advisor or robo-system is one of the results generated in the context of automation. By fully leveraging and understanding machine learning and artificial intelligence, financial planning services might be replaced by these smart systems and tools. Ideally, financial service providers can provide more accessible financial calculation tools and tailor-made services to cater to the various needs of customers.

Besides financial planning consultation services under PM and financial robo-advisors under TM, HM sits in the middle of the spectrum offering people another type of financial services with more flexible possibilities. People can better consider their financial situations by having more flexibility to choose the financial tools including visible products to invisible service models and experience that they want to curate from in-person consultation services to digital financial planning assistance tools.

However, the intention of mapping 13 potential service models from automobile industries, financial planning industries, hospitals, hospitality services, and others in Fig. 4.3 is to better help us think about where are the future service design and experiences that people desire that we cannot find on the market to identify potential business implications. The observation and research can effectively assist us to design accessible services with sound business models and qualities for our target audience and continue to explore future research opportunity areas for service design and social innovation.

## *People-Centered Model*

The term "people-centered" was inspired by the term "human-centered design." IDEO, an international design and innovation company, has promoted the concept of human-centered design (HCD) and design thinking since the 1990s, even though the HCD concept started in the 1960s (IDEO, 2022). It is one of the design processes that starts from users' desirability and also considers the feasibility of technology and viability of the business to come up with innovative HCD solutions focusing on building empathy for users or potential consumers (IDEO, 2015).

In the study, we put emphasis on the service touchpoints provided by people, including staff, leadership teams, interns, and other service providers in the service system.

PM is constructed based on the customer-centered service guidance (Cook et al., 2002). The majority of service currency/cost that we assume in the study is the labor cost. The service providers deliver their service to users/customers through their staff, for example, waiters/guides/nurses. For example, in banks, financial planners/advisors or bank tellers can give customers consultation advice; in restaurants or hotels, waiters serve customers; in museums, tour guides give tours to visitors; in hospitals, doctors and nurses help patients.

The benefit of PM for customers is that they are entitled to the service packages that service providers designed for them because they can feel a sense of being served and enjoy the privilege, whereas service providers offer tailor-made premium services to cater to customers' unmet needs and designers through providing labor-intensive services.

The downside of PM is that the high cost of labor service fees may not attract customers. From service providers' perspectives, they will need to consider the cost of training their staff and also think about service updates to make the seasoned customers feel not only premium, but also fresh and inspired.

## *Technology-Centered Model*

Prior to discussing the concept of TM, we want to introduce our definition of the term technology in the study. What is technology by definition? There are several explanations of this context-driven term. However, our definition is modified from de Weck's book: *Technology Roadmapping and Development: A Quantitative Approach to the Management of Technology*. We view technology as an enabler to assemble processes and objects in the different stages from conception, to design, execution, and operation to realize the functional aspect of technological artifacts that are also associated with knowledge to achieve goals that people set and that generate value for people (De Weck, 2022).

When technology is the center of service design, the control of service shifts from people to technology (De Weck, 2022). Service providers use the emerging technologies to replace the majority of labor costs. Thus, the service currency/cost that we assume in the study is the technology development and maintenance cost.

The service providers deliver their service to users/customers through technology, for example, financial planning toolkits, software, financial robo-advisors, autonomous vehicles, autonomous delivery services, and automatic manufacturing/production lines. The consumer-facing touchpoints are driven by technology. For example, customers can take autonomous vehicles to their destinations without driving themselves. They can have more flexible space and time to work or relax during transportation.

The benefit of TM for customers is that they can rely on systems driven by technologies, for example, AI and machine learning, to do the tasks for them. Therefore, people can focus on other more important things, whereas service providers do not need to pay much attention to people's side. Instead, they can focus on designing optimized mechanisms or trained algorithms to adapt to various situations.

For service providers, the downside of TM is the challenge of how to use the emerging technologies, for example, AI and machine learning, to develop tailor-made customer services without making people feel distant or cold due to automation. Service providers also need to consider the services and systems safety of the users when operating or controlling by machines or smart devices.

## *Hybrid Model*

HM is a mix of PM and TM. Service providers use the idea of self-service to empower customers to make their own decisions, which enables more emphasis on experience-based design (Perrott, 2013). Some of the service touchpoints of customer journeys might have offered staff or machines to help customers finish their tasks. The estimated service currency/cost that we assume in the study is the learning time and effort for customers to understand the self-service model.

There are many self-models on the market and it has become very common. For example, customers order their preferred house through the Airbnb house rental service on digital devices. Amazon marketplace provides an interactive platform to make connections with communities (Patrício et al., 2018a). They probably do not necessarily need to meet the house owner, since they have already dropped the keys in the mailbox.

Many airports have set up lots of self-check-in kiosks to help travelers process their tickets and luggage check-in to save more time and labor costs. One analogous example is that many financial planning services, online banking or banking apps (e.g., personal capital), contain both in-person consultation services and financial toolkits or software that allow customers to understand the financial products better and maintain the transparency of the conversation and process.

Another example is a shared lab space on campus. We can also view it as a great HM case study, since lab scientists, staff, or students can work flexibly either in the actual physical lab space or digital space via Zoom or other online software to contribute their works (Lee, 2022b). The research works and services can be delivered in multiple channels, because either people or technology is one element within the service systems.

The benefit of HM for customers is that they will have a certain level of freedom to spend their time and control the frequency and approaches to using services. Service providers offer maximized flexibility and adaptability to satisfy customers' needs and might save the cost between labor/staff training and technological development and maintenance.

However, in HM, the self-service experience design will play a critical role for service providers, since customers can navigate themselves directly through the process and outcome of service design and they might participate in part of the creation of service design. Due to the complexity of the self-service, system design, and sophistication of customers, the design and development of HM might cost much more money than the other two models.

## Discussion and Next Step

Based on Fig. 4.1 and Table 4.4, we divided Fig. 4.4 into nine opportunity areas marked by numbers from one to nine. We are especially interested in exploring circles three, five, and seven for further discussion. What will the scenarios look like if we consider a PM connecting with experience design (circle seven)? How do we envision applying a TM to reshape the product design process and frame of reference (circle three)? What if we use a HM by combining two models of PM and TM to create and explore future service design approaches, models, and applications (circle five)?

**Table 4.4** Three types of macro-trend models through the lens of service systems

| Macro-trend | People-centered model | Technology-centered model | Hybrid model |
|---|---|---|---|
| Explanation | Service providers use labor or people services to have customers served by people, for example, waiters/guides/nurses, and make them feel a sense of privilege | Service providers use technologies to replace the majority of labor costs and services through using autonomous systems or AI powered by big data | Service providers might use self-service to empower the customers to make their own decisions. Some of the service touchpoints might have staff or machines to help customers finish their tasks |
| Service currency/cost | Labor/staff training and mentoring cost | Technology development and maintenance cost | Customer service learning and education cost |
| Examples | Financial planning consultation service (e.g., financial planners/advisors and bank tellers)<br>Traditional hospitality service<br>Hospital service<br>Tour guide service | Financial planning toolkits and software (e.g., financial robo-advisors)<br>Autonomous vehicle<br>Autonomous delivery service<br>Automatic manufacturing/production line | Financial planning service with multiple touchpoints (e.g., online banking or banking apps)<br>Airbnb house rental service<br>Uber service<br>Airport self-check-in service<br>Museum audio guide system |
| Advantages | Customers can fully enjoy the service packages that service providers designed for them<br>Service providers offer labor-intensive services to cater to customers' unmet needs<br>It is a "high-touch" service for both service recipients and service providers. It is also a model with a high potential to develop tailor-made services and systems with flexibility and adaptability | Customers can rely on the service system driven by technologies or smart devices, for example, AI and machine learning<br>Service providers do not need to pay much attention to people's side. Instead, they can focus on designing optimized mechanisms or trained algorithms to adapt to various situations<br>Service providers can relatively easy to maintain service quality | Customers will have a certain level of freedom to spend their time and control the frequency and approaches of using services<br>Service providers offer maximized flexibility and adaptability to satisfy customers' needs<br>Service providers might save the cost between labor/staff training and technology development and maintenance |

(continued)

**Table 4.4** (continued)

| Macro-trend | People-centered model | Technology-centered model | Hybrid model |
|---|---|---|---|
| Weakness | The high cost of service fee and other labor fees may not be suitable for customers<br>The service providers, for example, a hotel, will need to consider service updates to make the seasoned customers still feel fresh and inspired<br>It might be relatively hard to maintain the quality of services if the service providers do not control the quality of people's training | The service providers might find it hard to offer tailor-made services due to the limitation of emerging technologies or smart devices<br>The service might make customers feel distant due to automation and machine interfaces<br>Service providers consider the services and systems safety of the users when operating or controlling by machines or smart devices | The self-service experience design will play a critical role for service providers, since customers will navigate directly through the outcome of service design<br>Due to the complexity of the self-service, system design, and the sophistication of customers, the design and development of HM might cost much more money than the other two models |

## *People-Centered Model and Experience Design*

Envision one design scenario that is about applying PM to experience design. Even though the majority of service is delivered by people and may heavily rely on people or labor-intensive tasks, how do we smartly leverage people's talents and creativities or design a set of toolkits or platforms to empower them to curate better experiences and services in the future? When we celebrate the incredible outcome that originates from "high-tech" advances in our society seamlessly connecting to our lives and work, we also need to consider another layer: high-touch to improve the quality of life.

High-touch can be interpreted in many different ways based on contexts and purposes. The term was coined by John Naisbitt, the author of the book *Megatrends* in 1982 (Naisbitt, 1984). It helps us reflect on the roles and responsibilities of service providers (including individuals and companies) in the context of creating or curating human-centered experience.

One obvious example is to apply a human-centered approach to our design and research process. We need to put more emphasis on our users, customers, and key stakeholders' perspectives to not only design for them but also design with them. While we enjoy the outcomes and use the convenient, efficient services and experiences provided by high-tech innovation, for example, autonomous vehicles, home social robots, and other high-tech products, we need to consider all these great design solutions, apart from providing more convenient services for people: what are other critical high-touch service touchpoints that can enable us, as well as service providers and designers to offer more human-centered experience.

## *Technology-Centered Model and Product Design*

Consider the concept of TM connecting to product design. In this study, we define product designs as physical products with tangible interfaces and features. Obviously, most product designs in the early-stage design and development process naturally fall into TM, since users/customers tend to view technology as part of product features. How does this technology work on this type of product or what are the benefits that users/customers will gain by applying the technology? In addition, the outcome of using technology in product design is relatively obvious and foreseeable.

In further discussion, we might need to reconsider the effects in different stages of product design and development, and the position and value of applying TM. We might need to think about not to overemphasize the importance of TM across all stages of product design and development without thinking about other aspects of product design, for example, the voices of potential users, the cost of the manufacturing process, time and labor, the whole product life cycle within the systems, and the sustainability issues of the product design.

Especially, product design now has become more complicated and powerful than ever to satisfy the needs of sophisticated users/customers. Emerging technologies only serve as one component among the rest of the complex product design system, and its goal is to deliver great value and services to its users/customers. For further studies, we can start by asking one question: How do we find the balance between using TM and PM during the process of product design and development to integrate considerations from users, manufacture, business, cost, and other critical factors into the product-service design process within the dynamic system of business, social impact, and emerging technologies?

## *Hybrid Model and Service Design*

Envision one of the future conditions by applying HM to the service design domain. When we think of operation in hybrids, it is a combination of multiple systems and potentially many subsystems. People as users of systems have options to switch between automated, manual, or in-between/hybrid mode of controlling systems. Thus, it naturally makes the hybrid process itself complicated. We need to consider it comprehensively with more angles prior to making critical decisions. The discussion of applying HM in the context of service design will generate even more complicated systemic challenges, even though, in general, the nature and the process of HM connecting to service design is relatively challenging. And the service outcome and value are still very impactful in terms of influencing people's behavior, business models, and social impact.

For example, innovative shared economic services like Airbnb and Uber have fundamentally transformed the way we view these services. The concept of "ownership" has disrupted and shaped people's perceptions from having automobiles to sharing their mobility services: from owning the house to providing extra room for rental services. These services use both PM and TM during the whole user journey including service providers, service recipients, and other service-relevant stakeholders in the system.

For further studies of HM and service design, we can explore how to identify the critical service touchpoints to make PM and TM smooth and seamless and how to evaluate the service design quality both through a qualitative and quantitative approach.

## Conclusion

We live in the world of experience economics (Lai et al., 2022), and our work and lives are full of services that are not only useful, usable, and desirable but also efficient and effective (Holmlid & Evenson, 2008). We consume various services to make our life more convenient and improve our quality of life; meanwhile, we might generate more consumers' needs by asking for more new types of services. For example, when we travel abroad, we look for effective and efficient services to enjoy the trip. Regardless of other financial issues, if people can bike, they will not walk; if people can drive an automobile, they will not bike; if people can hire drivers, they probably will not drive themselves. People always tend to choose an option to achieve their goal in a most convenient way by instinct.

So it makes us reflect on the meaning of the term service design in today's society. In the study, we want to provide a new perspective to understand service design in three macro-trends: people-centered model, technology-centered model, and hybrid model, with three service systems: product, service, and experience design, which can help us better understand, reframe, or even solve social-technological challenges (Amatullo et al., 2022; Vink & Koskela-Huotari, 2021). Thus, we identified and suggested three opportunity areas in Fig. 4.4.

### *High-Touch Is a Critical Catalyst in Service and Experience Design*

In the people-centered model, service providers use labor or people services to have customers served by people and make them feel a sense of privilege. For example, in a five-star hotel, the servants or waiters should know your name, prepare check-in material in advance, and recognize your car before you enter the lobby. We want to leverage the "high-touch" of the people-centered model across the services to amplify the importance of the human-centered experience design.

We break down one general user experience by five critical touch points using the 5E experience design model: entice, enter, engage, exit, and extend (Sontag, 2018). Each key moment is created and curated not only through the lens of service functions and rational angles but also through the emotional and human-centered design aspects. People actually want to purchase neither products nor services; instead what consumers think about buying is the offerings that render services that bring value to them (Perrott, 2013).

## *Technology Is Viewed as a Vehicle to Deliver Value to People*

In the technology-centered model, service providers use technologies to replace the majority of labor costs and services through using autonomous systems or AI powered by big data. How do we design products to provide a better user experience to people and with people? Sometimes, we overemphasize the term technology while introducing or learning about new products. Obviously, technology is one of the components of product design, and technology can be considered a vehicle to realize the functions of products. What we need to put more emphasis on is the services and experiences around the products to generate benefits for our users.

Apple is a great well-known example not only to launch a world-class product design but also to establish an accessible platform, subscription business models, and services, for example, Apple Pay and iTunes, to enhance the overall user experience and service.

## *Sophisticated Considerations for Users Are a Key to Creating Service Innovation*

In the hybrid model, a mix of a people-centered model and a technology-centered model, service providers might use self-service to empower the customers to make their own decisions. Some of the service touchpoints might have staff or machines to help customers finish their tasks. This model is a relatively complex one, since we need to consider many components with multiple layers of considerations across the user journey. Do we have the right staff/experts to help users when the system goes wrong? Do we have adequate technologies to empower users to pursue their tasks? How do we connect people and technology in the service system by designing service models following rigorous and flexible protocols?

Tsutaya Books, an international bookstore established in 1983 in Osaka, Japan, has now opened over 1400 stores nationwide and is a convincing example to discuss the hybrid model with a service design focus (Yang, 2018; Culture Convenience Club Co., Ltd., 2010; Slywotzky & Wise, 2003). They offer a T-card membership service with collaborations with many brands across different industries, which share similar visions and goals, to form a strong network to bring benefit to consumers with more competitive prices and more options.

Meanwhile, their omnichannel business strategy has seamlessly interconnected the physical (e.g., bookstore space, commercial items) and digital (e.g., membership, Tsutaya brand curation with the seasonal recommendation) service touchpoints (Sugiyama et al., 2015). Ultimately, Tsutaya Books positions itself as a lifestyle brand to bring more value to its customers.

## *Roadmapping of Macro-Trend Models and Service Systems*

Service design is an integrated domain of knowledge to build, translate, and communicate the interaction, connections, and values between the product design and the experience design to meet the desirability of users, the sustainability of business, and the feasibility of emerging technologies to improve our society (van der Bijl-Brouwer, 2022).

The study uses a preliminary experimental research methodology to explore and understand the definition of service design in today's context and use three different service system lenses: product, service, and experience design to analyze three macro-trend models: the people-centered, technology-centered, and the hybrid to help us better envision the roles and value of service design attached to these complex systemic social-technological challenges. Therefore, we summarized critical questions around service innovation for future research shown in Table 4.5.

In the era of transformation of organizations and society, service design is also under a paradigm shift to adapt to new changes and build its capabilities to solve these complicated and systemic social-technological challenges (Wizinsky, 2022; Telalbasic, 2021; Patrício et al., 2018b). For example, how do we consider the transition phase of the creation of new service concepts from the stage of service prototyping and service implementation (Perrott, 2013)? How do we apply service engineering methodologies to reframe, ideate, solve, refine, and implement service-relevant issues in a more evidence-based engineering approach (Tomiyama, 2000, 2001)?

In the foreseeable future, in the midst of a new global paradigm, we look forward to establishing and reshaping services and service designs as evolving dynamic platforms prepare for the next-generation challenges and create more contemporary service design innovations driven by emerging technologies, business models, and people's motivation to embrace the frontier possibilities, as well as build impactful ecosystems and culture.

**Table 4.5** Macro-trend models and service systems

| Macro-trend | People-centered model | Technology-centered model | Hybrid model |
|---|---|---|---|
| Product | How do we integrate people-side of services, policies, structures, and even culture into high-tech products? | How do we design products powered by emerging technology to provide better user experience for people and with people? | Do we have adequate technologies to empower users to pursue their tasks in the service system? |
| Service | How do we re-emphasize the human-centered culture and safe environment in a people-centered model to establish a sustainable service system? | When we launch a new product with new technologies, how do we design the service system accordingly in order to enhance the user experience? | Do we have the right staff/experts/professional team to help and support users when the service system goes wrong? |
| Experience | How do we amplify the human-side of the service model for users to create and curate a high-touch experience and address people's needs? | What will future scenarios influenced by emerging technologies look like that will transform service systems and people's behaviors dramatically? | How do we connect people and technology in the service system by designing service models following rigorous and flexible protocols? |

**Acknowledgments** Thank you to Mari Suoheimo, Professor of Service Design at The Oslo School of Architecture and Design (AHO), for her invitation and encouragement for this book chapter contribution and the great support from Dr. Umar Zakir Abdul Hamid, Lead of Strategic Planning, CEVT AB, and a co-editor of this book. We also want to express our appreciation to the MIT AgeLab team for their insightful suggestions and ideas for this work, MIT Ideation Lab, MIT Engineering Systems Laboratory (ESL), Susan Spilecki, and Xuan Zhou.

## References

Almossawi, H. (2022). *The innovator's handbook: A short guide to unleashing your creative mindset*. Mossawi Studios LLC. https://www.theinnovatorshandbook.com/

Amatullo, M., Boyer, B., May, J., & Shea, A. (Eds.). (2022). *Design for social innovation: Case studies from around the world*. Routledge, Taylor & Francis Group.

Anderson, S., Nasr, L., & Rayburn, S. W. (2018). Transformative service research and service design: Synergistic effects in healthcare. *The Service Industries Journal, 38*(1–2), 99–113. https://doi.org/10.1080/02642069.2017.1404579

Bethune, K. G. (2022). *Reimagining design: Unlocking strategic innovation*. The MIT Press.

Bitner, M. J. (1992). Servicescapes: The impact of physical surroundings on customers and employees. *Journal of Marketing, 56*(2), 57–71. https://doi.org/10.2307/1252042

Brown, T. (2008, June). Design thinking. *Harvard Business Review*. https://hbr.org/2008/06/design-thinking

Catalanotto, D. (2018, August 19). *Book: A tiny history of service design*. https://service-design.co/book-a-tiny-history-of-service-design-368ed603797c

Chuang, P.-T. (2007). Combining service blueprint and FMEA for service design. *The Service Industries Journal, 27*(2), 91–104. https://doi.org/10.1080/02642060601122587

Cook, L. S., Bowen, D. E., Chase, R. B., Dasu, S., Stewart, D. M., & Tansik, D. A. (2002). Human issues in service design. *Journal of Operations Management, 20*(2), 159–174. https://doi.org/10.1016/S0272-6963(01)00094-8

Crawley, E. F., Cameron, B., Selva, D., & Augustine, N. R. (2016). *System architecture, strategy and product development for complex systems* (Global Ed.). Pearson.

Culture Convenience Club Co., Ltd. (2010). *CCC Group consolidated, annual results for year ending Mar 31, 2010*. https://web.archive.org/web/20101227140151/http://www.ccc.co.jp/eng/fileupload/pdf/news/100528FY09-4Q_eng.pdf

De Weck, O. L. (2022). *Technology roadmapping and development: A quantitative approach to the management of technology*. Springer.

De Weck, O. L., Roos, D., & Magee, C. L. (2012). *Engineering systems: Meeting human needs in a complex technological world*. MIT Press.

Duan, Z., Vink, J., & Clatworthy, S. (2021). Narrating service design to account for cultural plurality. *International Journal of Design, 15*(3), 18.

Engel, A. (2018). *Practical creativity and innovation in systems engineering*. John Wiley & Sons.

FJORD. (2017, August 31). *What is service design a tale of two coffee shops*. https://youtu.be/HNOY8GLVy_8

Furrer, O., Sudharshan, D., Tsiotsou, R. H., & Liu, B. S. (2016). A framework for innovative service design. *The Service Industries Journal, 36*(9–10), 452–471. https://doi.org/10.1080/02642069.2016.1248420

Gheerawo, R. (2022). *Creative leadership: Born from design*. Lund Humphries Publ.

Goldstein, S. M., Johnston, R., Duffy, J., & Rao, J. (2002). The service concept: The missing link in service design research? *Journal of Operations Management, 20*(2), 121–134. https://doi.org/10.1016/S0272-6963(01)00090-0

Goodwill, M., & Bendor, R. (2021). Beyond good intentions: Towards a power literacy framework for service designers. *International Journal of Design, 15*(3), 15.

Griffin, A., & Hauser, J. R. (1993). Marketing Science, *12*(1), 1–27

Haberfellner, R., de Weck, O., Fricke, E., & Vössner, S. (2019). *Systems engineering: Fundamentals and applications* (1st ed.). Springer International Publishing.

Hegeman, J. (2017, January 27). *So you want to be a service designer*. https://youtu.be/bA_gnQqjmz4

Hevner, A. R., March, S. T., Park, J., & Ram, S. (2004). Design science in information systems research. *Management Information Systems Research Center, University of Minnesota, 28*(1), 32.

Holmlid, S., & Evenson, S. (2008). Bringing service design to service sciences, management and engineering. In B. Hefley & W. Murphy (Eds.), *Service science, management and engineering education for the 21st century* (pp. 341–345). Springer US. https://doi.org/10.1007/978-0-387-76578-5_50

IDEO (Ed.). (2015). *The field guide to human-centered design* (1st ed.). Design Kit.

IDEO. (2022). *IDEO design thinking*. https://designthinking.ideo.com/

Jones, P., & Van Ael, K. (2022). *Design journeys through complex systems: Practice tools for systemic design*. BIS Publishers B V.

Kankainen, A., Vaajakallio, K., Kantola, V., & Mattelmäki, T. (2012). Storytelling Group—A co-design method for service design. *Behaviour & Information Technology, 31*(3), 221–230. https://doi.org/10.1080/0144929X.2011.563794

Kim, M. (2021). A study of dignity as a principle of service design. *International Journal of Design, 15*(3), 14.

Kim, M., Srinivasan, D., & Zhou, X. (2019). The morphology of dignity: Service storytelling and prototypes for a service design tool. *The Design Journal, 22*(6), 793–812. https://doi.org/10.1080/14606925.2019.1662633

Kuang, P. H., & Chou, W. H. (2017). Research on service blueprint of food banks. *The Design Journal, 20*(sup1), S3425–S3435. https://doi.org/10.1080/14606925.2017.1352846

Lai, M., Huang, J., & Pine, B. J. (2022). *X thinking: Building better brands in the age of experience*. X Thinking Institute.
Lee, S.-H. (2022a, July 5). *Transformative service innovation and design: From touch points to data points*. MIT AgeLab. https://agelab.mit.edu/home-logistics-and-services/blog/transformative-service-innovation-design-touch-points-data-points/
Lee, S.-H. (2022b, September 8). *Service innovation in space design: How to build an immersive and interactive lab space experience in a post-Covid era*. DesignWanted. https://designwanted.com/space-design-service-innovation/
Lusch, R. F., & Vargo, S. L. (2014). *Service-dominant logic: Premises, perspectives, possibilities*. Cambridge University Press.
Mager, B. (Ed.). (2004). *Service design: A review*. International School of Design.
Mager, B. (2008). Service design. In M. Erlhoff & T. Marshall (Eds.), *Design dictionary* (pp. 354–357). Birkhäuser. https://doi.org/10.1007/978-3-7643-8140-0_244
Mager, B. (2009). Service design as an emerging field. In *Designing services with innovative methods* (p. 17). University of Art and Design Helsinki.
Mager, B. (2013, June 6). *Introduction to service design—What is service design?* [Interview]. https://youtu.be/f5oP_RlU91g
Maglio, P. P., Vargo, S. L., Caswell, N., & Spohrer, J. (2009). The service system is the basic abstraction of service science. *Information Systems and e-Business Management, 7*(4), 395–406. https://doi.org/10.1007/s10257-008-0105-1
Naisbitt, J. (1984). *Megatrends: Ten new directions transforming our lives* (Pbk. ed). Warner Books.
Parasuraman, A. P., Zeithaml, V. A., & Berry, L. L. (1988). SERVQUAL: A multiple-item scale for measuring consumer perceptions of service quality. *Journal of Retailing, 64*, 12–40.
Parasuraman, A., Berry, L. L., & Zeithaml, V. A. (1991). Perceived service quality as a customer-based performance measure: An empirical examination of organizational barriers using an extended service quality model. *Human Resource Management, 30*(3), 335–364. https://doi.org/10.1002/hrm.3930300304
Patrício, L., de Pinho, N. F., Teixeira, J. G., & Fisk, R. P. (2018a). Service design for value networks: Enabling value cocreation interactions in healthcare. *Service Science, 10*(1), 76–97. https://doi.org/10.1287/serv.2017.0201
Patrício, L., Gustafsson, A., & Fisk, R. (2018b). Upframing service design and innovation for research impact. *Journal of Service Research, 21*(1), 3–16. https://doi.org/10.1177/1094670517746780
Penin, L. (2017). *An introduction to service design: Designing the invisible*. Bloomsbury Publishing.
Perrott, B. E. (2013). Including customers in health service design. *Health Marketing Quarterly, 30*(2), 114–127. https://doi.org/10.1080/07359683.2013.787882
Rechtin, E. (1991). *Systems architecting: Creating & building complex systems*. Prentice Hall.
Rodrigues, V., Holmlid, S., & Blomkvist, J. (2021). A designerly approach to exploring disruptions in service: Insights from employing a systems perspective. *International Journal of Design, 15*(3), 12.
Sangiorgi, D. (2009). *Building up a framework for service design research* (p. 7). 8th European Academy Of Design Conference
Seravalli, A., & Witmer, H. (2021). (Service) Design and organisational change. *Balancing with Translation Objects, 15*(3), 14.
Shostack, G. L. (1984). Designing services that deliver. *Harvard Business Review, 62*(1), 133–139.
Simonse, L., Albayrak, A., & Starre, S. (2019). Patient journey method for integrated service design. *Design for Health, 3*(1), 82–97. https://doi.org/10.1080/24735132.2019.1582741
Slywotzky, A., & Wise, R. (2003). An unfinished revolution. *MIT Sloan Management Review, 44*(3), 94–95.
Sontag, A. (2018, January 26). *The 5E experience design model: A step-by-step guide to designing meaningful experiences*. Andy Sontag. https://medium.theuxblog.com/the-5e-experience-design-model-7852324d46c

Spohrer, J., Maglio, P. P., Bailey, J., & Gruhl, D. (2007). Steps toward a science of service systems. *Computer, 40*(1), 71–77. https://doi.org/10.1109/MC.2007.33

Sugiyama, D., Shirahada, K., & Kosaka, M. (2015). Elements to organize the third place that promotes sustainable relationships in service businesses. *Technology in Society, 43*, 115–121. https://doi.org/10.1016/j.techsoc.2015.05.013

Sun, Q. (2020). Towards a new agenda for service design research. *The Design Journal, 23*(1), 49–70. https://doi.org/10.1080/14606925.2019.1694808

Sun, Q., & Park, H. (2017). The map as an object of service design. *The Design Journal, 20*(sup1), S4101–S4119. https://doi.org/10.1080/14606925.2017.1352911

Sung, T.-J. (2014). The essence of service design and process tools. *Journal of Design, 19*(2). https://www.jodesign.org.tw/index.php/JODesign/article/view/1130

Telalbasic, I. (2021). A conceptual framework for social currency innovation: A service design perspective. *Strategic Design Research Journal, 14*(2), 407–422. https://doi.org/10.4013/sdrj.2021.142.03

Tomiyama, T. (2000, April 18). Knowledge intensive engineering towards sustainable products with high knowledge and service contents. In *TMCE 2000, Third international symposium on tools and methods of competitive engineering*. https://cir.nii.ac.jp/crid/1570291225325672576

Tomiyama, T. (2001). Service engineering to intensify service contents in product life cycles. In *Proceedings second international symposium on environmentally conscious design and inverse manufacturing* (pp. 613–618). https://doi.org/10.1109/ECODIM.2001.992433

Tomiyama, T., Shimomura, Y., & Watanabe, K. (2004). A note on service design methodology. In *Volume 3a: 16th international conference on design theory and methodology, 2004* (pp. 251–259). https://doi.org/10.1115/DETC2004-57393

Trischler, J., Dietrich, T., & Rundle-Thiele, S. (2019). Co-design: From expert- to user-driven ideas in public service design. *Public Management Review, 21*(11), 1595–1619. https://doi.org/10.1080/14719037.2019.1619810

Urban, G. L., & von Hippel, E. (1988). Lead user analyses for the development of new industrial products. *Management Science, 34*(5), 569–582. https://doi.org/10.1287/mnsc.34.5.569

van der Bijl-Brouwer, M. (2022). Service designing for human relationships to positively enable social systemic change. https://doi.org/10.57698/V16I1.02

Vargo, S. L., & Lusch, R. F. (2004). Evolving to a new dominant logic for marketing. *Journal of Marketing, 68*(1), 1–17.

Vargo, S. L., & Lusch, R. F. (2008). From goods to service(s): Divergences and convergences of logics. *Industrial Marketing Management, 37*(3), 254–259. https://doi.org/10.1016/j.indmarman.2007.07.004

Vargo, S. L., & Lusch, R. F. (2016). Institutions and axioms: An extension and update of service-dominant logic. *Journal of the Academy of Marketing Science, 44*(1), 5–23. https://doi.org/10.1007/s11747-015-0456-3

Vink, J., & Koskela-Huotari, K. (2021). Social structures as service design materials. *International Journal of Design, 15*(3), 15.

von Hippel, E. (1986). Lead users: A source of novel product concepts. *Management Science, 32*(7), 791–805. https://doi.org/10.1287/mnsc.32.7.791

Wizinsky, M. (2022). *Design after capitalism: Transforming design today for an equitable tomorrow*. The MIT Press.

Wolfe, K. (2020). Service design in higher education: A literature review. *Perspectives: Policy and Practice in Higher Education, 24*(4), 121–125. https://doi.org/10.1080/13603108.2020.1792573

Yang, X. (2018). *A localized book sharing service: Create a new scenario for physical bookstore*. Politecnico di Milano. https://www.politesi.polimi.it/handle/10589/140740

# Chapter 5
# A Proposed Transformation Service Design Research Framework for Underserved Settings

**Retha de la Harpe and Oluwamayowa Ogundaini**

## Introduction

Sub-Saharan Africa boasts of being a gateway to several opportunities on the continent for the development of emerging industry 4.0 technologies, but parts of the region still experience socioeconomic issues associated with less-than-adequate service delivery. The reality on ground is that service providers are often under-resourced and stretched to the limit due to the dearth in infrastructure, proximity from urban areas, and lack of holistic understanding to address the complexities of an underserved setting. Government and non-state actors usually do not have the luxury to design services that would address the needs of their clients who rely on these services to continuously improve their well-beings and socioeconomic statuses.

Without a proper understanding of the lived realities of the people in underserved settings, the situation will not change. Underserved settings in sub-Saharan Africa are characterized by socioeconomic difficulties characterized by varying degrees of poverty, unemployment, high crime rates, and limited infrastructure development. It is imperative to carefully understand the contexts, how services can be designed and deployed to transform the lives of people in these underserved settings.

Lessons learnt from field work while implementing the Double Diamond service design framework developed by the UK design council suggests that the first diamond is not sufficient to obtain a deeper understanding of the lived experiences

R. de la Harpe (✉)
Cape Peninsula University of Technology, Cape Town, South Africa
e-mail: delaharper@cput.ac.za

O. Ogundaini
University of South Africa Graduate School of Business Leadership, Midrand, South Africa
e-mail: ogundoo@unisa.ac.za



U. Z. A. Hamid, M. Suoheimo (eds.), *Service Design for Emerging Technologies Product Development*, Springer Series in Design and Innovation 29,
https://doi.org/10.1007/978-3-031-29306-1_5

in underserved settings of sub-Saharan Africa. At the initial time of our field works, the service design network (SDN, 2022) was not known. They proposed a compulsory third diamond to deal with the complexities experienced when designing in response to public challenges across agencies and sectors. Their additional diamond presented as the first diamond provides an opportunity to better understand the problem situation to focus on the "what" and "why" of root causes to design services that are implementable in complex situations. They combined system and design thinking to propose a systemic approach.

Although our proposed diamond is not yet aligned to systemic thinking, it already addresses the need to better understand the problem situation in complex settings. However, in our case, the context is typical of an underserved setting. The work of Bijl-Brouwer and Malcolm (2020) focuses on framing complex problems in society. They also use a systemic approach to adapt to dealing with complex societal challenges. Jagtap (2022) suggests that the nature of input provided by multiple stakeholders for integrate design solutions when co-designing in resource-limited societies, with similar conditions to our concept of underserved settings. Therefore, it is necessary to spend more time on understanding the context of the lived realities using empathizing methods and encouraging stakeholder and community participation.

Hence, the authors have proposed a transformation service design research framework derived from related literature through a conceptual analysis approach integrated with the experiences and reflections of conducting service design research studies in similar settings. The proposed framework has been used as a lens to gain insights in a study to design a patient-centered information service in maternal healthcare in an underserved setting (Nyatuka & de La Harpe, 2021). The research question to guide the discussion in this chapter is as follows: How can service design research be conducted in underserved settings with a focus on the lived experience of the people in the setting to commit to a positive transformation?

This chapter is structured to firstly provide an overview of service design research from a design and research perspective followed by why the need to transform is an important consideration. Next, the proposed framework is grounded in the literature and based on reflections from conducting service design research studies in such settings where the need for the proposed framework was established. This chapter is concluded with a reflection on the proposed framework and with suggestions for further research.

## Service Design Research

### *Designing Services*

A service is conceptualized as an activity that is designed and provided by one or a group of individuals to meet the needs and benefit of another group or individuals.

There are two main characteristics of a service which are evident in its tangibility and ownership of the provider or receiver (Williams et al., 2008). While tangibility is attributed to the ability of an entity to be perceived by the sense of touch, a service is usually not tied to a physical product. In other words, a service could be intangible as observed in digital systems. The ubiquitous nature of computing devices like smart mobile phones and personal computers and the pervasiveness of the internet have influenced the creation of services enabled by technology (Valderrama Bahamondez et al., 2011). In the current industry 4.0 era, underserved settings are plagued by wicked problems.

Digital or technology-enabled services are obtained and delivered through a digital transaction involving information or consumer product exchange facilitated by software modules over the Internet Protocol (IP) infrastructure (Wulf et al., 2017). The service delivery describes how the design of a service meets the requirements for the consumer or client to participate at different levels and the peculiarity of the context that requires a new or improved service.

The strength of design is in the use of specific methods to translate ideas into tangible products or services. The design of a service is informed by a pressing need to address complex or wicked problems considering the diverse stakeholders and their multiple realities, and the end goal is an optimal solution to meet supra-functional needs (Weightman & McDonagh, 2003). The logic of design thinking is captured by (Easterday et al., 2018):

> If you want to design intervention <X> for the purpose <Y> in a context <Z>, then you are best advised to give that intervention the characteristics <A, B, and C> (substantive emphasis), and to do that via procedures <K, L, and M> (procedural emphasis), because of arguments <P, Q and R>.

With the logic of design as described above, designing a service is prescribed beyond creating a tangible activity for a specific purpose but as a process toward situational transformation in the sense of positive impacts on public and community services, toward a progressive society. To tackle the complexity of tangibility of a service, it is imperative to adopt an interdisciplinary approach such as service design research (Holmlid & Evenson, 2008). Service design research process is explained in the next section.

## *Service Design Research*

Research is about creating new knowledge based on an identified gap supported by existing studies. In design research, new knowledge can be generated by regarding the design process as moving from a problem space to a solution space and to what extent the solution addresses the problem (Vom Brocke et al., 2020)—in other words, transitioning from an existing situation to a desired state, with measurable evidence. They also propose the aspects in both spaces that need to be considered during the design process. These include the context and goodness criteria for the

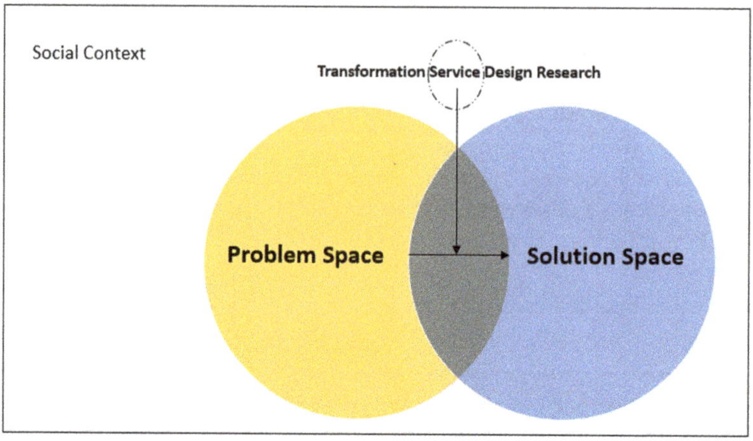


**Fig. 5.1** Transformation design process space of a service

problem space and the representations and process of the new solution space as shown in Fig. 5.1.

The concepts associated with the problem space, in our case the current situation, are needs, goals, and requirements, enabling institutions and stakeholders as highlighted by Maedche, Gregor, Morana, and Feine (2019). Considerations for conducting research as part of the design process means that both design in practice and research processes need to be considered simultaneously as proposed by Hevner (2007), that is, moving between the relevance (application domain); design (design in practice), and rigor (theorizing and grounding) cycles. These cycles were adapted by Thuan, Drechsler, and Antunes (2019) to assist with the formulation of research questions.

The service design and research process are equally human-centered and participatory to enable the co-creation of a new or improved service based on a deep dive into people's realities and adopting a transformative approach to improve the well-being and interest of stakeholders (Patricio et al., 2020). Embarking on a simultaneous yet collective transformative service design and research process should be facilitated using tools that enable flexibility when applied to underserved contexts, as in this chapter, if the need for transformation is to be addressed as intended (Sengewald et al., 2021). The next section presents the need for transformation and the role of service design to achieving a changed situation.

## *The Need for Transformation*

There are already limited or inadequate services in an underserved context. The reality is that the peoples' needs are often not addressed with the available but limited services. The people in such resource-restricted contexts are often indigents, poor, unemployed, and have to deal with the many social ills resulting them being in a state of powerlessness (Walsh, 2008; Curtin et al., 2016). The design of a new or improved services needs to incorporate a deep understanding of the service provider and consumer needs in an underserved setting to commit to the new or improved services by not only addressing the needs but in fact lead to a better situation for the people. Brereton et al. (2014) highlight that often times, indigenous communities are engaged by external researchers without the adequate knowledge of the settings, culturally appropriate methods. Consequently, the resulting benefits are hampered or not fully realized.

Transformative service research considers the well-being of people in underserved settings as well as how engaging with them and co-creating knowledge can be improved toward a positive transformation (Islam et al., 2022). Vulnerable people may find it difficult to express their lived experiences, often not knowing what is possible. A service design research approach provides a pathway for communities and researchers to generate knowledge within the context and cultural settings for which a service is being designed. At the same time, transformative service research has the methods to achieve the goal of a positive transformation (Hurley et al., 2018).

The aim of transformation service research is to create services that lead to uplifting changes and improvements by considering the well-being and quality of life of individuals and communities (Ostrom et al., 2010). The focus of service design for transformation is to create an intervention in the form of a new or improved service through collaborative human-centered action-oriented design (Koskela-Huotari et al., 2021).

## Transformation Service Design Research

In this section, a framework is proposed to guide service design research when transformation is needed to improve the situation and capabilities of people in an underserved setting. The rationale for transformation service research is to create sustainable solutions either a product or service, toward improving the well-being of directly affected persons especially vulnerable and marginalized dwellers in underserved areas (Ostrom et al., 2010). The focus of a transformation service research is optimizing the value clients get from a service, promotion of service co-creation, improved access to a service, and enhanced quality of life. The process of service design and underpinning research is explained as simultaneous and parallel activities.

### *Proposed Transformation Service Design Framework*

The proposed transformation service design research framework is presented in Fig. 5.2 and discussed in this section. There are two main processes, namely, the

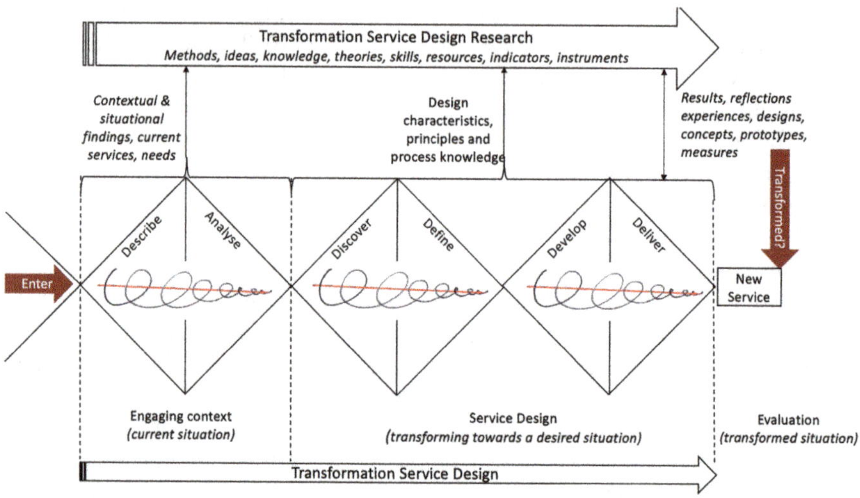


**Fig. 5.2** Proposed transformation service design research framework

service design and research processes. The processes span across three situations to depict moving from a current situation to a transformed situation. The outcome of the service design process is a new or improved service, and from the research process, new knowledge is created about the characteristics of the designed service (representation knowledge); the process of designing it (process prescriptive knowledge); situational knowledge (descriptive knowledge); and transformation knowledge (indicators; transformation insights and measures) (Panwar & Khan, 2021).

The research process provides methods, ideas as the current status of the situation, theories, instruments, indicators, etc., to inform design process. From the design process, new knowledge is created using a theorizing process based on the needs, insights, reflections, and experiences of the people living in underserved settings. As such, transforming of a person's or population's lives and situation through a service is underpinned by ensuring the protection of human rights and dignity (Anderson et al., 2013).

Three situations are considered. First, a current situation is established by understanding the lived experiences of the service provider and people in their reality of the situation. Second, is the immediate state of change when the new or improved service has being designed and introduced towards the desired transformed situation. The service design process is guided by the double diamond with iterative cycles using methods that encourage divergence followed by methods toward convergence of ideas, concepts, and designs. During the design, all the stakeholders of the proposed service including the client of the service are actively involved in the design process. The third situation is the transformed state attributed to the new service available to the people of the setting. Insights gained during the research process will guide similar studies.

It is important to note that the research (transformation service design) and design (service design) processes are not separate processes but are intertwined, where the design is guided by the research and the design provides information for the research. For example, information about the characteristics of the design solution, the new service, and information about the design process can be analyzed to gain insights as part of the research to then in turn guide further design activities.

Although the first diamond of the proposed framework is to provide for a deeper understanding of the situation, it is necessary to prepare and plan for the transformation service design processes, and this is discussed next as entering the field.

## Entering the Field

As part of entering the field, it is important to first prepare for engaging with the situation. From our practical experience in conducting research studies in underserved settings, we found that conducting research in such settings requires meticulous planning, and for that purpose, we consulted with community-based research practices by Kerstetter (2012) and Olshansky and Zender (2008) and adapted these to our own practices informed by the situation (de La Harpe et al., 2015; Debrah et al., 2017; Nyatuka & de La Harpe, 2021; Ogundaini & de La Harpe, 2022).

The first consideration is the importance to conduct ethical research where the participants in such situations are often regarded as being vulnerable. This means that the recruitment of and engaging with the participants need to be carefully considered. It is important to ensure that informed consent is sufficiently understood by the participants. Once the setting has been identified for the study, it is important to identify the right persons who represent the target group for whom the new service is intended. A continuing relationship should then be established and sustained, to ensure participation and that the interest of the target group is considered.

A clear research intention should be communicated, for the target group to feel comfortable with the service design process. This is especially the case when a service is introduced by outside stakeholders, for example, health and well-being services where the need was identified by the government. Although the intention for such a service is to address a real need, the direct stakeholders in that setting still need to agree to participate in the design process. When researchers and designers who are not familiar with the setting are involved in the process, it is important to get acquainted with the cultural practices and contextual factors to know how to engage with participants during the design and research processes. This step is demonstrated by Brereton et al. (2014) and Wyche et al. (2015).

It is also important to use appropriate methods and data collection instruments to ensure equal and active participation and to capture an insightful understanding of the underserved setting. Examples of methods will be the use of design probes to stimulate a dialogue to give opportunity for all persons to express themselves and to get some indication of intention to participate and commit to the design process. Finally, it is important to provide regular feedback to all the stakeholder groups as a way to reflect. Researchers have to continuously reflect on their practices and engagement to be able to adapt to the situation and unexpected encounters. Once the above aspects are considered, the design and research team can proceed with the research and design processes. The first part of the proposed new first diamond is discussed next.

## *Discussion*

The discussion section deals with the proposed two halves of the proposed novel diamond to firstly describe the current situation followed by a comprehensive analysis. The section highlights the tasks that should be done and the techniques to achieve the intended purpose of the tasks.

### Describing the Current Situation

The need to understand the current situation requires a research approach that focuses on situational inquiry and the lived experiences of persons in the situation for which a new service is being designed. Examples of strategies that can be adopted to understand a lived situation are ethnography, interpretive phenomenology, and postcolonial approaches as suggested by Wyche et al. (2015). An ethnographic research strategy allows for empathetic understanding of the context and people's living or working practices, while an interpretive phenomenological research approach focuses on the lived experience of people in a situation (Frechette et al., 2020; Neubauer et al., 2019; Teherani et al., 2015). In addition, postcolonial approaches acknowledge the input of local or indigenous actors as informed by their lived experiences and cultural relevance to settings for which a transformation-oriented service or product is being designed. These strategies create a balanced power dynamic between the service clients and the service designers or researchers.

Frechette et al. (2020) position interpretive phenomenological research as being orientated toward understanding or uncovering lived experiences of individuals as a constant being with others. Baskerville and Myers (2015) propose design ethnography to engage with the context as part of the design process at the beginning to incorporate ethnography as part of the design. Goldkuhl (2011) proposes situational inquiry and theorizing to study local practice to then contribute toward general practice and the body of knowledge for practice research. Design practice loops can be used for renewed practices by addressing problem situations (Goldkuhl & Sjöström, 2018).

During this phase, the context of the current situation is unpacked to establish the factors that need to be considered during the design of the new or improved service. This is done through a situational inquiry to consider situated actions in practice that are typically influenced by the lived experiences of the people in the situation. The situation is bound by place and time in the context of particular and concrete circumstances, and these do not often happen as planned (Ghajargar & Bardzell, 2019). The designer and researcher interpret the current situation with its associated problems, needs, experiences, and opportunities as insights to plan the new service (Dalsgaard, 2014; Maedche et al., 2019).

It is important to determine the situation and stakeholder roles in relation to the current service delivery as a reference point of departure for the new or improved service to be designed. Stakeholder involvement is important to ensure that the new or improved service is desirable, viable, and feasible. During this phase, the relevant stakeholder groups are identified, recruited, and engaged at the start of the design process. The engagement process helps the designer and researcher to have an idea about each stakeholder's interest and power in terms of the proposed service. Once the stakeholder's interests are aligned and agreed as the collective issue at stake, or point of view, with an associated change objective, their participation will be more focused during the design process.

Since transformation research commits to an improved situation through the introduction of a suitable intervention, in the case of a new or improved service, a measure is needed to determine the level of change that will hopefully happen as a result of the new service. An important consideration is the identification of suitable indicators and instruments to use for the baseline measures at the beginning of the service design process. The indicators should be relevant to the situation and the measures be agreed upon by the relevant stakeholders' consensus. The same instruments should then be used to measure the level of change at different timelines once the service is infused into the lives of persons in the changed situation, to establish if transformation has happened and to what extent.

During this phase, it is the researcher or designer who should pay as much attention to details as possible that could describe the situation, and hence the nature of the methods at this stage is to diverge. The divergence would ensure that sufficient details about the needs, experiences, and opportunities that are required to analyze the situation are captured and understood toward the design process of a new or improvement of an existing service.

## Analyzing the Current Situation

Once sufficient details to describe the situation are collected, the data is analyzed to ensure that the change objective are synthesized and achievable. It may also be necessary to scope the expected service at this stage to ensure that it is possible within the boundaries associated with the situation and issue at stake within an underserved setting. Often, resources are limited, and participants may not have sufficient time to commit for their participation in the design process.

Other technological factors typically associated with underserved contexts such as internet connectivity, availability of technology, and unstable electricity need to be identified for the viability of the new service. The research process during this stage is focused on understanding the situation using strategies typically associated with interpretive research and application of tools and instruments that would capture qualitative data required for designing the new service. The analysis process is considered to be convergent as it focuses on the most relevant data to derive findings for the next stage within the constraints of the situation.

## *Researching and the Design of Services*

After the introduction of the new diamond, the UK Design Council's Double Diamond (2005) for designing services is followed. Although the diamonds appear sequentially, the design process follows several iterative cycles. However, the outcome of each cycle is closer to the desired outcome where divergent methods assist with exploring different possibilities. Convergence methods on the other hand refine the identified possibilities to consider a few prioritized possibilities. The overall design process can be regarded as converging toward the design solution with cycles of divergence and convergence to explore more or less possibilities. The outcome of the discover and define diamond could translate to be a proposed service concept, which can be evaluated with different case studies.

During the Discover phase, new ideas are considered, and this serves as an opportunity for the service providers and clients to propose service solutions from their perspectives. The methods used stimulate discussions to encourage participants to propose unprecedented ideas. Then, in the Define phase, the proposed ideas are reviewed jointly by the service providers and client, and prioritized to define the service concepts based on the selected options. The outcome would be a feasible service concept in response to the issue at stake. To realize the change objective, in the development phase, different possible solutions for the defined concepts are considered, prototyped, and tested. This is again achieved through different iterative cycles of trial-and-error process (build it quickly, break it, fix it). At the end of this phase, a demonstrable prototype service should be available.

The final phase of the design process is the Deliver phase. In this phase, the new or improved service is implemented in practice to evaluate how it impacts the situation and the persons' lived experiences of providing or using the service. It is important to emphasize that during the design process, all the stakeholders are involved with the designer facilitating the process by using suitable and relevant methods to encourage participation. The design process produces the new or improved service as happens all the time in practice. However, by adding research to it, research about the design components and design process is also done and adds insights to the design of services and the process that could be used to guide general practices of designing services. The situational insights guide the design process of services in difficult situations or peculiar underserved settings.

### *After Delivering New or Improved Services*

When the aim is to transform a situation, one cannot stop at the delivery of the new or improved service but need to consider to what extent the initial situation has changed and whether the desired transformation was achieved. The same instruments used during the baseline study should be used to measure the same indicators to make a comparison possible. However, the evaluation of the changed situation should be done with caution because introducing a new service may produce unintended consequences or positive results that may change, as the novelty of service wears off. Therefore, the use of the new service should be evaluated over time to establish to what extend it has infused in the daily practices of the people in the situation.

We believe that there is a need for another diamond after the original Double Diamond framework, to investigate the use of the new service beyond adoption. The additional diamond would assist researchers to establish if the new service has become infused in the practices of the people and whether the situation has reached a sustainable transformed state based on the new or improved service. Such a fourth diamond could consider the deployment of the service in practice, the new capabilities evolved, followed by establishing to what extend the service is incorporated as part of daily living. During this final phase of the design process, it is important to determine the level of change toward transformation. In this case, the focus is on an ex-post evaluation after the design process is completed and the research changes its focus to an evaluation approach of the service in practice (Baskerville et al., 2009).

## Conclusion

This chapter has described a proposed service design framework that can be applied by service providers, product designers, and academics to conduct research in underserved settings with a focus on positively transforming the lived experiences of the setting dwellers. The contribution is evident in the introduction of an additional diamond to the original service design Double Diamond framework by the UK design council. The additional diamond highlighted a need to understand the current situation of the underserved settings with particular focus on situational inquiry, enabled participation, and the lived experiences of persons affected toward achieving the sustainable development goals.

However, more research is needed to align our proposed first diamond to other triple diamond frameworks. Aligning our thinking to a systemic thinking approach could also help dealing with complexities associated with societal challenges and long-term implications of designed solutions. The results and insights gained from our work can be of benefit by aligning it to the work, for example, of the SDN,

Bijl-Brouwer, Jagtap, and other colleagues contributing to the development and refinement of the service design research framework.

The limitation of this work is that the proposed transformation service design research framework has not been tested. Therefore, we propose that future service design studies in the underserved setting of sub-Saharan Africa should deductively evaluate the framework and refute or support the arguments for the additional diamond and establish evolving capabilities of both service users and designers. Our proposed transformation service design research framework adds the novelty of engaging stakeholders, encouraging community participation as co-designers to get a holistic understanding of the underserved context and the need to measure the impact of a new or improved service for sustainable development and transformation.

## References


Anderson, L., Ostrom, A. L., Corus, C., Fisk, R. P., Gallan, A. S., Giraldo, M., Mende, M., Mulder, M., Rayburn, S. W., Rosenbaum, M. S., & Shirahada, K. (2013). Transformative service research: An agenda for the future. *Journal of Business Research, 66*(8), 1203–1210.

Baskerville, R. L., & Myers, M. D. (2015). Design ethnography in information systems. *Information Systems Journal, 25*(1), 23–46.

Baskerville, R., Pries-Heje, J., & Venable, J. (2009, May). Soft design science methodology. In *Proceedings of the 4th international conference on design science research in information systems and technology* (pp. 1–11). Association for Computing Machinery.

Bijl-Brouwer, M. V. D., & Malcolm, B. (2020). Systemic design principles in social innovation: A study of expert practices and design rationales. *She Ji, 6*(3), 386–407.

Brereton, M., Roe, P., Schroeter, R., & Lee Hong, A. (2014, April 26). Beyond ethnography: Engagement and reciprocity as foundations for design research out here. In *Proceedings of the SIGCHI conference on human factors in computing systems* (pp. 1183–1186). Association for Computing Machinery.

Curtin, K. A., Schweitzer, A., Tuxbury, K., & D'Aoust, J. A. (2016). Investigating the factors of resiliency among exceptional youth living in rural underserved communities. *Rural Special Education Quarterly, 35*(2), 3–9.

Dalsgaard, P. (2014). Pragmatism and design thinking. *International Journal of Design, 8*(1), 143–155.

de La Harpe, R., Korpela, M., & Zyl, I. V. (2015, August). Co-design for development: Lessons learnt from an information systems project in underserved communities. In *Scandinavian conference on information systems* (pp. 61–74). Springer.

Debrah, R. D., de La Harpe, R., & M'Rithaa, M. K. (2017). Design probes and toolkits for healthcare: Identifying information needs in African communities through service design. *The Design Journal, 20*(sup1), S2120–S2134.

Design Council UK. (2005). The Design Process. Available at https://www.designcouncil.org.uk/our-work/skills-learning/tools-frameworks/framework-for-innovation-design-councils-evolved-double-diamond/ Accessed 4 April 2023.

Easterday, M. W., Rees Lewis, D. G., & Gerber, E. M. (2018). The logic of design research. *Learning: Research and Practice, 4*(2), 131–160.

Frechette, J., Bitzas, V., Aubry, M., Kilpatrick, K., & Lavoie-Tremblay, M. (2020). Capturing lived experience: Methodological considerations for interpretive phenomenological inquiry. *International Journal of Qualitative Methods, 19*, 1609406920907254.

Ghajargar, M., & Bardzell, J. (2019). What design education tells us about design theory: A pedagogical genealogy. *Digital Creativity, 30*(4), 277–299.

Goldkuhl, G. (2011). The research practice of practice research: Theorizing and situational inquiry. *Systems, Signs & Actions, 5*(1), 7–29.

Goldkuhl, G., & Sjöström, J. (2018, June). Design science in the field: Practice design research. In *International conference on design science research in information systems and technology* (pp. 67–81). Springer.

Hevner, A. R. (2007). A three cycle view of design science research. *Scandinavian Journal of Information Systems, 19*(2), 4.

Holmlid, S., & Evenson, S. (2008). Bringing service design to service sciences, management and engineering. In *Service science, management and engineering education for the 21st century* (pp. 341–345). Springer.

Hurley, E., Trischler, J., & Dietrich, T. (2018). Exploring the application of co-design to transformative service research. *Journal of Services Marketing, 32*, 715.

Islam, S., Muhamad, N., & Sumardi, W. H. (2022). Customer-perceived service wellbeing in a transformative framework: Research propositions in the area of health services. *International Review on Public and Nonprofit Marketing, 19*(1), 219–245.

Jagtap, S. (2022). Codesign in resource-limited societies: Theoretical perspectives, inputs, outputs and influencing factors. *Research in Engineering Design, 33*(2), 191–211.

Kerstetter, K. (2012). Insider, outsider, or somewhere between: The impact of researchers' identities on the community-based research process. *Journal of Rural Social Sciences, 27*(2), 7.

Koskela-Huotari, K., Patrício, L., Zhang, J., Karpen, I. O., Sangiorgi, D., Anderson, L., & Bogicevic, V. (2021). Service system transformation through service design: Linking analytical dimensions and service design approaches. *Journal of Business Research, 136*, 343–355.

Maedche, A., Gregor, S., Morana, S., & Feine, J. (2019, June). Conceptualization of the problem space in design science research. In *International conference on design science research in information systems and technology* (pp. 18–31). Springer.

Neubauer, B. E., Witkop, C. T., & Varpio, L. (2019). How phenomenology can help us learn from the experiences of others. *Perspectives on Medical Education, 8*(2), 90–97.

Nyatuka, D. R., & de La Harpe, R. (2021). Service design as a catalyst for patient-centered eHealth innovation: An architectural design framework for cloud-based maternal health information service in underserved setting. *International Journal of Information System Modeling and Design (IJISMD), 12*(3), 62–85.

Ogundaini, O., & de La Harpe, R. (2022). The interplay between technology performativity and health care professionals in hospital settings: Service design approach. *JMIR Formative Research, 6*(1), e23236.

Olshansky, E., & Zender, R. (2008). The use of community-based participatory research to understand and work with vulnerable populations. In *Caring for the vulnerable: Perspectives in nursing theory, practice and research* (pp. 269–275). Jones & Bartlett Learning.

Ostrom, A. L., Bitner, M. J., Brown, S. W., Burkhard, K. A., Goul, M., Smith-Daniels, V., Demirkan, H., & Rabinovich, E. (2010). Moving forward and making a difference: Research priorities for the science of service. *Journal of Service Research, 13*(1), 4–36.

Panwar, T., & Khan, K. (2021). Integrating design thinking in service design process: A conceptual review. *Journal of Design Thinking, 2*(1), 23–36.

Patricio, L., Sangiorgi, D., Mahr, D., Čaić, M., Kalantari, S., & Sundar, S. (2020). Leveraging service design for healthcare transformation: Toward people-centered, integrated, and technology-enabled healthcare systems. *Journal of Service Management, 31*, 889.

SDN. (2022). Service design network. https://www.service-design-network.org/community-knowledge/combining-service-and-systemic-design-in-norways-public-sector. Accessed 20 Nov 2022.

Sengewald, T., Jalowski, M., & Schymanietz, M. (2021). A software ecosystem for the development of digital service design tools: A conceptual framework. In F. Ahlemann, R. Schütte, & S. Stieglitz (Eds.), *Innovation through information systems. WI 2021* (Lecture notes in information systems and organisation) (Vol. 46). Springer. https://doi.org/10.1007/978-3-030-86790-4_15

Teherani, A., Martimianakis, T., Stenfors-Hayes, T., Wadhwa, A., & Varpio, L. (2015). Choosing a qualitative research approach. *Journal of Graduate Medical Education, 7*(4), 669–670.
Thuan, N. H., Drechsler, A., & Antunes, P. (2019). Construction of design science research questions. *Communications of the Association for Information Systems, 44*(1), 20.
Valderrama Bahamondez, E. D. C., Winkler, C., & Schmidt, A. (2011, May 7). Utilizing multimedia capabilities of mobile phones to support teaching in schools in rural panama. In *Proceedings of the SIGCHI conference on human factors in computing systems* (pp. 935–944). Association for Computing Machinery.
Vom Brocke, J., Winter, R., Hevner, A., & Maedche, A. (2020). Special issue editorial–accumulation and evolution of design knowledge in design science research: A journey through time and space. *Journal of the Association for Information Systems, 21*(3), 9.
Walsh, D. (2008). Helping youth in underserved communities envision possible futures: An extension of the teaching personal and social responsibility model. *Research Quarterly for Exercise and Sport, 79*(2), 209–221.
Weightman, D., & McDonagh, D. (2003, June). People are doing it for themselves. In *Proceedings of the 2003 international conference on designing pleasurable products and interfaces* (pp. 34–39). Association for Computing Machinery.
Williams, K., Chatterjee, S., & Rossi, M. (2008). Design of emerging digital services: A taxonomy. *European Journal of Information Systems, 17*(5), 505–517.
Wulf, J., Mettler, T., & Brenner, W. (2017). Using a digital services capability model to assess readiness for the digital consumer. *MIS Quarterly Executive, 16*(3), 171–195.
Wyche S., Dillahunt T. R., Simiyu N., & Alaka S. (2015, September 7). "If god gives me the chance I will design my own phone" exploring mobile phone repair and postcolonial approaches to design in rural Kenya. In *Proceedings of the 2015 ACM international joint conference on pervasive and ubiquitous computing* (pp. 463–473). Association for Computing Machinery.

# Chapter 6
# Service Prototypes as the Setting for Product Innovation Agenda

**Aguinaldo dos Santos, Johan Blomkvist, and Alessandra Caroline Canfield Petrecca**

## An Epistemological Perspective on Product, Service and System Requisites

Services can be the main source for a product design brief. Ryd (2004) defines briefing as the communication of instructions on intentions and objectives, with the result regarded both as a product (the brief) and a process (the briefing). An ideal briefing process must create a platform for a clear understanding of user needs and ensure that the final product/service meets these needs. From a strictly design perspective, service is a new type of design material (Blomkvist et al., 2016) that is often made tangible using different visualisation techniques (Segelström, 2013). One common technique is the customer journey map, where all interactions between firms and their customers are captured in the so-called touchpoints: visual representations with the main information about each such interaction.

In an integrated process involving both products and services, the identification of requisites and design proposals cannot be seen solely as a technical and rational problem-solving task (Barret et al., 1999; Luck et al., 2001) but as a social process based on interaction, learning and the joint creation of meaning. The inner characteristics of a service impose a high complexity on this process: high intangibility, high variability/heterogeneity, inseparability of production and consumption and perishability (a service cannot be stored) (Lovelock, 1983; Zeithaml et al., 1985). These characteristics affect the scope and nature of the service design

A. dos Santos (✉) · A. C. C. Petrecca
Federal University of Paraná (UFPR), Curitiba, Brazil
e-mail: asantos@ufpr.br

J. Blomkvist
Linköping University (LiU), Linköping, Sweden
e-mail: johan.blomkvist@liu.se

 
U. Z. A. Hamid, M. Suoheimo (eds.), *Service Design for Emerging Technologies Product Development*, Springer Series in Design and Innovation 29,
https://doi.org/10.1007/978-3-031-29306-1_6

requisites/requirements and will have a direct impact on the translation into product requisites. In such a process of joint meaning making, elements of service such as customer interactions, behaviours, technical solutions, products and motivations emerge. This can then be translated into an overarching strategy of coherent, consistent and appropriate resource integration in service systems. That strategy then informs not only choices on the service level but also on the level of products.

Requirements for service design address the functionality and form of services, in issues such as usefulness, usability and desirability from a client's perspective, contemplating the service provider needs on effectiveness, efficiency and differentiation (Mager, 2008). The most widespread categories for service requisites include accessibility (also availability), competence (accuracy, knowledgeability), communication (transparency), reliability, courtesy, credibility (integrity, reliability), empathy (attention, care), responsiveness (readiness, promptness, flexibility, speed, timelessness), safety (privacy) and tangibles (cleanliness, appearance, thermal/acoustic/light comfort, functionality) (Parasuraman et al., 1985; Silvestro, 2005). On the aesthetic dimension of services, Löwgren (2009) presents four propositions that are relevant for the requisite definition of services: pliability (flexibility); rhythm (of interaction); dramatic structure (dramatic tension along the time) and fluency (the grace with which multiple tasks are fulfilled). Other concepts include pleasure, fun (Blythe & Hassenzahl, 2005), surprise and charm (Mccarthy & Wright, 2003). Cho (2013) includes in the quality roll respect, consideration, friendly contact, credibility and politeness.

In contrast to the content of service requirements categories, Garvin (1987) proposed eight dimensions of quality by which consumers judge products: performance, features, reliability, conformance, durability, serviceability, aesthetics and perceived quality. Alternatively, Lin and Cheng (2017) propose that products have instrumental (e.g. utility, functionality or usability) and noninstrumental (e.g. aesthetic, symbolic or motivational aspects) qualities, with interaction and perception affected by the transient internal state of the user him/herself. Product attributes/requisites can be also classified as hard-attributes, which are objective and measurable (e.g. strength, speed, weight and price) and soft attributes, which are subjective and emotional (e.g. attractive/ugly, modern/old, masculine/feminine). Another common classification for product design attributes/requisites is the distinction between 'functional design requirements' and 'aesthetic design requirements'. The first provides a direct indication of what the product should do, and it is expressed in a functional language (e.g. easy to hold) (Van De Poel, 2009), whilst the second deals with the perceptive qualities such as balance, colour, style, movement, pattern, scale, shape and visual weight.

Although there are clear epistemological differences among service and product requirements, it is also clear that products that represent 'touchpoints' on a customer journey need to be harmoniously aligned with the service concept. Their functional and aesthetic characteristics as well as their position in space and time do affect the quality of the customer experience. The issue is quite complex, as various artefacts representing touchpoints throughout the customer journey are often not controlled by the primary service provider and quite frequently are owned/produced by a variety of different organisations. Different brand strategies among the product manufacturers might present conflicting goals with the requisites intended for the service touchpoints. Without such alignment between product design and service design, touchpoints might present reduced contributions or even opposing impacts in the service proposition (e.g. a service concept based on high flexibility whilst the artefact does not allow any functional or aesthetic flexibility).

There is a limited amount of studies regarding the conversion of requisites from services to products. Liu (2000) exemplifies the complexity of this conversion by showing that a service might present counterintuitive intents: be healthy but 'unattractively displeasing' (e.g. rehabilitation service for drug addiction) or even reckless thrills and potentially 'harmful' (e.g. the full experience of a roller coaster). In order to bring light on this issue, the next section explores the alternative directions that such conversion can take in a design project.

## Alternative Approaches for Flowing Down System, Service and Product Requirements

### *Flowing Down Requisites from System to Services/Products*

Ideally, the requisites should scale down from a system design standpoint into the mix of products and services, as illustrated in Fig. 6.1. The holistic and strategic nature of a system perspective can result in more effective definitions for product and service requirements, resulting in more efficiency by promoting their synergy and more effectiveness by searching for higher coherence on the production of value for the end customer.

Baines et al. (2007), in his proposal of an evolution of the product-service system concept, argues that the 'servitization of products' and 'productization of services' imply moving away from isolated design of products or services into products designed to services or services designed to products and, on the more ideal situation, an integrated design of a product-service system (Fig. 6.2).

System requisites are by definition much broader in scope and present a much richer embedded information. They involve issues such as the required profile

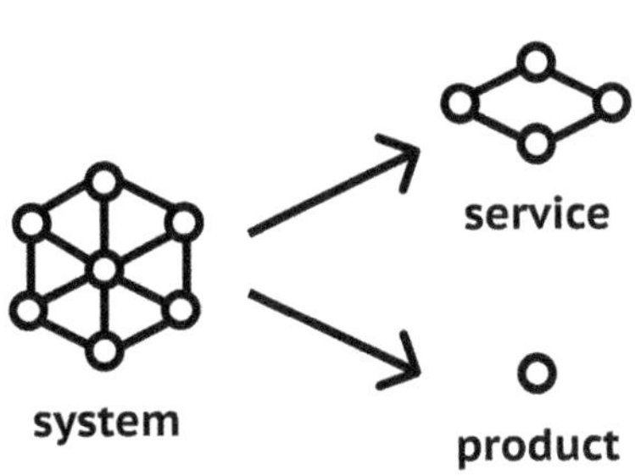


**Fig. 6.1** Flowing down requisites from the system to services and products

**Fig. 6.2** Evolving integration of products and services. (Based on Baines et al. (2007))

and roles of key system stakeholders, general characteristics expected for their interactions, the direction and content of their flows (e.g. material, information and capital) as well as the general characteristics of the mix of products and services that they should provide to a given user. Furthermore, requirements at the system level do consider wider issues such as innovations in the business processes and, also, strategies for holistic transition towards sustainability. Such requirements should be generated by inclusive approaches that involve customers and relevant stakeholders, or they will fail to identify perspectives that are integral to both service and product development.

The research of Silva (2010) illustrates a breakdown of requisites from the system to service and product requirements. Her research focused on systems of products and services in Brazil for enabling remote work. It was motivated by the growing demand for home offices in some particular markets in Brazil, enabled by legislation that provided the legal framework for virtualisation of the workspace (ten years later the impact of the pandemic became the central driver for home office expansion around the world). Her fieldwork involved the definition of system requirements, with a focus on the environmental dimension of sustainability. One of the identified requirements dealt with enabling eco-efficiency on the overall system infrastructure. The translation of this system requisite to products in the home office (e.g. furniture) resulted in requirements such as 'enable functional upgradability of furniture', 'facilitate easy assembling/disassembling of parts and components' and 'adoption of atoxic/biocompatible materials'. Also, the conversion of system requirements into service requirements resulted in issues such as 'readiness on furniture maintenance', 'provision of traceability on furniture waste destination' and 'providing safety by background checking on maintenance personnel'.

However, as stated before, the flow of requirements from the system to products can encounter barriers as usually depends on the availability of services and product offers by the organisation, its partners or other stakeholders on the market. For instance, Silva (2010) stated that it was difficult to find furniture or electronic manufacturers around her case study that reached the basic system requirements.

## *Flowing Down Requisites from Products to Service/System*

This approach is relevant particularly when technological innovations are involved. In such instances developing a (meta) product, an initial product prototype may be a

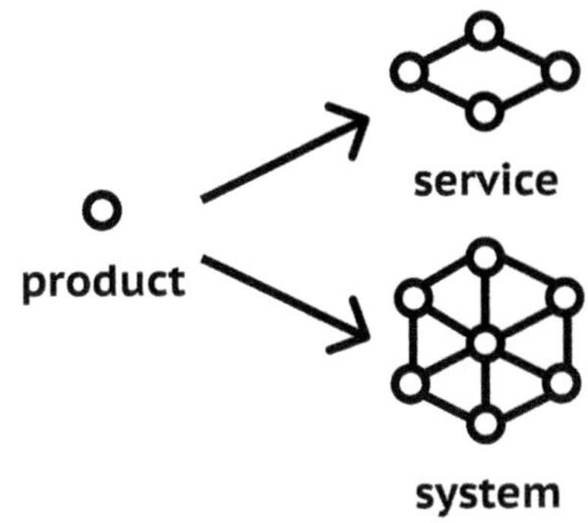


**Fig. 6.3** Flowing down requisites from the product to the service and system

necessary phase in order to fully understand the affordances of the new technology and its implications on value co-creation. In a way, such a physical artefact can act as a probe, through which users interact, record their perceptions and express their thoughts and ideas (Mattelmäki, 2006). Another possible approach is 'Situated and Participative Enactment of Scenarios' (SPES) where the artefact, for example, a rough prototype, is placed in a real environment. Then, a potential user is provided with this artefact, and it is followed/observed by designers in daily life activities. The user is asked to perform a possible use of the new artefact in their everyday environment, envisioning ideas of services and product features (Iacucci & Kuutti, 2002). From a sustainability perspective, such pilot products are also useful to understand the potential positive and negative environmental, economic and social impacts of the new offer. Hence, in such situations, the flow of requirements for both system and services has on the physical artefact the starting point to requirement capture, as illustrated in Fig. 6.3.

This approach is relevant when knowledge is scarce on radically new technologies associated with production and consumption. An example is the research of Serbena (2013) which had as the main research object a 'product-service system for light provision on low-income households in Brazil'. It specifically involved LED technology which, at the time of the research, was widely absent in low-income households in the country. Thus, Serbena (2013) had to develop a LED lamp that took into consideration the dynamics of light usage in Brazilian poor households. Among the product requirements was the need for high flexibility regarding the position of the lamp on the house as the daily activities in those homes are highly dynamic.

In the same room, for instance, the daily activities might involve work, sleeping, studying and social gatherings. He designed two alternatives: one manufactured in metal, aiming for centralised production, and one in ceramic, aiming for distributed production. The prototype was installed in seven low-income households, and the contact of users with the product provided key information to understand system requirements such as the need to integrate small local actors for the assembling and maintenance services of the LED lamps and to consider the provision of local solutions to the end of the life cycle services for LED lamps. It also enabled the identification of key service requirements such as energy consumption monitoring (as energy represents an important portion of the family budget expenditures) and safety services that benefit from the product movement sensors since the target users often work long hours far away from their homes. Although nowadays LED products are widely available, including to low-income consumers, it is worth emphasising that most product and service requirements identified on Serbena's (2013) research remain relevant and are still widely absent on the low-income Brazilian market.

## *Flowing Down Requisites from Service to System/Product*

As stated at the beginning of this chapter, services can be the starting point to device the system and product requirements, as illustrated in Fig. 6.4. This is often the case in organisations that offer services as their core business and adopt a user-centred design to define their business characteristics and search for continuous improvement. Adopting service design practices in these companies implies the design of the user experience and, also, the business process itself required to deliver such experiences (Mager, 2008). Thus, the requirements of service should point to some key elements of the expected directions of the system, as well as the main characteristics of the tangibles that constitute the touchpoints along the customer journey.

This service approach to understand system and product requirements can be illustrated by the undergraduate class assignment developed by Leme and Frare (2022). The challenge presented to them was to conceive a solution that generates income for immigrant women coming from Venezuela to Curitiba, Brazil. They come up with a concept based on the notion of a pop-up cinema, integrating a street food market with gastronomic features of Venezuelan street food. Hence, their concept merges the provision of cultural entertainment via projections of Venezuelan movies, Venezuelan street food and the opportunity for customers to interact with the Spanish language. The value provided to the client would be to get a more profound encounter with the Venezuelan culture whilst tasting Venezuelan

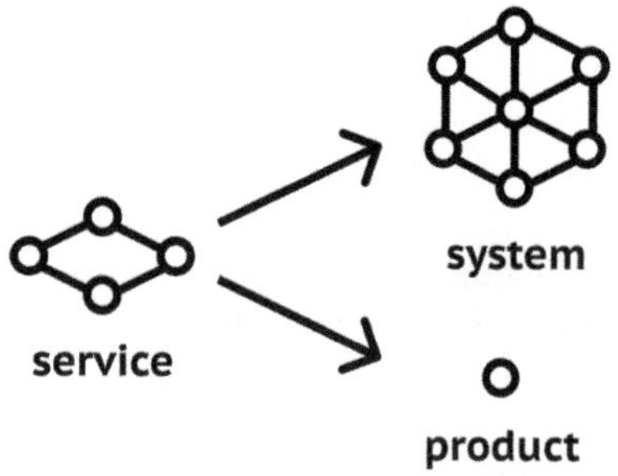


**Fig. 6.4** Flowing down requisites from the service to the system and its products

street food. Therefore, the service concept has an underlying social contribution regarding the reduction of xenophobia against immigrants.

Once the service concept was established, it was possible to set requisites for the products encountered during a customer journey, for example, the structure of the pop cinema itself, the furniture characteristics, the interface between the pop cinema and the food truck and the signalling system – as well as the system-wide dimension – for example, defining who would produce the food; which stakeholder would select and rent the movies, which stakeholders would be involved to provide access to free urban spaces required by the pop cinema and how to involve the immigrants into the system.

## Service Prototypes: A Tool to Gather and Refine User's Requisites

We have illustrated how requisites can flow in systems of products and services with examples. We now look specifically at service prototyping and how it can help gather useful knowledge in design projects.

### *Defining Service Prototypes*

Blomkvist (2014) defines service prototyping as an activity that can explore, evaluate and communicate future service situations based on shared representations of design ideas. A shared representation means a prototype that manifests service and makes it physical and tangible. Representing service gives them form and makes them available for design. As precursors of the final design (Sanders & Stappers, 2012), prototypes can manage existing knowledge and present possible futures (Kimbell & Bailey, 2017). They can simulate parts of the project without the cost of final implementation (Houde & Hill, 1997). Also, it assists in thinking, problem-solving, communication, understanding and testing possibilities (Plattner, 2010).

Whilst designing a service, it is possible to prototype in different levels: isolated artefacts, moments of interaction, contexts or entire customer journeys. Prototyping entire journeys is the most holistic perspective which allows designers to balance requirements of different elements of service, aiming for a more coherent and cohesive experience (Blomkvist, 2016). One technique that allows exploration on the service level is service walkthrough, where key touchpoints are identified and then represented in a tangible way that enables relevant stakeholders to walk through each step of a service (Blomkvist & Bode, 2012). A service walkthrough involves enacting different roles, like customers, salespersons and domain experts, based on different assumptions.

Service prototypes often involve a variety of artefacts throughout the customer journey, distributed in time and space (Blomkvist, 2014). Hence, service prototypes can simulate and test materials, environments or person-to-person/person-to-artefact/person-to-organisation relationships that represent one or more aspects of a service (Hoss, 2014). The interaction with the designed service can be observed when representing the place, situation and condition (Diana et al., 2009), providing direct insights on user requirements associated with isolated products, a group of products and the impact of synesthetic interactions in a given servicescape. The next section presents the roles of service prototypes and how they can assist the requirement capture for a given product design project.

## *Different Roles of Service Prototypes Throughout the Design Process*

Prototypes can be applied for more instructional or communicative purposes (Blomkvist, 2014). In this process of capturing product, service and system requirements, a service prototype can play a number of roles. Highlighted below are those roles and how they aid in gathering requirements for the design:

- *Support the 'understanding' phase*: Service prototypes can be used to understand a given problem even before the requisites of a system, product or the service itself is defined. Indeed, a provocative prototype can be used, for example, to enable physical interactions and, by doing so, facilitate problem definition as well as work as a bridge to foster dialogue regarding the plausibility, viability, barriers and attractiveness of newer solutions. The Smart Textile Services project, by Kleinsmann and Bhömer (2020), uses a 'provotype' as a boundary object in an initial session to aid users and designers to understand the limitations and opportunities of new technology and services for people with dementia in an elderly care organisation. From this initial session, the participants could understand the integrations with different parts of the service (user, products, actual context, a possible mix of technologies, etc.). Prototypes with a role focused on enabling understanding allow the designer to think from a personal experience (immersion), empathising and getting acquainted physically with the situation, grounding acknowledged requirements and learning what is possible in that context.
- *Assisting the 'ideation' phase*: Service prototypes can provide requirements for the ideation process. When converting user requirements into initial service concepts, for example, the prototype acts as a more effective communication channel for the user to express his/her agreement of and with the defined requirements, suggesting new solutions and pointing to new requirements, not only for the service itself but also for the artefacts involved and system-wide issues. In this manner, the prototypes can act as a filter, as proposed by Lim et al. (2008), consolidating and integrating essential elements, contributing to refine

the project scope. In an initial representation, one can manifest functionality and appearance variables to be mandatory in the final project. As an example, Alves et al. (2018) present a project aimed at an elderly residence called NurseAid, developed in an undergraduate class of service design. In this project, a bodystorming session was used as the strategy to conceive service prototypes. The session considers a list of requirements, grouped in mandatory and desirable, all of them raised during the problem definition research phase. From this creative session, the required touchpoints for the service were identified, and it was possible to initiate the product creative process already on the bodystorming session. This study has shown that it is possible to simultaneously carry out a creative session for both product and service design by take advantage of the characteristics of service prototype.

- *Enabling the test of concepts*: Service prototypes for testing usually contain more elaborate ideas, presenting probable solutions rather than just fundamental requirements (Blomkvist, 2014). Ideas can be tested and new requirements may be identified, helping to anticipate risks and barriers, contributing to refine requisites for both system and products. This role might be the most common application of service prototypes, fulfilling the need to validate the project concepts in a holistic manner. A service prototype described by Hoss (2014), focused on the customisation of light solutions to low-income households, adopted a high-fidelity approach, with a realistic representation of the service. In this prototype, Hoss (2014) identified that operational factors such as opening period, logistics constraints to reach the service, including characteristics of the service personnel may influence the service effectiveness. Moreover, for the touchpoints, it was identified that the space was too small and some of the physical artefacts adopted on the prototype were not suitable for such an environment. Hence, product requirements derived from this service prototype included a review of form and functionalities in order to enable its operation on small spaces. In another example, a service walkthrough approach was used as an evaluation method of a new technology-enhanced service experience (Arvola et al., 2012). The aim was to enhance the experience of visiting Astrid Lindgrens Näs using augmented reality. The service walkthrough used mobile phone mock-ups and separate transparent screens to illustrate the interaction with the AR solution. The walkthrough evaluated both the service and the product elements in its rough, early stages and provided valuable knowledge on both levels.

According to Stickdorn et al. (2018), high-fidelity service prototypes help to consolidate more structured and finished ideas, and it is more about refinements as final requirements. Also, they are usually applied at the end of projects since they can be costly. Low-fidelity (rough) prototypes such as cardboard representations can also be used to test service/product ideas and their integration. In this case, more disruptive requirements and significant changes for the products can be evaluated, and, like any other form of service prototyping, it can result in the identification of new service and product requirements. Prototypes of low fidelity can encapsulate the main designer ideas and, at the same time, let the concept and its fine details be revised or completed using the imagination of the user him/herself.

As it has been shown so far in this chapter, although epistemologically different, services, systems and products, the definition of their requisites should follow an integrative and collaborative approach in order to reach coherent and more effective propositions. Whilst there is a relatively reduced amount of knowledge regarding service prototyping when compared with the highly consolidated topic of product prototyping, the evidence from the recent studies point to the great benefit of considering service prototypes as a necessary phase on the definition of product requisites.

## The Potential of Emerging Technologies for Requirement Capture and Refinement During Service Prototyping

Emerging digital technologies (e.g. IoT, Big Data, data mining, blockchain, augmented reality, virtual reality, AI/deep learning/machine learning) offer new possibilities for a more effective and efficient assessment of service prototypes. These technologies can be either embedded into the service concept or provisionally adopted solely for assessing the service prototype, converting conventional touchpoints into interactive touchpoints and allowing a higher level of user participation in the co-creation process. Their implementation on the service can adopt both a visible and invisible interaction approaches (Gkouskos & Linde, 2016).

Among these digital technologies perhaps, the most relevant for the context of service prototyping is IoT (Internet of Things). IoT can be defined as intelligent artefacts connected via the Internet, where artefacts exchange information with each other, with their users and with their database (Wortmann & Flüchter, 2015). An example comes in the form of wearable technologies and automatic lifelogging. Arvola et al. (2017) used a wearable camera and heart rate monitor to collect data about people in everyday situations. The data was then used to create an experience timeline together with self-reported key experiences, heart rate, decisions and valence. The outcome was very detailed and rich data that creates multifaceted images of lived experiences. Approaches like these facilitate the identification of new opportunities for innovation on service offers, supporting changes towards more sustainable behaviours, including those involving the automatisation of decisions on issues that may not be apparent, natural or habitual for users (Bocken et al., 2019).

IoT can enable service design assessment on the temporal and spatial dimension of customer flows, with sensors embedded into touchpoints, with or without requiring human intervention to create or supply data content. For instance, specific IoT applications such as eye tracking allows the development of maps with heat islands on interaction, which can later be adopted as criteria to select the most critical touchpoints on the customer journey; face recognition devices can enable real-time evaluation of user mood throughout the customer journey, with a high level of detail and speed that would be impossible through conventional analogic tools such as emotional wheel. When combining stationery sensors with wearables (e.g. smart glasses, smart watches, activity monitors, heart rate monitors), IoT can further leverage the quality and volume of data that is collected from a service prototype. Very importantly, it can increase the speed of decision-making during the service prototyping and enable designers to revisit details from the pool of Big Data produced by IoT.

IoT and other digital technologies benefit from their integration with artificial intelligence (AI). AI refers to systems designed by humans that act on the physical or digital world by perceiving and interpreting the environment, reasoning on the knowledge derived from structured or unstructured data and, from this cumulative learning process, deciding the best course of action to achieve a given goal (EC, 2018). It can help designers to deal with the growing amount of data gathered at increasingly high speed (Nasirian et al., 2017).

AI's potential functions on the design process fall into one of three groups: 'assess' (e.g. to record the flow of people in a hotel lobby by the second), 'infer' (e.g. define in real-time probable scenarios regarding the amount of people and behaviours) and 'respond' (e.g. redistribute the task priorities on the hotel check-in service in response to sudden rise in the amount of clients) (Vockea et al., 2019). On its various dimensions like thinking humanly, thinking rationally, acting humanly or acting rationally (Russell & Norvig, 2003), AI can be used to carry out specific tasks during a service prototyping evaluation as well as interacting with users and other information systems. Frick et al. (2019) alert that the use of AI in service design must be authentic, particularly regarding communication. Furthermore, since it can be used to generate recommendations and decisions or even fulfilling tasks autonomously, it has to be as trustworthy and transparent as possible (Wüenderlich et al., 2012; Wüenderlich & Paluch, 2017; Frick et al., 2019).

## Conclusion

This chapter has demonstrated the multiple directions on which a briefing can be carried out when navigating the orbits of systems, services and products. The authors argue that ideally the system requisites should be cascaded down into products and services requisites. Similarly, in order to achieve system coherence, service requisites need to be translated into product requisites and vice versa. This is not an easy task as the nature and scope of system, service and product requisites are quite different, and yet there are no consolidated guidelines or tools to support such translation.

Due to the higher complexity and scope of service design, with its range covering multiple artefacts throughout the customer journey, its briefing should have precedence over a product design briefing whenever a project intends to obtain a coherent mix of products and services. In order to facilitate and enrich the flowing down of the complexity of service design requisites into product design, this chapter introduces the potential roles of some of the emerging technologies, understanding that such aspects also represents a gap of knowledge in the field of design and, thus, a great opportunity for research endeavours.

## References


Alves, M. C., Müller A. G., & dos Santos, A. (2018). Product service system development for transportation of elderly people. 13° Congresso Brasileiro de Pesquisa e Desenvolvimento em Design.

Arvola, M., Blomkvist, J., Holmlid, S., & Pezone, G. (2012). A service walkthrough in Astrid Lindgren's footsteps. In *Service design and service innovation conference, ServDes*. Espoo, Finland.

Arvola, M., Blomkvist, J., & Wahlman, F. (2017). Lifelogging in user experience research: Supporting recall and improving data richness. *The Design Journal, 20*(sup1), S3954–S3965.

Baines, T. S., Lightfoot, H. W., Evans, S., Neely, A., Greenough, R., Peppard, J., Roy, R., Shehab, E., Braganza, A., Tiwari, A., Alcock, J. R., Angus, J. P., Bastl, M., Cousens, A., Irving, P., Johnson, M., Kingston, J., Lockett, H., Martinez, V., Michele, P., Tranfield, D., Walton, I. M., & Wilson, H. (2007). State-of-the-art in product-service systems. *Proceedings of the IMechE, 221*(B). https://doi.org/10.1243/09544054JEM858

Barret, P. S., Hudson, J., & Stanley, C. (1999). Good practice in briefing: The limits of rationality. *Automation in Construction, 8*(6), 633–642.

Blomkvist, J. (2014). Representing future situations of service: Prototyping in service design. Doctoral Dissertation, Linköping University.

Blomkvist, J. (2016). Benefits of service level prototyping. *Routledge, Design Journal, 19*(4), 545–564.

Blomkvist, J., & Bode, A. (2012). Using service walkthroughs to co-create whole service experiences: A prototyping technique for service design. In *Proceedings of ISIDC 2012*, Tainan, Taiwan.

Blomkvist, J., Clatworthy, S., & Holmlid, S. (2016). Ways of seeing the design material of service. In *Proceedings of ServDes: Service design geographies* (pp. 1–13). Linköping University Electronic Press.

Bocken, N. e. I., & Emilia e Gonzalez, D. (2019). Designing sustainable business models: Exploring IoT-enabled strategies to drive sustainable consumption. In AAGAARD, A (Ed.), *Sustainable business models* (pp. 61–88). Palgrave Macmillan.

Blythe, M., Hassenzahl, M. (2005). The semantics of fun: Differentiating enjoyable experiences. In.: Funology. Springer.

Cho, E. J. (2013). Designing for sociability: A relational aesthetic approach to service interaction. Doctoral Thesis, Politecnico di Milano.

da Silva, J. S. G. (2010). Diretrizes para o design de sistema produto-serviço voltado ao trabalho remoto. MSc Dissertation, Universidade Federal do Paraná.

Diana, C., Panceti, E., & Tassi, R. (2009). Visualtiles: Communication tools for (service) design. *DeThinkingService ReThinkingDesign*, 112.

European Commission. (2018). High-level expert group on artificial intelligence. A definition of AI: Main capabilities and scientific disciplines, Brussels.

Frick, N. R. J., Brünker, F., Ross, B., & Stieglitz, S. (2019). Towards successful collaboration: Design guidelines for AI-based services enriching information systems in organisations. In *Australasian conference on information systems*.

Garvin, D. A. (1987). Competing on the eight dimensions of quality. *Harvard Business Review, 65*(6), 101–109.

Gkouskos, D., & Linde, P. (2016). Designing for IoT multi-touchpoint UX. In *Experience design for multiple customer touchpoints workshop in conjunction with NordiCHI'16*.
Hoss, M. J. (2014). Prototipagem de Serviços: um Estudo Exploratório com Foco na Iluminação de Habitações de Interesse Social. MSc Dissertation, Universidade Federal do Paraná.
Houde, S., & Hill, C. (1997). What do prototypes prototype? In *Handbook of human-computer interaction* (pp. 367–381). Elsevier Science B. V.
Iacucci, G., & Kuutti, K. (2002). Everyday life as a stage in creating and performing scenarios for wireless devices. *Personal and Ubiquitous Computing, 6*(4), 299–306.
Kimbell, L., & Bailey, J. (2017). Prototyping and the new spirit of policy-making. *CoDesign, 13*(3), 214–226. Taylor & Francis. https://doi.org/10.1080/15710882.2017.1355003
Kleinsmann, M., & Ten Bhömer, M. (2020). The (New) roles of prototypes during the co-development of digital product service systems. *International Journal of Design, 14*(1), 65–79.
Leme, F. D. P., & Frare, L. G. (2022). A product-service system concept for income generation to Curitiba residents immigrant women from Venezuela. Sustainability and Design Course, Bachelor Program in Design, Universidade Federal do Paraná.
Lim, Y. K., Stolterman, E., & Tenenberg, J. (2008). The anatomy of prototypes: Prototypes as filters, prototypes as manifestations of design ideas. *ACM Transactions on Computer-Human Interaction, 15*(2), 27.
Lin, C. J., & Cheng, L. (2017). Product attributes and user experience design: How to convey product information through user-centered service. *Journal of Intelligent Manufacturing, 28*, 1743–1754. https://doi.org/0.1007/s10845-015-1095-8
Liu, Y. (2000). The aesthetic and the ethic dimensions of human factors and design. Proceedings of the 5th industrial engineering conference on industrial engineering theory, applications, and practice (CD-ROM). Taiwan.
Lovelock, C. H. (1983). Classifying service to gain strategic marketing insight. *Journal of Marketing, 47*(3), 9–20.
Löwgren, J. (2009). Toward an articulation of interaction esthetics. *New Review in Hypermedia and Multimedia, 15*(2), 129–146. https://doi.org/10.1080/13614560903117822
Luck, R., Haenlein, H., & Bright, K. (2001). Project briefing for accessible design. *Design Studies, 22*(3), 297.
Mager, B. (2008). Service design – Definition. In T. Marshall & M. Erlhoff (Eds.), *Design dictionary* (pp. 354–357). Birkhäuser.
Mattelmäki, T. (2006). *Design probes*. Publication Series of the University of Art and Design Helsinki.
Mccarthy, J., & Wright P. (2003). Technology as experience. http://www-ist.m.ac.nz/plyons.pdf. Accessed 19 Dec 2015.
Nasirian, F., Lee, O. K. D., & Ahmadian, M. (2017). *AI-based voice assistant systems: Evaluating from the interaction and trust perspectives*. Twenty-third Americas Conference on Information Systems.
Parasuraman, A., Zeithmal, V. A., & Berry, L. L. (1985). A conceptual model of service quality and implications for future research. *Journal of Marketing, 49*(4), 41–45.
Plattner, H. (2010). *Introduction to design thinking process guide*. Design School Stanford, Hasso Plattner Institute of Design.
Russell, S., & Norvig, P. (2003). *Artificial intelligence: A modern approach* (2nd ed.). Pearson Education.
Ryd, N. (2004). The design brief as carrier of client information during the construction process. *Design Studies, 25*(3), 231–249.
Sanders, E. B. N., & Stappers, P. J. (2012). *Convivial toolbox: Generative research for the front end of design*. BIS Publishers.
Segelström, F. (2013). *Stakeholder engagement for service design: How service designers identify and communicate insights*. Linköping Electronic Press.
Serbena, H. J. (2013). Plataforma de luminária LED para habitação de interesse Social, 2013. MSc Dissertation, Universidade Federal do Paraná.

Silvestro, R. (2005). Applying gap analysis in the health service to inform the service improvement agenda. *International Journal of Quality and Reliability Management, 22*(3), 215–233.

Stickdorn, M., Hormess, M. E., Lawrence, A., & Schneider, J. (2018). *This is service design doing: Applying service design thinking in the real world a practitioners' handbook.* O'Reilly Media.

Van De Poel, I. (2009). Values in engineering design. In *Philosophy of technology and engineering sciences. Handbook of the philosophy of science* (pp. 973–1006). Elsevier B.V.

Vockea, C., Constantinescu, C., & Popescu, D. (2019). Application potentials of artificial intelligence for the design of innovation processes. In *29th CIRP Design.*

Wortmann, F., & Flüchter, K. (2015). Internet of things: Technology and value added. *Business and Information Systems Engineering, 57*(3), 221–224. https://doi.org/10.1007/s12599-015-0383-3

Wüenderlich, N. V., & Paluch, S. (2017). A Nice and friendly chat with a bot: User perceptions of AI-based service agents (p. 011).

Wüenderlich, N. V, Wangenheim, F. V., & Bitner, M. J. (2012). High tech and high touch: A framework for understanding user attitudes and Behaviors related to smart interactive services. 16(1), 320.

Zeithaml, V. A., Parasuraman, A., & Berry, L. L. (1985). Problems and strategies in service marketing. *Journal of Marketing, 49*(2), 33–46.

# Chapter 7
# Service Design for Medical Devices

## How the Service Design Approach Can Empower Medical Device Development to Reach Its User-Centered Goals

**Ilkka Juuso and Tapio Seppänen**

## Introduction

Modern medical devices have come a long way from the time field surgeons placed the first wooden twig between the jaws of an amputation patient – such as may have been the case in the first century according to Celsus (Sachs et al., 1999) and during the American Civil War over 1700 years later. Medical devices have since evolved from bandages and surgical instruments to programmable electronic devices and, today, devices approaching a level of artificial intelligence.

Medical devices have, in one way, always acted as an extension to the limbs of the physician enabling the experts to have an extra pair of hands, extra time, or just the right instrument to treat patients. The advent of the microchip in the late 1950s and its entry into medical devices in the subsequent decades enabled devices to achieve increasingly smart functions. The present-day interest in leveraging artificial intelligence and machine learning (AI/ML) technologies in healthcare applications is just the most recent step on this evolutionary path, although the gradual embracing of AI is a marked step. With artificial intelligence, we are for the first time contemplating medical devices that are not just inanimate devices but also include some small part of the skillset and autonomy of a medical professional. For the first time, the device may be promoted to the level of an assistant medical staffer who gets to prep the patient or close up the wound after a master surgeon has done their work. It is thus no surprise that AI has raised both vast interest and cautious opposition in the medical profession and in the regulatory and standardization bodies overseeing such devices (Alcini et al., 2021).

I. Juuso (✉) · T. Seppänen
Center for Machine Vision and Signal Analysis, University of Oulu, Oulu, Finland
e-mail: Ilkka.Juuso@oulu.fi



U. Z. A. Hamid, M. Suoheimo (eds.), *Service Design for Emerging Technologies Product Development*, Springer Series in Design and Innovation 29,
https://doi.org/10.1007/978-3-031-29306-1_7

As devices are about to again take a leap forward in complexity and attempt to help users in new ways the importance of understanding the user as well as the disease is only growing in importance. The medical device industry has an ingrained appreciation of, but not always an understanding of, the user. A stated appreciation of the user is expected by every stakeholder in the field from patient to manufacturer, but the methods to obtain or ensure that understanding vary. The resources available to a manufacturer will vary and so too will the risk profile of their medical device, but the user-centered design, usability engineering and the overarching validation of medical devices remain hot topics for all in the field. Against this backdrop of activity, the discipline of service design holds great promise in streamlining manufacturer activities and ensuring the focus is appropriately placed on the user and the use of the medical device.

In this chapter, we will briefly discuss medical device development, the role emerging AI-based technologies may have in it, and the standards and regulations that govern such work. We will then look at how service design fits together with the expectations placed on medical devices and how the discipline may provide welcome structure and assistance to meeting the key medical device requirements. Figure 7.1 presents a conceptual mapping between the core tenets of service design and the standard life cycle model for medical devices. Note that while the figure shows the stage in medical device development where the tenet is seen to have the most impact, the tenets maintain their relevance throughout the development, deployment, and use medical devices all the way from initial ideation to the eventual decommission of the finished device.

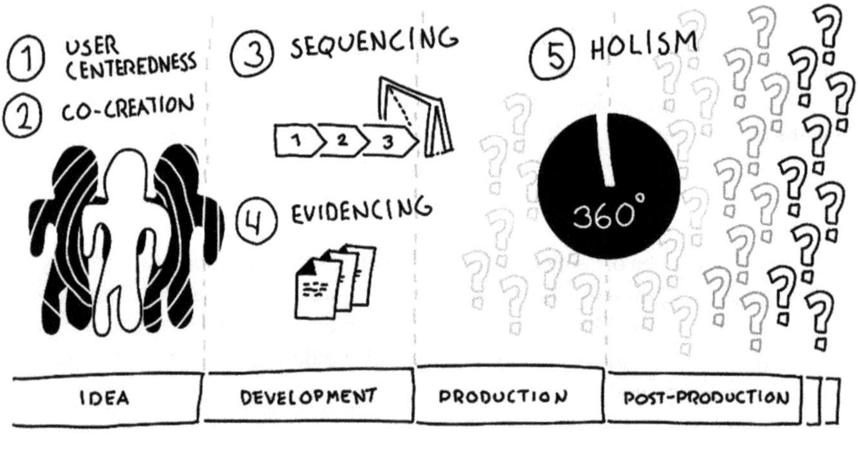


**Fig. 7.1** Service design tenets mapped on to the life cycle of medical devices

## Medical Devices as a Regulated Industry

The field of medical devices is an example of a highly regulated environment where the development of products comes with great expectations to cure diseases and alleviate medical conditions. Great advances are expected of the product's developed, and at the same time, significant fears are attached to any serious malfunctions encountered with their use. Errors may, for example, be caused by misunderstandings between the user and the manufacturer, or the misalignment of expectations across this boundary. The industry and its standards acknowledge this by talking of use errors instead of user errors. Any manufacturer will find it hard to blame the user. The burden of identifying and meeting all requirements, too, is placed squarely on the shoulders of the manufacturers by such key industry standards as the international ISO 13485 standard (International Organization for Standardization, 2016) on the quality management of medical devices. Within its first few pages, the standard already defines its scope as follows:

> This International Standard specifies requirements for a quality management system where an organization needs to demonstrate its ability to provide medical devices and related services that consistently meet customer and applicable regulatory requirements.

From the definition above, it is clear that the active party who needs to ensure all the applicable requirements are met is the manufacturer. The products offered by the manufacturer need to be safe and effective, as demanded by the regulations and the customer, and they need to remain so every time and over their entire lifetime. The burden of determining the requirements is on the manufacturer who will then use the requirements to build the product, or service, requested. Note that this does not give the manufacturer the freedom to define the product as anything they like, but instead the requirement is to see the manufacturer understand all the applicable requirements for the product they are contemplating building.

The user requirements, or the definition of needs for the device, may come from patients, from doctors, from nurses, or from scientists, engineers, and business folks of all backgrounds. The pathway from each of these beginnings to a finished medical device sold to real users is in each instance governed by the same strict regulations. Standards and regulations, too, place demands on the behavior of the manufacturer in designing, realizing, and maintaining products on the market. At the same time, standards and regulations will, most of the time, balk at placing requirements on the laymen users of a medical device. There are some exceptions to this rule, as demonstrated by the current draft for the Artificial Intelligence Act under preparation in the EU, but mostly the manufacturer is the active subject in defining requirements, not the user.

The consideration of benefits and risks related to each product is an essential part of the design of each product, and also its later maintenance once in the market. The validation of each product is today a part of the assessment before the release of any product. As a key part of the validation, the manufacturer is asked to ensure that the product meets the expectations of its customer prior to providing the product, or service, in question. Obtaining feedback from clinical experts is often a major part of this validation, and a big-ticket item to fund from the point of view of the manufacturer. Depending on the product, the size of the clinical studies required may range from a few people, for a low-risk product perhaps dealing with a rare disease, to tens of thousands of people across geographically disparate locations, as was the case with the COVID-19 studies before those revolutionary mRNA vaccines were cleared for sale.

## *The Role of Artificial Intelligence in Medical Devices*

Artificial intelligence (AI) is not a new invention, but it is seeing new increasingly advanced applications in healthcare (Qadri et al., 2020). The overarching field of AI in healthcare may be roughly divided into two subfields: radiological imaging and the use of electronic health records.

Radiological imaging and the use of machine vision technologies in this field has led to numerous applications in healthcare that have also received regulatory approvals around the world. The identification of cancer cells, cell structures, and other regions of interest in biopsies or magnetic resonance imaging (MRI) is a typical example. Failures here include deep learning prototypes that have latched onto notes and marking made by doctors in the images to identify regions of interest to the doctors.

In the case of electronic health records (EHR), the use of AI holds great promise in identifying patients for preventative screening of some disease, and the streamlining of hospital care paths based on the multitude of available information. Failures here include a deep learning prototype that identified patients requiring complex or lengthy treatment based on the accumulated cost of their treatments.

At the same time, it is not just the amount of data that is accumulating but also the types of data, as it is increasingly straightforward to retain not just the background information of a patient collected via forms but also the various results of imaging and other exams. The data available for a patient may now also include bio signals (e.g., ECG, EEG, blood pressure) and digitized tissue samples. The concept of a "digital twin," a digital representation of a person, has been suggested as a means to studying the health and physiology of both the individual and the populations they comprise. The challenge for the user is how to interpret this vast body of data. Traditionally this has been done by the physician who has compared test results to widely accepted reference values and thresholds. The use of AI may enable such comparisons to be analyzed using several parameters at the same time. The combination of increasingly comprehensive datasets and AI technologies thus holds great promise in identifying, predicting, and treating ailments.

It is also notable that the user of an AI-based product may be a patient, a physician, or even a healthcare authority. The expectations that can be placed on a layman user of an AI product versus a medical professional using the same product will differ.

In both abovementioned fields, radiological imaging and the use of EHR data, AI has been investigated to assist medical professionals in identifying regions and patients of interest in terms of some usually well-defined symptoms of a disease. In both fields AI has provided both triumphs and spectacular failures to the extent that providing practical transparency over the actions of the AI, and achieving sufficient explainability of the underlying models, is now often cited as a major area to address when bringing an AI product to the market. The current state of explainable models in the context of healthcare is examined in (Ahmad et al., 2018).

The task of modeling data using machine learning is today a major area of activity inside the development of AI technologies. This toolbox contains both traditional white-box technologies that readily provide the user with a view into the workings of the model and the outcome it provides, and black-box technologies that do not. The newest cutting-edge group of AI applications utilizes the latter, where explainability does not come naturally. Black-box models are in essence mathematical models based on vast numbers of parameters that make it difficult or even impossible to know what is taking place inside the box. In using such tools, the user is forced to trust the inner workings of the tool and the correctness of the results they provide. Significant research effort is currently funneled into providing mechanisms for seeing into, at least, parts of such black boxes, and thus turning them into gray-box models. Unsurprisingly, the adoption of black-box models in the healthcare context has been slower than that of white-box models.

Similarly, the task of providing the analysis over large amounts of data is a considerable challenge. The choice between displaying a row after row of numbers or some visual summary of the results may come down to the amount and type of data, as well as the analysis needs of the user. Presenting the user with a large amount of raw data may in some cases be warranted, but often some form of aggregated analysis or the calculation of derived measures is called for. The development of such measures then depends heavily on the user's ability to understand the measures and perform further analysis based on them.

In terms of developing products and services based on AI technologies, the field is currently in a state of considerable change. The intense activity around the application of AI technologies, both old and new, and the solutions for keeping users abreast of the pertinent factors involved may mean that the continuing evolution of almost any AI-based product is to be expected going forward. This evolution applies to both new products becoming available but also existing products receiving updates, and the users of the products needed to be kept in the loop from early development phases to later post-market surveillance phases. The need to provide actionable information without burdening the user with superfluous details or unnecessarily high requirements for prior knowledge are major points of interest for developing AI-based medical devices regardless of the specific technologies used. Identifying practical solutions here will go a long way toward facilitating a meaningful dialogue between manufacturers and the users of their devices, and in this a discipline such as user-centered design may prove invaluable.

## *The Role of the User During Development*

The medical device industry standards place requirements on the manufacturer, not the user. This does not absolve the user from keeping their wits about them when using the products, but it is a wake-up call for any manufacturer who thinks it is enough to just build it and let the users come.

The above is evident in the key quality management standard for medical devices, ISO 13485, which in its Clause 5.2 places the burden of determining and meeting both customer requirements and applicable regulatory requirements squarely on the manufacturer's top management. The implication is that the very top management must ensure that appropriate appreciation of the above requirements and the adequate resources to meet them are in place throughout the organization. The standard thus elevates customer requirements to an almost parallel standing with applicable regulatory requirements and insists the organization appreciates both at all relevant levels and functions. Notably this also applies when an organization outsources some process or a part of a process: the work itself may be outsourced, but the responsibility over it may not if it can affect product conformity to the above requirements (see Clause 4.1.5). Responsibility here extends beyond mere legal responsibility to the monitoring of the work and the appropriate actions being taken if that work falls short.

Furthermore, the identification of the requirements is the responsibility of the organization (Clause 7.2.1). Here the manufacturer is required to consider any requirements specified by the customer (including delivery and postdelivery activities) but also those requirements left unsaid by the customer but necessary for the specified use. User training, too, is specifically called out by the standard as something that may need to be ensured for reaching the specified performance of the device and its safe use.

As a result, stringent clinical evidence is expected of any new device or technology that goes beyond the status quo and what earlier predicate devices have perhaps attempted. Rightly, so as we need to be convinced, our trust in any new technology is justified, and the case for its reliability sufficiently made.

Obtaining this clinical evidence may mean lengthy clinical trials conducted for the individual products. The cost of arranging clinical trials is often cited as one of the major hurdles to cross when bringing a new innovation to market. This cost in turn may hinder the arrival of new technological advances by leading to both aborted development projects and the curbing of development ambitions to incremental products.

The roles of the patient, the physician, and the healthcare facility may at times merge in different ways to answer to the stakeholder of a user. The user of an at-home device may intuitively be the patient themselves, but at the same time also the doctor monitoring the obtained results and the wider health operator, too, may be considered as a user. To keep the standards practical, they most often talk of the organization adopting the standard (e.g., the manufacturer) and their customer. The chain of organizations involved in the development of a particular device may be several organizations deep, but in each case, the link between organizations may be described in terms of a provider and a customer. Depending on the product, your network of stakeholders may be intricate and your elicitation of requirements equally complex, but it may also be straightforward in other cases. That is the beauty of medical device development: it is all standardized and governed by regulations, but what your user (customer) actually wants will guide you in selecting the correct requirements and processes. The role of standards is first and foremost to help align expectations, improve compatibility, and remove unwelcome surprises.

## *The Role of the Manufacturer During Use*

It is intuitive to think that once a device is built and leaves the factory floor, it is in the hands of the user, and the manufacturer may turn their thoughts to the manufacturing of the next device.

All the key medical device standards, for example, ISO 13485 quality management, IEC 62304 software life cycle, and ISO 14971 risk management, drive home the fact that you must monitor how your device performs in the field and react to any changes in that performance. Similar demands are made by the regulations, too, such as the European Medical Device Regulation (EU 2017/745). These requirements often concern the so-called post-market surveillance of your product and involve keeping track of how your users use the device, what issues they report, how your competition is faring, and how the science or literature around your type of device is evolving. Note that in the years past, it may have been enough to keep a phone line open to receive complaints and reports of adverse events, but today the manufacturer is expected to perform active surveillance of the above sources of information. The information may, for example, be reported as periodic safety update reports (PSUR) that are then made available to regulatory authorities. A new subclass of safety is the cybersecurity and information security of the device. Here too it is not enough to pass some check once and sail smoothly from thereon, but instead the requirements extend across the whole lifetime of the device.

The above approached the topic of post-market activities from the point of view of intelligence gathering, but the manufacturer will also need to keep the products safe and effective via appropriate maintenance. When developing the product, the manufacturer is expected to define, among other critical pieces of information, the appropriate installation, servicing, and training information for the device. In the case of physical products, this may mean testing once installed, recalibration thereafter, and the replacement or repair of parts as described. In terms of software, this will often mean the provision of software updates to address some discovered shortcomings, security threats, or new features to the device. All of this is maintenance, and it must take place according to what the manufacturer has described in the interaction with their customers and regulatory authorities. In this and other respects, the responsibility of the manufacturer extends greatly beyond the mere initial production of a device.

## Application of Service Design to Medical Device Development

Medical device standards and regulations acknowledged the need to maintain efficient and effective communication between the user of a device and its manufacturer. The prime objective is to ensure that the user's expectations are met by the device and that no dangerous misalignment may occur in, for example, how a device functions or what it actually does. Despite this ready realization, the practical assistance given by standards and regulations is largely limited to high-level requirements and good intentions. A more rigorous model is needed to help the manufacturer design their own processes so that the development work can proceed in a practical, productive, and efficient manner.

Service design holds the promise of being just the right approach here. It is a design methodology that attempts to provide structure to how the conversation between the user and the manufacturer can unfold, and how it can be utilized as fruitfully as possible in the development of services and products. The five key hallmarks of service design, as listed by Stickdorn and Schneider (2011), are centered around the topics of user-centrality, co-creation, sequencing, evidencing, and a holistic approach. In the following, these aspects are discussed from the point of view of medical devices.

### *User-Centrality*

The central tenet in service design is a focus on the user. This is manifested in a stated desire to see all the developed services (products) through the eyes of the customer and in the context they naturally inhabit. This drive to understand the user and continually meet their requirements is engraved deep into the development of medical devices, too. The most fundamental standard in medical devices, the ISO 13485 standard on quality management, expressly mentions the customer a total of 51 times and devotes a whole clause, Clause 7.2, to defining customer-related processes. The standard cannot place requirements on the customer, but it does place requirements on how the manufacturer interacts with the customer. In fact, the standard often speaks of customer requirements on a level on par with regulatory requirements and insists that the manufacturer also identifies those customer requirements that go unsaid but are known to be necessary for meeting the intended purpose of the device. The impetus is clearly on the manufacturer here, as shown in ISO 13485 Clause 4.1.5 below:

> The organization shall retain responsibility of conformity to this International Standard and to customer and applicable regulatory requirements for outsourced processes.

In the example quote, it can be clearly seen that the standard makes no separation of standing between its own requirements, those given by applicable regulations, and those sourced from the user. In general, regulations are thought to come first, be implemented via adopting standards, and then together set up the framework for meeting user needs, but this is not highlighted in the text of the standard.

Furthermore, the ISO 13485 standard moves away from the old ISO 9000 notion of a product as being something produced by the manufacturer independently of any conversation with the customer to something that is quite the opposite produced after interaction with the customer. The present-day ISO 13485 therefore considers the customer very much a part of the requirements elicitation process for any product and the number one audience for any definition of purchasing information and postdelivery activities.

In general, the ISO 13485 standard speaks of customers in, for example, the following contexts:

- It is the responsibility of the manufacturer's top management to ensure that customer requirements are observed (Clause 5.2).
- The top annual management review must look at the improvement of the product in light of customer requirements (Clause 5.6.3).
- Product requirements must be reviewed before a commitment is made to supply product to the customer (Clause 7.2.2).
- The manufacturer must have arrangements for communicating with the customer on, for example, product information, order handling, feedback, and advisory notices (Clause 7.2.3).
- Customer property is to be protected (Clause 7.5.10).
- Monitoring must look at how the manufacturer has met customer requirements (Clause 8.2.1).

Based on the above list, it is clearly the case that the standard places great emphasis on understanding and interacting with the customer. Thus, the first tenet of service design is right on the mark for addressing the design of medical devices. In fact, the only way service design could be more directly applicable to the standard was if it made explicit the role of the management in ensuring customer-focus is maintained.

## Co-creation

Continuing from the previous topic of user-centrality, service design instructs that all stakeholders should be involved in the design of a service (or product). This is called co-creation and is the second tenet of service design. The aim here is to ensure that all relevant points of view are heard when developing the design.

The ISO 13485 standard is in agreement although it does not name the stakeholders that might be in question. In addition to the user of the device, this might, for example, include the patient, their family, the customer, the distributor, the notified body, the regulatory authorities, and even health policy officials (Juuso, 2022). The standard does, however, introduce the concept of "personnel affecting quality" within the manufacturer. Here it is required that the responsibilities, authorities, and interrelation of such personnel are documented and understood (Clause 5.5.1). Thus, the medical device world goes one large step beyond service design to understand how co-creation may take place. The difference here does not prevent the adoption of service design to assisting medical device development, but it is significant enough to suggest that some overhead or tweaking of the service design approach may be called for to obtain the optimal fit.

## *Sequencing*

The third tenet of service design is sequencing. The goal here is to visualize the service (or product) as a sequence of interrelated actions. Traditionally service design sees sequencing as an act involving the use of a product: how the product is discovered, how it is interacted with, and what type of a memory residue it leaves.

The development of medical devices goes even further. Here according to the ISO 13485 standard, the sequencing starts even earlier during the early development of a product. The whole process from an idea to a finished product is seen as a process consisting of distinct design and development stages, some of which may be internal but a few of which de facto require the involvement of the user. The standard sees the involvement of the customer as critical, for example, during the requirements elicitation phase and the final validation of the product before its launch and provision into use. This way the development process is at the very least bookended by user involvement: the customer gets to say both what should be done and if that something was done to their satisfaction.

The standard does not stop there, either. After the product, or service, has been provided to the customer, the manufacturer must stay attentive toward whether any issues are discovered during its use. This is called vigilance and post-market surveillance of the product. If issues are detected the manufacturer must respond to them within the expectations of the standards and the regulations. This may mean developing an update fixing the issue. Similarly, new features may be added to the products over time even if no issues are detected. All of these incremental changes must make the product better, and must thus pass the same expectations as before. Supporting medical device products over extended periods of time, while perhaps performing incremental updates on them over time, is perhaps not addressed by service design theory head-on. The present theory is, of course, compatible with sustained support and incremental improvements, but no clear-cut or efficient mechanism is yet presented on how to apply the original full development path to perform smaller repeated changes. This is a key facet to consider when pitching service design as an approach to improve the development and maintenance of medical devices.

## Evidencing

The fourth tenet of service design is evidencing. This is where service design and the expectations of medical device development really hit it off. The aim service design has in mind is to visualize intangible services in terms of physical artifacts. The idea is to let the user appreciate even the intangible outcomes of the service (product) via some concrete token or other evidence. The medical device world needs no prod to appreciate evidence: in fact, the whole existence of medical devices, the pedigree given to them via the application of processes based on standards and regulations, relies heavily on the notion of retaining evidence in the form of documentation. This concept of evidence is, of course, somewhat different from that called for by service design, but both concepts apply to medical devices.

Evidencing in service design is used to refer to, among other things, how intangible or behind-the-scene work can be presented to the user in an unobtrusive way. The user should be able to appreciate what goes into the product or service they use without being sunken into overwhelming detail. An example here given by (Stickdorn & Schneider, 2011) is that of placing a placard in a hotel washroom pointing to environmental considerations for not using cloth towels. The implication is that the user's attention is directed toward the thinking that has gone into making a certain service choice.

In medical devices, the equivalent of the previous example might be a label on the recycling properties of a device or indeed the regulatory approval of the device before it can even be purchased. A further example, and indication of the increasingly complex nature of medical devices possible to manufacture today, is the evidence of how that device actually works and why the user should be able to trust it. Such evidence has been a part of the medical device business since at least 1938 when the Federal, Food, Drug, and Cosmetic Act was passed in the USA, after which the Food and Drug Agency (FDA) was given the powers to oversee the safety of such devices. In the century since, the importance of evidence provided ahead of market clearance has only increased.

Recently, with the advances in artificial intelligence and other technologies ushering in new generations of devices that may evolve once already in use, the need for sufficient evidence has only been heightened. In the case of artificial intelligence and deep learning, for example, the need to understand what a device is doing and what data it is basing its conclusions on is even more important than before. Such advanced products today work under the supervision of humans who are in the loop before any impact may be felt by the patients, but nonetheless such devices are taking over some of the tasks of the medical professionals using the devices. The million-dollar question is what tasks can be taken over in this way, and what evidence from the tasks and the development of the devices should be provided to the user. Providing all of the data and source code to the user, and expecting them to spend the weeks needed to analyze the material, is not feasible or helpful. Providing enough practical evidence for the user to adopt and trust a device is key, and here the concept of evidencing and its workflows from service design may provide valuable insight.

Evidencing as keeping receipt of the work done during development is very much appreciated by medical devices. The idea of building in evidencing into the use of medical devices is not, but it holds great promise for designing in usability into both the devices and the processes leading to the development of those devices. This is especially true with devices incorporating intangible emerging technologies such as artificial intelligence, where facilitating transparency into the workings of the device and explaining its output is often difficult. If the manufacturer cannot explain a device to the user without a master's degree on the device and going through stacks of documents and data, the user is not likely to adopt the device and realize the benefit it offers.

## *Holistic Approach*

The last tenet of service design is that it aims to be a holistic approach capturing the entire environment of the service (product) for investigation. The objective here is to point out that every angle should be considered and every rock turned to end up with the best possible design for a product.

In the development of medical devices, the same concept is approached from the point of view of stakeholders, manufacturer's internal teams and personnel, and the product realization process which is covered from ideation, requirements elicitation, development, and testing all the way to delivery, installation, training, and maintenance – and indeed the customer-facing processes from there on as part of post-market surveillance. The standards, such as ISO 13485, devote much column space to defining the requirements for the processes in order to ensure all transitions occur predictably and all the important activities run until perpetuity, or at least the end of life for the device and its decommissioning.

Other standards such as the IEC 62304 standard on medical software development add further points of view to consider during product realization. Here equal importance is placed on understanding the entirety of the requirements – be those from the customer, the applicable regulations, or the standards themselves – but this is then elaborated on to describe how the units comprising the finished software should be designed, implemented, tested, and finally integrated into a whole. The IEC 62304 thus takes an engineering view to developing medical devices, whereby it wants to see the components adequately defined and then the whole assessed in terms of also its parts. The ISO 13485 by comparison looks at whether the final product meets the expectations you had for it at the start but also whether it matches with the expectations of your customer (Juuso, 2022). This former part is often referred to as verification and the latter validation.

The calls for a holistic approach in service design thus map on to the expectations of medical device development. The needs of medical devices are more complex and at present also more evolved via the introduction of standards than perhaps understood by service design, but the goals appear to be aligned. If service design can assist in the development of medical devices, this may mean that less pertinent points of view are overlooked in the development which is certainly a worthy outcome and reason for adopting the approach.

## Service Design Modified to Medical Devices

The core tenets of service design appear readily applicable to the development of medical devices. The focus on the user, the appreciation of all stakeholders involved in the realization of a product, the sequencing, evidencing, and holistic understanding of the use of products and services speaks to the core of the needs involved in designing and producing medical devices. Many of the lessons gained from service design may thus be of immediate value in medical devices and lead to better products that behave more closely as the user expects both now and over the long term. A more predictable and streamlined development process will not only lead to cost savings in the development that can be transferred to the final product but also the availability of more solutions to perhaps previously unmet needs.

The five tenets of service design can be utilized throughout the product realization process of medical devices. This is particularly true for the second tenet, co-creation, which can be conveniently expanded to appreciate the whole team both within and without the manufacturer. Medical device standards and regulations see this dialogue between the manufacturer and the customer as critical in order to ensure all relevant points of view are heard before the appropriate functions provide permission to move forward. The standards and regulations, however, lack practical tools to facilitate this dialogue and could thus benefit from the approach offered by service design.

The tenet that holds the greatest promise for improving the development of medical devices is, however, the concept of evidencing. This concept, too, can be brought to enhance all the phases of product realization from requirements elicitation to training, use of the device, and learning more through using the device. If the tools and best practices available on evidencing from service design can improve these activities by making the outcome safer and the path there more efficient, it will have affected a small revolution in medical devices. If these solutions can assist the manufacturer in providing the necessary information to the user in a practical, user-friendly, and perhaps self-guiding manner over time, the solutions offered by service design may be just what the doctor prescribed for adopting new emerging technologies as part of medical devices. How to best support transparency, explainability, and the practicality of information supplied to the user is at the moment a key topic in enabling such advanced technologies via regulations and standards.

There is one omission in service design, and this is also less-than-perfectly covered in the present medical device standards. There could be a sixth tenet to highlight that a service (or a product) does not just need to meet user expectations once but also over the long term. The notion of maintenance is built into many of the standards, but here too the continued interaction of the manufacturer and the customer deserves more practical tools. The advent of evolving technologies, such as artificial intelligence, has heightened the need to allow products to evolve also once released. The bullpen on truly autonomous self-learning AI in healthcare remains firmly shut, but the frequency of software updates appears to be growing, and the need to streamline the maintenance release process is growing as well. The development of new releases and patches for previous devices also means having to reassess or reperform the testing, verification, and validation activities related to those changes and perhaps the entire product, too. Here service design could facilitate activities by instructing how a continuous long-term dialogue might proceed between the manufacturer and the customer. The holy grail would be to safely and efficiently perform piece-wise validation of changes to medical devices. Until then, service design can add best practices and practical tools to shape this important interface between the manufacturer and the user. As with the rest of the world, some form of agile service design tied with the life cycle model of medical devices would be a very welcome addition to the toolbox of medical device manufacturers.

## Conclusions

The paradigm of service design touches on many of the core requirements of designing, developing, and maintaining medical devices. The resolute focus on understanding the user, the drive to reliably meet the needs of the user, and goal of completing the user experience by making all the critical aspects that go into a product or a service tangible to the user are all aligned.

The need to keep up the service, meet all the requirements continuously – regardless of whether those come from the patient, the doctor, the standardization bodies, or the regulatory authorities – and perform incremental updates to the product or the service are not specifically addressed by the present service design vocabulary. The current service design approach is compatible with such needs, but the approach might benefit from an in-depth consideration of how to best push the envelope toward a continuous, incremental process of product provision. This way the fifth tenet of service design, holism, would extend over the entire life cycle of the medical device and provide both welcome structure and user focus to the activities during this increasingly important post-market phase.

## References

Ahmad, M. A., Eckert, C., & Teredesai, A. (2018). Interpretable machine learning in healthcare. In *Proceedings of the 2018 ACM international conference on bioinformatics, computational biology, and health informatics (BCB'18)*. Association for Computing Machinery. https://doi.org/10.1145/3233547.3233667

Alcini, P., Baird, P., Williams, P., Bhattacharyya, S. B., Cooper, T., de la Cruz, R., Kumar, C., Datta, G., De Marco, D. A., Franken, A., Geierhofer, R., Goldschmidt, P. G., Juuso, I., Klimenko, H., Kung, A., Laroche, F., Lewelling, J., Meyer, M., Morozov, S., Orlova, A., Panagiotis, T., Penzel, T., Ritz, D., Sander, G., Shin, S.-Y., Värri, A., Vladzimerskiy, A., & Lebedev, G. (2021). Artificial intelligence in healthcare: Directions of standardization. In *Handbook of artificial intelligence in healthcare* (pp. 231–257). Springer Publishing. ISBN: 978-3-030-83620-7.

International Organization for Standardization. (2016). Medical devices – Quality management systems – Requirements for regulatory purposes (ISO 13485:2016). https://www.iso.org/standard/59752.html

Juuso, I. (2022). *Developing an ISO 13485-certified quality management system – An implementation guide for the medical-device industry*. Productivity Press/Routledge. ISBN 9781032065748.

Qadri, Y. A., Nauman, A., Zikria, Y. B., Vasilakos, A. V., & Kim, S. W. (2020). The future of healthcare internet of things: A survey of emerging technologies. *IEEE Communications Surveys & Tutorials, 22*(2), 1121–1167. https://doi.org/10.1109/COMST.2020.2973314

Sachs, M., Bojunga, J., & Encke, A. (1999). Historical evolution of limb amputation. *World Journal of Surgery, 23*(10), 1088–1093. https://doi.org/10.1007/s002689900628

Stickdorn, M., & Schneider, J. (2011). *This is service design thinking*. BIS Publishers. ISBN 9789063692797.

# Chapter 8
# AI Service Model for an Airline Ecosystem: A Systemic Design and Thematic Approach to Service-Dominant Logic: Examining State-of-the-Art Technologies for Service Centricity

**Vássil Rjsé, Taiba Sadeq, and Satu Miettinen**

## Introduction: Service

In a previous study, the framing of a conceptual model where AI is designed to play a vital role to the creation of necessary channels, digitally integrating service actors within the ecosystem, has been established for the domain of airline mobility service systems. Referred to as AI-augmented services (AIAS), it is a model that signifies the relationships amongst service providers and the user (Rjsé et al., 2022). This is made possible by rendering high-tech capabilities onboard an aircraft, to interactions that co-create values and service quality amongst the passenger and cabin crew (Grönroos, 1984). The model goes further by linking accessibility to external stakeholders such as ground mobility services, hotel reservations and flight data information from other affiliated airlines within the given cabin experience (e.g. Akaka & Vargo, 2014; Lusch et al., 2016). Hence, AIAS attempts to provide both functional and emotional values by the deployment of AI machine learning and processing within cabin services, opening channels that connect more stakeholder competencies, placing more control within the reach of the user and allowing the personalisation of aesthetics throughout the cabin journey. Essentially, when technology can assist the user to co-create values, 'high-tech' is turned into 'high-touch' (Rjsé et al., 2022). The best example would be when AIAS can give the right kind of solution at the exact time of need (Google, 2016), especially when service failures occur and the user needs quick solutions within the limited time and space of the cabin. The provision or even access to such functional values may naturally lead

---

V. Rjsé · T. Sadeq · S. Miettinen
University of Lapland (uLapland), Service Design, Rovaniemi, Finland

Public Authority for Applied Education and Training (PAEET), Kuwait City, Kuwait
e-mail: vesaal.rajab@ulapland.fi

 
U. Z. A. Hamid, M. Suoheimo (eds.), *Service Design for Emerging Technologies Product Development*, Springer Series in Design and Innovation 29,
https://doi.org/10.1007/978-3-031-29306-1_8

to 'emotional values', some of the most sought-after value groups in any industry (Almquist et al., 2016). AIAS strives towards such values through the provision of access that may lead to better service integration which ultimately leads to a more personalised journey, referred to as 'high-touch' services (Rjsé et al., 2022; Vargo & Lusch, 2010). Here, technology is seen as any device or hardware system capable of computing multimodal data inputs within the cabin service system. The role of AI is to thusly utilise such data by learning and providing quick and meaningful solutions at any given time of need (e.g. Bland, 2017). In other words, AI is an actor that converts data into 'high-touch' values comprehensible to the user (Jylkäs et al., 2018).

The co-creation of values through any channel during service consumption leads to 'value-in-use' (Finne & Grönroos, 2017; Grönroos, 2008); hence, value-in-use is the proposal and co-creation of values amongst system actors that are justified by the user, that is, providing grounds where the user is an active participant in the development of the enterprise's value propositions, and all stakeholders are seen as beneficiaries (Vargo & Akaka, 2009). Here, value-in-use is achieved by giving beneficiaries interactive access to reflect 'need' and service provider to propose 'solutions' in a timely or rather instantaneous manner; thus, value-in-use increases probabilities of emotional value, leading to an enhanced user perception of service quality (Grönroos, 1984) which also reflects on the brand image as whole (Almquist et al., 2016; Grönroos, 2007).

Providing the user with integrated communication to a wider range of service providers allows value-in-use to build relationships amongst the user and stakeholders (Akaka et al., 2014). Moreover, user 'control' is a major aspect in AI-enabled services that improves 'trust' towards digital interactions for high-touch values (Rjsé et al., 2022); here, the user must be in the frontline of data sharing and the decision-making process. AIAS is fundamentally initiated through a 'service-centred' approach where values and service gaps are determined by the user (Vargo & Akaka, 2009). In this sense, service centricity is further exercised through service design thinking (Miettinen & Koivisto, 2009; Stickdorn & Schneider, 2012; Stickdorn, 2018). Therefore, AIAS aggregates data from multiple sources within the services ecosystem to give prominence for value-in-use, analysing magnitudes of data through AI processing for the provision of a frictionless user-friendly service (e.g. Weller, 2019). This cycle between what the customer needs and what values the enterprise can propose creates a balance where needs are met with just the right number of resources (Dodge et al., 2012). To achieve wellbeing, service-dominant (SD) logic is employed through high-tech advantages of AIAS (e.g. Burr et al., 2020), where it can reach out further into the service ecosystem (Vargo & Lusch, 2004), enabling the model to have more resource at hand for that one critical moment the customer is mostly in need of (e.g. Grönroos, 1990). In this sense, values are expanded to reach multiple stakeholders, stretching further than the microservice system, that is, cabin services, to connect more providers that have traditionally been unreachable during a typical flight. This allows the model to utilise available technology capable of centralising the service, that is, human centricity, achieving a variety of co-created values (Grönroos, 2008; Miettinen et al.,

2014). Hence, technology is always seen as a tool to seek value-in-use for a high-touch experience. This approach is generally regarded as service design thinking (e.g. Miettinen & Alhonsuo, 2019); the user co-creates 'determines' values, and stakeholders are at better reach to 'propose' and provide benefits (Finne & Grönroos, 2017; Vargo & Akaka, 2009). This essentially expands the co-creation of values to reach out further into the macro level of service ecosystems (Godsiff et al., 2018; Layton, 2008; Rijn & van der Burgt, 2021). The airline industry can be one of the best examples where digital transformation is witnessed in almost every single stage of the journey; nevertheless, with that comes consequences where digital service interactions are continuously updating (ACI, 2021), and the user can likely be left out from such fast past evolutionary process, hindering service centricity logic (Vargo & Lusch, 2016, 2017). Thus, service design thinking is seen as the necessary bridge that empowers stakeholders to co-create value-in-use, supporting the necessary relationships of a 'sustainable' service ecosystem through technology (Mele & Russo-Spena, 2022).

In this chapter, data is collected through a series of observations and interviews, 'listening' to the airline ecosystem, observing state-of-the-art components by benchmarking to 'find' what constructs an airline ecosystem; consequently, data is then analysed through a systemic design approach by 'understand the system' and then 'defining the desired future' (Jones et al., 2021). Theoretical approach of user centricity, value co-creation and relationships encompass the major grounds from which the interviews are analysed—SD logic and service quality. User's service centricity, also known as human centricity, is approached through SD logic and service design thinking (Stickdorn & Schneider, 2012; Vargo & Akaka, 2009). Stakeholders' relationships, competencies and technical service skills are approached though SD logic and the Nordic model of service quality 'technical dimension' (Grönroos, 2007). Perception of the user's service quality during service consumption is approached through the Nordic model of service quality 'functional dimension' (Grönroos, 1984). The type of values these approaches may produce are evaluated according to 'the elements of values pyramid' (Almquist et al., 2016). Hence, SD logic and service quality are filtered through Jones et al.' (2021) systemic design toolkit to produce the framework of this study, namely, values that are co-created throughout the central service journey are referred to as 'value-in-use'. In this sense, state-of-the-art airline services are explored in terms of the viable application of AIAS for future services where greater value-in-use is facilitated by the employment of AI as an actor (Jylkäs et al., 2018), adding more layers of value proposals especially during critical service moments (Davey & Grönroos, 2019; Grönroos, 2008). Data analysis has produced a new rendition to the AIAS model where 'value-in-use' plays a very important part in the implementation of AI-enabled services. The AIAS model 2.0 is further elaborated in the Discussion section of this chapter.

Hence, this chapter seeks to 'evaluate' the service ecosystem in terms of its latest technologies and its effect on stakeholder relationships, and how such relationships may facilitate value-in-use through a high-touch cabin experience— what is the predominant service system used in the airline industry today? And how does this type of system affect value co-creation in terms of SD logic?

Nine company representatives were contextually interviewed through service and product demonstrations during a major airline interior expo. The goal is to generate thematic patterns where the dynamic connection between service design and technology is recognised and 'evaluated' (Braun & Clarke, 2006). Moreover, the research design observation process is to recognise signs of 'mechanism' (Craven, 2021; Tysk, 2021). A mechanistic system is a service process that was developed through bases of conventional 'goods-dominant logic' rather than service centricity 'human centricity' (Jylkäs, 2020; Vargo & Akaka, 2009; Vargo & Lusch, 2004). In other words, how does the current cabin experience, that is, microservices, enable resource integration towards the 'macro level' of services (Flint et al., 2014)? An analytical discussion highlights opportunities where AIAS values can be applicable within the airline's ecosystem.

## Background: System

It is necessary to see any service as a fluid and ever-changing system that allows improvement, intervention and complete reconstruction when necessary (Jones, 2014; Jones et al., 2021). Therefore, AIAS does not attempt to fit itself into a crippled system to mend some parts of it in hopes of balancing the system's feedback loop. A 'balancing feedback loop' is regarded as the heart of the system where system actors are well integrated for the user to adjust and tweak system parts in order to achieve a desired value (Sweeney, 2017). When service actors and parts fail to integrate resources within a system, it causes an imbalance to the system feedback loop, perhaps overcompensating with the remaining functional parts but eventually depleting the system's overall 'stock' (Sweeney & Sterman, 2007). This leads to an unsustainable service where system flaws and failures become more probable (Chapman, 2004; Sweeney, 2017). Unsustainability often occurs when service systems are hierarchically or linearly constructed, placing the enterprise in the position of creating and delivering value on behalf of the user 'goods-dominant logic' (Flint et al., 2014; Greer et al., 2016). Such systems can also be referred to as 'mechanistic' where the user is often the receiving part of values and lacks the channels to adjust and tweak a balancing feedback (Chapman, 2004; Craven, 2021; Tysk, 2021). Thus, wellbeing as a 'balanced' experience is unattainable without the provision of value-in-use (Dodge et al., 2012).

In this sense, when systems are based on the above mechanism, it becomes inefficient and not cost effective to replace some parts of it with high-tech devices in hopes of achieving high-touch values. The goal is not to give the user the aesthetics of a digital experience. The goal is for technology to open more ways of access, enabling the service to integrate its resources well enough for the user to regard them as value stocks, from which the user can deploy to sustain a balancing feedback loop (Jones, 2014; Rijn & van der Burgt, 2021; Sweeney & Sterman, 2007). This can more accurately define how high-touch services are achieved. In other words, high-touch is achievable when digital transformation in mobility (e.g. ACI, 2021) is service centric, capable of expanding relationships across the service ecosystem as whole (Akaka & Vargo, 2014; Rjsé et al., 2022). Thus, a service is comprised of an ecosystem that stretches beyond what is experienced at the time of service consumption 'functional dimension' (Grönroos, 1984). When taking service design mentality in view (Stickdorn & Schneider, 2012), services must always be built throughout the contextual observation of the user. This means that ideation, user pain points and solutions are often generated during the time of service consumption at the micro level (Stickdorn, 2018). However, value-in-use through SD logic is achieved through the development of co-created values to infinite possibilities—underlining and reinforcing relationships amongst system actors (Mills et al., 2010). That is why a zoom-out viewpoint of the ecosystem, where situations that are linked to the meso- and macro levels are visualised through 'mapping' relationships (Clarke, 2022), allowing evaluation and perspectives to build upon strategies of integration (Akaka et al., 2014; Gummesson & Grönroos, 2012). Hence, service design thinking, regarded here as a major axiom of SD logic, is a necessary force that feeds into a sustainable high-touch service system (Vargo & Lusch, 2016). To place this into context, developing a digital service system is initiated through service design tools well within the service consumption period (Stickdorn & Schneider, 2012), observing user behaviour through one of the most critical touchpoints of the service journey—'moment of truth' (Grönroos, 1984); subsequently, SD logic promotes the integration of system actors to run the moment of truth more smoothly, providing various resources, 'skills and competencies' that increase the chances of system sustainability (Field et al., 2021; Sweeney et al., 2022). Thus, service centricity (i.e. service design) and SD logic are factors that are interchangeably used throughout the analysis process as means of expanding AIAS implications towards a 'macromarketing' domain (Layton, 2008).

This study, however, is mostly a zoom in to that micro level of service consumption where service design thinking is pinnacle. The airline industry's major function is to transport passengers. Even though in recent times, more airlines have also shifted towards cargo services (Huileng, 2022), mitigating around the industry's volatile marketplace by utilising emerging selling opportunities (Frost, 2019), the transportation of passengers remains to be the predominant function of airline business models (Kökény et al., 2022). Thus, human centricity through service design remains to be an essential factor of airline ecosystem sustainability from which SD logic enables scalability to the business model (Greer et al., 2016; Vargo & Akaka, 2009). To achieve such human (service) centricity, design digs deeper into various aspects that form a 'psychographic' profile of the user (Zins, 1998). Psychographics is regarded as a human-centric marketing approach where the user is seen from behavioural and identity-association perspectives, constructing what is referred to as 'user archetype' (Saunders, 2016). This includes the user's initial state of mind which predicts the user's expected service (Grönroos, 1984, 2007). Psychographics and user behaviour are further accessible today through technologies that collect the user's historical data more actuality (Siering et al.,

2018). Having such tools empowers enterprises to implement value-in-use through the consent of the user, improving service quality and brand image (Kujur & Singh, 2019). Thus, a psychographic profile that showcases customer pain points, goals, desires and dreams can be utilised to create enhanced 'high-touch' experiences throughout the extended service journey, for example, pre- and post services (Rjsé et al., 2022). Consequently, these profiles humanise the traveller's identity through 'archetype' categorisation to which each customer typically belong to, for example, explorer, retreater, hero, enthusiast and pilgrim (Pihlajamaa et al., 2018). This attitude towards user and environmental behaviour is a key factor towards service centricity, allowing a wider understanding of the human context of service interactions, from which the user shares similar behaviour and social traits with other like-minded users (Gad & Strandsbjerg, 2019). Mapping out these relationships gives evidence that may provide additional tools to improve a user's initial or expected service (Clarke, 2022; Pihlajamaa et al., 2018). Thus, service design on the micro level domain is a necessary steppingstone for resource integration and the co-creation of value-in-use within the entire scope of the airline marketplace (Grönroos & Gummerus, 2014; Vargo & Akaka, 2009) (Fig. 8.1).

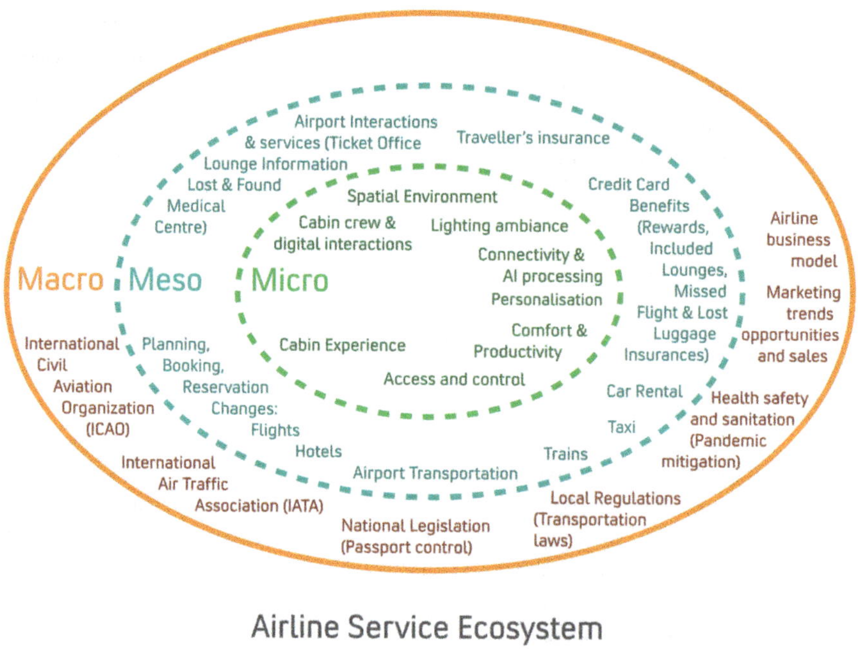


**Fig. 8.1** The macro, meso-, and micro levels of the airline service ecosystem. These system parts and actors have been founded on basis of relationships amongst stakeholders that directly and indirectly effect the cabin experience. This study alongside its predecessor (Rjsé et al., 2022) have defined the above airline service ecosystem

## Approach: Design

The main function of a systemic design process is to allow empirical research analysis to go through iterative steps in which technology and service centricity are continuously identified. This is applied through SD logic and service quality—and the process produces 'value-in-use' (Almquist et al., 2016; Grönroos, 1984, 2007; Jones et al.'s, 2021; Stickdorn & Schneider, 2012; Vargo & Akaka, 2009). Systemic design is initiated by attempting to frame, listen and understand the service system in practice, for example, observing demonstrations of products and services. Consequently, this guides data analysis to examine the type of service system at hand from which AI-augmented services (AIAS) can possibly be integrated. In essence, the approach allows necessary steps to be taken to define the respective type of an ecosystem, what elements are needed to empower services with value-in-use, and how to 'implement' AI for such cause. Hence, this systemic scheme or framework attempts to address system complexities (Akaka et al., 2013), accommodating the progressive nature of technology within the AIAS criterion cycle (later elaborated at Fig. 8.3 in the Discussion section).

In the predecessor study of 'AI Enabled Airline Cabin Services' by Rjsé et al. (2022), to 'frame a system' capable of personalising the user's cabin experience, we cross analysed key components of AI technology and user's needs to describe how 'high-tech' can transform into 'high-touch', that is, personalising the digital experience. As for this study, it continues by empirically 'listening to the system' to further discover system parts and actors, 'defining' their relationship with one another and how that impacts future 'implementations' of value-in-use; ultimately, it is to 'explore possible opportunities' through the discussion section of this chapter. Nonetheless, defining relationships amongst system actors can be a 'messy and complex situation' but is most notably one of the most crucial steps towards 'understanding' what values can be co-created and how solutions can be offered (Clarke, 2022); as such relationships are expected to grow within value-in-use interactions, complexities are addressed through 'system thinking' (Jones, 2014). Thus, 'seven steps to tackle complex issues' by Jones et al. (2021) is used to zoom out and observe service systems, whilst SD logic and service quality are used to zoom into system components to observe system parts, actors and relationships.

Current airline services are often evaluated according to a 'sustainable service system' (Gad & Strandsbjerg, 2019). To clarify this, a service 'feedback loop' is an integration of power that provides the user with necessary components to achieve a goal; a system is made of various parts that are either naturally connected or consciously constructed to achieve a particular 'value' (Sweeney & Sterman, 2007; Sweeney, 2017; Sweeney et al., 2022). The main goal is to allow the customer to determine what values should be provided for (Vargo & Akaka, 2009). In principle, this goal translates into service quality for the customer, value-in-use across the entire ecosystem and brand image for the company (Grönroos, 1984, 2007), that is, suitability. An ideal scenario of system sustainability is placed into context through the following example of a cabin experience from which the collected data will be analysed: the main actors involved at the microservice are the customer, service provider and technology. Passengers are regarded as the fundamental part of the system because their senses are what determines gaps, and such gaps can be expressed through the use of available onboard technologies such as Wi-Fi-enabled information and communication systems capable of integrating enough actors to propose values (solutions). The user interacts with these digital outputs which feed the user's permitted data into the system to determine the archetype of the user, that is, behavioural patterns. This interaction toggles through several tailored options according to who the user is and what resources are available for value-in-use to occur. Furthermore, user interactions enable the digital system to recognise weaknesses (system flaws) that may be improved upon by stating such touchpoints to the passenger, provider and enterprise.

Hence, as the user's cabin journey becomes more personalised, value-in-use is taking place across an integrated network capable of delivering system goals (Lusch et al., 2016). Nonetheless, streamlining such process requires an intensive amount of digital processing in the background capable of multimodal analysis (Bland, 2017, 2018). This means that both high-tech technology and service design must ensure the simplification of a rather complicated service system (Rjsé et al., 2022), enabling value-in-use through a user-friendly manner (Weller, 2019). With that comes two possibilities where AIAS may be functional, both of which are approached through the systemic design approach of Jones et al. (2021). The first is to observe the system through the examination of microservices at the cabin experience, determining the possible integration of AIAS within the current system. The second is to regard AIAS as an intervention model when signs of mechanistic systems are present (Chapman, 2004; Craven, 2021; Tysk, 2021). It is hypothesised that AIAS integration within the current system will rather be more problematic to achieve since airline services are linked to many legal and external legislative bodies that may decrease service flexibility, that is, operating a mechanistic system. Thus, to increase the probability of a sustainable system design (e.g. Field et al., 2021), it is assumed that AIAS will be more effective as an intervention model.

## Methodology: Themes

The key method comprises of contextual interviews within nine booths (enterprises), in the span of three consecutive days, at the Aircraft Interior Expo (AIX, 2022). This expo brings the largest number of state-of-the-art services and technologies under one roof (e.g. Street, 2022). The expo was spread across three connected halls, divided into the following theme categories: (1) interior fuselage composite materials, windows material and exterior lighting; (2) food and beverage catering and galley equipment; and (3) technology (seats, interior lighting, Wi-Fi connectivity), cabin services (research-based firms for innovative cabin services). Hall 3 was the main premise of data input. Nevertheless, the topics discussed were often interconnected to service technologies offered by all three halls. These conditions are regarded as an ideal environment for 'benchmarking' (Stickdorn, 2018), a form of an 'industrial ethnography' where several enterprises are evaluated according to the latest standards, failures, future developments (Miettinen et al., 2014; Miettinen & Alhonsuo, 2019, p. 488) and systemic service patterns (Jones et al., 2021).

Interview style is approached phenomenologically and contextually; in this sense, the latest topics of technologies and services—offered by neighbouring booths—are brought to the attention of the interviewee. This phenomenological approach attempts to organically set a stage where interviewees can observe the given environment and reflect through their own expertise (Van Manen, 2014, p. 216). These interviews are performed through 'contextual interviews' (Stickdorn & Schneider, 2012), a service design tool that allows interviewees to demonstrate the experience of using a product or service whilst the interview dialog takes place. Each participant was appointed, by their management, as a representative of their company or department, with a level of expertise high enough to delegate service or product benefits. This meant that each participant was actively involved in the service development process with at least four years of expertise in the respective field (see Table 8.1). The final two booths represent major relevance; therefore, topics generated from previous booths were more explicitly brought to the attention of the final two booths as means of phenomenologically benchmarking the scene at hand. All interviews were conducted in person at the service demonstration point, and most were documented through audio recordings; some of them, however, were documented through written notes. Complete demographical data was not collected since it does not provide direct relevance to the research design and analysis criteria. Consent was taken by audio recordings before each interview. This includes the rights to use the acquired data for various studies related to research and development. Photos of each booth and product were included as author's supplements and referencing means. The entire process produced 31 pages of transcribed text.

To recognise patterns (codes) and generate themes of the transcription (Saldaña, 2014; Suoheimo & Toni, 2020), a major analysis report was produced according to the following thematic methods: familiarising with the data, generating preliminary codes, recognising themes, reviewing the themes and defining the themes (Braun & Clarke, 2006). Within the 'reviewing the theme' phase, a twofold validity process from each of the co-authors was performed to 'ascertain whether the themes "work" in relation to the data set' (Braun & Clarke, 2006, p. 91). The initial analysis report generated a set of major themes from the data set. Subsequently, an uncoded copy of the data set (transcription file) was shared with the co-authors alongside a description of the generated themes. The co-authors produced two additional analysis reports accordingly. The three reports were then cross analysed to assert validity: co-author reports have been able to confirm accuracy, and the themes have been consequently ratified.

**Table 8.1** Showcasing the types of equipment and services each of the nine booths provide, including interviewee role and gender. Chronologically ordered from the first to the final booth interviewed

| Booth | Field, expertise and gender |
|---|---|
| 1 | *Seating experience*: Modern and thinner seats designed for comfort and spaciousness. Executive level. Male |
| 2 | *Seating technology*: Seats that include personalised individual experience through sound and passenger's own control to lighting. Executive level. Male |
| 3 | *Lighting technology*: Hardware for cabin LED lighting. Colour and lighting effects through software development. Two male participants interviewed for this booth: founders/CEO |
| 4 | *Onboard hardware boxes*: Manufacturer of data recorder and processor machines (computation equipment) for video and audio in-flight entertainment (IFE). Machine learning and data processing computation. Portable wireless streaming services. Executive level. Male |
| 5 | *Digital interface design*: Software development and UI design for digital communication and catering services. Wi-Fi and 4G connectivity for personal mobile devices and cabin IFE systems. Personalised digital user experience. Executive level. Female |
| 6 | *Connectivity and lighting experience*: Leading manufacturer of assorted aircraft light structures and fixtures for general cabin and individual capsules (suites) seating areas. Executive level. Male |
| 7 | *Light, power and audio manufacturer*; LED mood lighting, in-seat power controlled by passenger and crew. Lighting fixtures and ambience control through (outsourced) software control. Executive level. Male |
| 8 | *AI-enabled in-flight services*: Research-based AI in-flight assistant and communication platforms. Ability to integrate within the IFE system. Development through practical testing. Founder and CEO. Male |
| 9 | *Digital cabin services*: Information and communication systems through satellite. Wireless cabin connectivity. Digital catering services. A major aircraft manufacturer. Executive level. Male |

## Findings: Intervention (Fig. 8.2)

Marketing opportunities can be regarded as a theme that shares potential values with existing technologies, having attributes that connect international competencies for service centricity (Akaka et al., 2013). Some issues were identified as hurdles to the application of AI-augmented services (AIAS), where no value was recognised for the integration of AIAS within the microservice system; these topics are clustered within the theme of AIAS blockers. As for the regulation and security, this theme considers legislative laws that govern the passenger's cabin experience and can be regarded as both hurdles and potentials to AIAS value implementation. Most notably, even though the theme 'emerging technologies' generated the highest number of topics, the topic of 'passenger values' remains the most integrated topic across all themes. In other words, technology seems to be dominant (e.g. lighting technology and connectivity); however, almost all discussions revolve around passenger values.

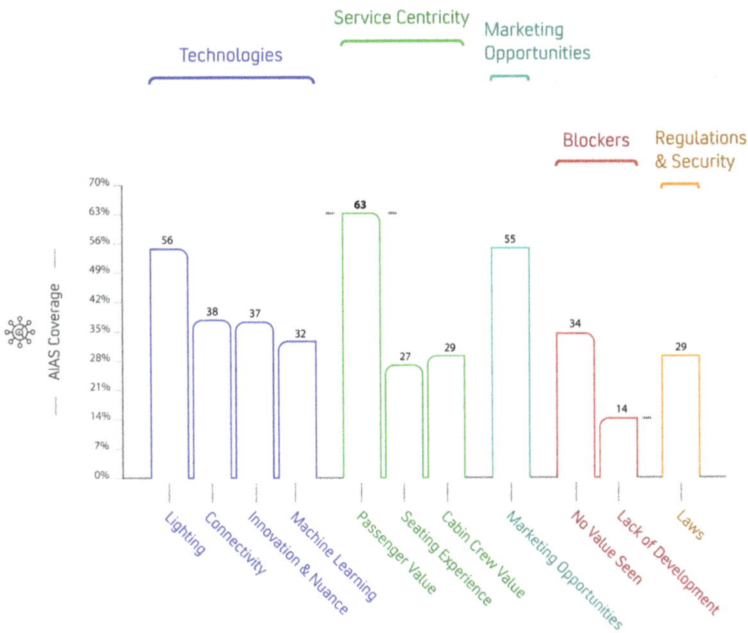


**Fig. 8.2** Thematic analysis results. The theme 'technologies' generated most widespread coverage amongst the 11 topics (i.e. codes) However, the topic of 'passenger values' within service centricity was consistently integrated within each of the five themes, and thus scored the highest integrated coverage of any other topic

All the themes are generated towards the potential of integrating the model AIAS into the current micro-system of airlines. Results indicate that AIAS properties are highly compatible with value-in-use (Grönroos, 2008), personalisation and opportunities in which co-created values are expanded through service-dominant (SD) logic to a wider network (Vargo & Akaka, 2009). Nevertheless, the current results primarily indicate favourable potentials towards personalising the user experience through digital communication channels within the microservice consumption period itself (Grönroos, 1990). This is evident by enhanced access to onboard food and beverage catering services through AI assistance and empirical testing (Booth, 5; 8; 9), primarily linked to the technical advancements of onboard Wi-Fi connectivity (Booth, 4), making it possible for faster interactive user interfaces to align with such developments (Booth, 5). On the other hand, there was less of an indication to the significance of expanding user access towards the greater ecosystem, for example, linking AI-enabled services to external stakeholders that may allow the quality of value-in-use to improve (Davey & Grönroos, 2019). In other words, these favourable results towards high-touch microservices are still limited because the full potential of AIAS is only possible when current high-speed communication stability is expanded beyond the cabin walls. Booths that were rather more customer oriented (e.g. software interface, lighting and seating environment) did not show much of an indication towards such extended SD logic (Booth, 1; 6); 'there is no visible pain point to fix' (Booth, 5). To the contrary, booths that are hardware oriented showed more indications of SD logic through intentions to further develop their technologies in such direction (Booth, 3; 4).

In essence, AIX (2022) expo is designed to create stronger bonds and integration of airline service competencies, viewing skills as an alternative to value-in-exchange (Vargo & Lusch, 2004; Grönroos, 2008) and inviting more external stakeholders within the industry's ecosystem, a criterion essential to value-in-use and SD logic (Hughes & Vafeas, 2018). In general, however, the expo seemed to focus on business-to-business relationships (Vargo & Lusch, 2011) and less towards direct customer service centricity. The number of booths providing user-centric research and services were considerably fewer than those showcasing products and technologies. AIX is geared towards linking large-scale competencies, but one cannot dismiss the lack of service-centric enterprises directly targeting the passenger or even cabin crew. Nonetheless, within these few enterprises, Booths 5 and 8 clearly exhibited service-centric understanding towards their service development, indicated by their empirical practices and the use of typical service design language (Stickdorn & Schneider, 2012); these two booths were consciously selected to investigate the relationship of service design in cabin services. Additionally, most topics generated from Booth 3 (lighting technology) were in fact human centric, but the enterprise's practical approach is rather mechanistic (Chapman, 2004); the two experts interviewed at that booth were indirectly aware of such systemic approach and elaborated that air traffic regulations and passenger security laws are largely what contributors to the slower service centric development of their business, 'but things are changing' (Booth, 3). They also explained that newer aircrafts are now equipped with more 'power supply' surrounding the passenger's seating area from various angles, unlike in older aircrafts, hinting that such development will allow lighting technology to evolve quicker towards service centricity whilst still abiding to the industries regulations (Booth, 3). Human-centric awareness of cabin services was echoed by every single interviewed personnel. This is because the airline industry holds a unique aspect where almost all service providers and product developers are effectively users/'passengers' themselves. This means that those proposing values have also experienced the service and can better relate towards the provision of solutions; even though the perception of analysis signifies that technology had overcome high-touch services, there was never a sense of complete disconnection from service centricity. Hence, the assimilation of 'passenger values' within all five themes is justified.

In this sense, emerging technologies within the microservices of airlines are often linked to service centricity. With advanced hardware capabilities (Booth, 4), data can be processed in unprecedented speeds through onboard computers. This reflects the user's increased demand for onboard mobile device applications and in-flight entertainment (IFE) to have better Wi-Fi connectivity, media streaming and catering services. These new onboard computers (Booth, 4) consume less power and are lighter in weight which in turn serves the airline company by decreasing fuel costs. On such note, Booth 8 is a software development firm for onboard passenger communications, conducting multimodal assessments through simulations and actual flight testing (Berggren, 2021). Whilst using personal mobiles or IFE devices, the passenger is connected to the flight attendant through AI-enabled 'chat services' (Booth, 8).

## Discussion: AIAS Opportunities

### *AIAS*

Results have produced an updated version to the model AI-augmented services (AIAS). In each study, AIAS attempts to zoom in and out to better observe the service system as whole, from the focal service consumption point, that is, 'micro' to a 'macro' view, where it is possible to inclusively map out stakeholders. The 'meso-' level is also observed to identify stakeholders who act as a bond between the micro and meso-regions, maintaining relationships and, thusly, allowing all stakeholders to be regarded as beneficiaries (Vargo & Lusch, 2010). Hence, through a systemic design framework, the AIAS model has now been reiterated to incorporate a service system recently identified through this study—mechanistic service system (Chapman, 2004). AIAS is designed to point out the necessary components of 'service logic' in which value-in-use though AI implementations is now supported for 'mechanistic' mobility services. Here, AIAS originates observations from the micro-service level; the cabin experience is seen as a place where service design tools can be applied to understand service-dominant (SD) logic implications, for example, the relationships technology can provide to alleviate possible service failures at the time of occurrence. AI is regarded as an augmenting instrument that offers a platform for service actors to be at better reach. AI also learns the behaviour and habitual nature of the user to offer quicker more high-touch solutions in real-time, and thus achieving service centricity. The microservice system is a delicate and critical stage where the customer's perception of service failures may not be easily repaired if propositions of fixes are not present at the spot (Grönroos, 2007; Vázquez-Casielles et al., 2007). In other words, when service failures occur, customers prefer 'immediate' service interactions rather than subsequent monetary compensation for future services (Xu et al., 2019). Accordingly, AIAS places digital operations to such criteria.

The previous AIAS model (1.0) was originally introduced to describe the types of 'technologies' needed in which human-centric user values through AI can be achievable (Rjsé et al., 2022). AIAS Model 2.0 (Fig. 8.3) is a latest synthesis that incorporates SD logic through the introduction of 'value-in-use', namely, when values are co-created through a 'relationship' between high-tech (technology) and high-touch (service centric design), it augments the user's ability to reach out towards further competencies—offered by the meso-system—that benefit the user during 'microservice' interactions. AIAS as a model is designed to continuously go

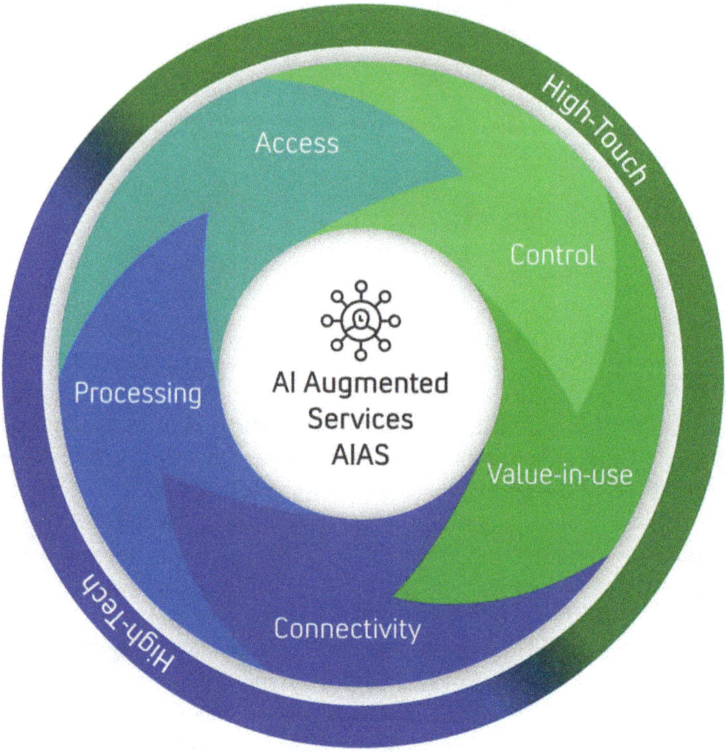


**Fig. 8.3** AIAS Model 2.0. By listening and understanding the ecosystem, 'value-in-use' is attainable by augmenting the user's ability through AI-enabled service interaction. AIAS 2.0 is the original outcome of a systemic design framework and a thematic analysis; thus, this model is unique to this study

through an iterative systemic design process, in each consecutive research project, allowing it maturity through either improving the service system or intervening with it to offer a new system transition. This study has revealed that future iterations of AIAS need to take an intervention structure (e.g. Jones et al., 2021) by identifying 'obstacles' hindering SD logic for value-in-use.

## *Opportunities*

Developments of technical hardware (Booth, 4) and interactive AI software (Booth, 8) are predicted to evolve, rendering high-tech for high-touch (Rjsé et al., 2022). Technology and service offerings from Booths 4 and 8 are substantial indicators that AIAS may be successful in future implementations, that is, offering to expand competency integration throughout the ecosystem to which a better cabin experience is provided for (e.g. Vargo & Lusch, 2017). However, a major finding reveals that the airline ecosystem is mostly governed by a mechanistic service system in which AI value-in-use through SD logic is a complicated mission to achieve.

Obtaining a better understanding of the type of technologies offered, wireless connectivity and LED lighting technologies emerged as two of the most exhibited emerging technologies. In this sense, the potential of integrating such technologies within the technical 'high-tech' part of AIAS was exercised thoroughly within the generated topics and themes, for example, providing tailored lighting aesthetics to the passengers' seating environment (Booth, 2; 6; 7), synchronising the lighting experience with the passenger's circadian rhythm through historic data and AI processing (Booth, 3), and utilising AI capabilities to give more user control through personal mobile devices and IFE (Booth, 8). Booth 5 indicated the strongest approach towards development through service design methods, echoing Booth 8 but with more emphasis towards cabin crew as a user for digital food and beverage catering interactions; nevertheless, Booth 5 development relays on improving a system that is already in place 'reliability', whereas Booth 8 is focused on innovative experimentation through AI-assisted software, often placing the passenger at the centre of research development 'service centric'. Booth 5 revealed less favourable indications towards an implementation of 'individually synchronised lighting environment' for all passengers through AIAS implications, stating that the efficacy of cost and development is not justified at this time. Even though Booth 5 suggested service design guidance, there was a stronghold on how AIAS would immediately be interpreted in the market rather than what potential AIAS would have on the market. In contrast, Booth 8 expressed a welcoming attitude towards AIAS rationality. Both enterprises can be considered well within SD logic since both are linking competencies within the microservice period but seem to be limited to that premise only.

Nevertheless, Booth 8 showed most potential to the application and testing of innovative AI-enabled services within the cabin, which can be regarded as a steppingstone towards a radical personalised cabin experience (e.g. Street, 2022). The expert expressed an experimental attitude towards services, using AI technology as a tool to reach greater user values. This firm has started to empirically test out AI-enabled services within selected actual flights: The passenger can place a request through digital channels where cabin crew can attend to in a timely and efficient manner. This can be regarded as an indication to significant AIAS foundations: AI background processing augments the capabilities of cabin crew to process more reliable and quicker services, improving wellbeing and allowing cabin crew to focus on other critical matters (Rjsé et al., 2022, p. 11). When asked about how user values and the cost of development is justified to the airline company, a subject that cannot be dismissed in such cost-conscious industry (Frost, 2019), the personnel expressed that during the testing phase—the user's feedback (passenger and cabin crew) is where the justification of such systems rely—signifying more representation towards user attitude rather than relying on 'mechanisms' already placed by a given marketing system (Chapman, 2004). In other words, Booth 8 facilitates radical approaches outside the normative context of cabin services which may potentially influence future technologies, allowing the user to independently express goals and create values (Holmqvist et al., 2020) and giving enterprises potential tools to explore relationships for value-in-use (Kelly et al., 2017; Vargo & Akaka, 2009). To interpret such possibility through AIAS thinking, when a passenger experiences a flight delay whilst onboard, AI multimodal analysis (Rjsé et al., 2022, p. 4–6), reliable connectivity, communication channels linking external stakeholders such as hotels and other airline companies would empower more actors to offer solutions in real-time whilst expanding beneficiaries across a network of 'relationships' (Clarke, 2022).

In this sense, AIAS aims to reach a prospective where the meso-services of planning, booking and problem-solving (Google, 2016) are as accessible to the user as the current onboard micro-communication services. Acknowledging the sheer number of companies at AIX already offering advanced satellite connectivity for high-speed cabin Internet (e.g. Booth, 4; 5; 8; 9) (e.g. OneWeb, 2022), greater accessibility can indeed be anticipated. For this to be fulfilled, onboard computers (Booth, 4) must demonstrate even larger capacities of data processing for AI to facilitate value-in-use meso-services. General results indicate that AIX emphasises more innovations towards technology than human centricity. On the one hand, this implies that technical equipment are likely to continue developing in which AIAS high-tech requirements are met; on the other hand, this may leave a noticeable gap where technology does not offer adequate high-touch cabin services, with the exception of Booth 8. That is not to say that SD logic is not applied within airline services because all experts from various booths showed a great understanding of service centricity; one of the expo's main objectives is to encourage networking and relationships that link skills and competencies. Nonetheless, participants' narratives indicated that external matters that affect the airline industry makes innovative high-touch services more difficult to implement, especially towards cabin services. As a result, technical advancements within the airline industry seem to be geared towards the implementation of a mechanistic system that is too rigid for radical changes. This means that the enterprise is at large the decision-maker of what values the user should get, leaning towards a goods-dominant logic (Vargo & Akaka, 2009). Nonetheless, there are also noteworthy efforts where technology empowers the user to co-create value-in-use within cabin services (Booth, 3; 8); how big of a role these companies are expected to play in future services will determine the type of service system to anticipate and its effect on the passenger experience and ecosystem as whole.

There are two scenarios where AIAS can prove efficacy. The first pertains to microservices (cabin experience). With experimental developments through testing (e.g. Booth, 8), access to further values of controlling the passenger's lighting environment and improved cabin crew communications throughout the critical period of service is conceivable; AIAS is capable of utilising technology to co-create values within the limits of the 'functional dimension' of services (Grönroos, 1984). Alternatively, in a scenario where AIAS has the capacity to co-create value-in-use that connects system parts and actors from meso- and macro spectrums (Rijn & van der Burgt, 2021), alleviating service failures that affect the passenger's journey beyond the cabin experience, it is far more complicated to expect such scenario to take place now. In general, the ecosystem of airlines indicates more signs towards a mechanistic nature to value creation. This is evident because airline regulations must abide by various entities from international associations (ICAO Structure, 2022), to local and international security measurements (IATA History, 2022), as well as external disruptions that majorly affect the passenger and industry as a whole (Huileng, 2022). This makes the pace of service centricity much slower, especially within the microservice expectations. Hence, the scenario of AIAS macro functionality is met with questionable results; no indication of pain points has been previously presented to justify development costs (Booth, 5). Unfortunately, this proves a major hurdle since AIAS is not just about finding solutions after pain points occur; it is about predicting service failures before they happen, enabling the system to offer necessary solutions at 'the moment of truth' (Google, 2016; Grönroos, 2007, 2008). Given that each passenger has a unique set of criteria, the application of service centricity through AIAS is more probable when there is a data flow from various competencies beyond the cabin's environment (Vargo & Akaka, 2009).

To initiate AIAS as an AI-capable service system, much of the model's AI technical functionalities and data sharing analogy was adopted through benchmarking the advances of AI for autonomous driving (Rjsé et al., 2022, p. 6–7), namely, a close collaboration with Bland, B (2017, 2018), at 'Sensum', a development lab for AI-assisted driving systems and autonomous driving has established many of the examples human and AI interactions can facilitate though an aircraft cabin journey. This study alongside its predecessor finds that the relationship between customer, AI and service centricity within a cabin journey is more extensively researched and practised through automotive mobility services in comparison to the airline industry. For example, in mobility as a service (MaaS), where mobility is regarded as a service rather than a vehicle or good (MaaS Alliance, 2022), passengers are aware of the type of data needed to allow MaaS digital platforms to function and offer more personalised values. The nature of MaaS is to join more transportation means such as busses, trains and ferries into one single streamlined journey. Hence, MaaS users are often more willing to engage with subscription schemes that permissibly collect data and payments because values are made visible through each digital and physical transition they interact with (Rajab & Rjsé, 2020), namely, 'trust' between the customer and AI is more probable because cost, and data sharing is continuously justified (Rjsé et al., 2022). This is especially applicable when vehicles with AI sensors (Bland, 2017) are integrated within a MaaS scheme that uses such vehicles as part of its service system parts. In this case, AI connects a larger network of transportation means whilst maintaining service competencies for a more streamlined and personalised journey.

Therefore, even though AIAS is geared towards the domain of airline services, its functional component and value proposition are closely related to any mobility service that views AI as an actor facilitating value co-creation (e.g. Jylkäs et al., 2018). Nevertheless, when it comes to AI implementation through the systemic design framework (Jones et al., 2021), the above scenario of AI-enabled 'vehicles' seem to integrate better within a MaaS system without the need of an intervention model. This is because such scenario already indicates more openness towards AI development and data sharing with less reliance towards past mobility habits; in this sense, it is not regarded as a mechanistic system (Chapman, 2004), unlike the case with AIAS implications for aircraft cabin services. Hence, the above example better illustrates how approaches of systemic design can assess the type of approach needed for value integration within various mobility systems.

## Limitations: Macro

An ecosystem of an established service economy such as the airline industry is often interlocked in a complicated web of system actors that most certainly impact that end user (passenger) as well as one another (service providers). The introduction of AI-enabled service dominate solutions within the cabin can prove to be a complicate affair with undefined values in terms of SD logic. This study was focused on the micro levels of the airline service (zoom in). More research is needed to analyse the relationships amongst a wider range of stakeholders from both the meso- and macro levels (zoom out). It is not possible to clearly identify the general efficacy of AIAS until broader practical testing is made. Future iterations of AIAS need to take an intervention method to which an in-depth analysis may highlight relationship links amongst stakeholders, possibly identifying obstacles hindering value-in-use implementations.

## Conclusion: Centricity

This study has produced two significant outcomes. (1) It has revealed the predominate nature of the airline ecosystem as 'mechanistic' and how this knowledge can assist in the integration of co-created values for better service quality. (2) AIAS as a model has gone through its second iteration, defining the general type of ecosystem airlines function upon. Across a future iteration, this brings the model closer to a full-fledged AI implementation blueprint for value-in-use in airline services.

Advancements in airline digital services within the micro level (cabin experience) show some signs of service centricity. Nevertheless, when taking factors that affect the industry's ecosystem as whole, a more mechanistic approach to service systems is evident. On a smaller scale, this is promising as far as enhanced cabin communications and aesthetics; it can be viewed as a benefit that improves what is already there within the cabin walls. However, it is rather difficult to expect AIAS to perform through means of value-in-use to demonstrate wider SD logic. As hypothesised, the observation and understanding of the current system suggests that AIAS must transform into an intervention model that initiates systemic design through human centricity as the most rational approach. In other words, AIAS would only be able to co-create values within a sustainable ecosystem when passenger needs and behaviour are placed as the main foundation of system design.

The evaluation process of benchmarking mobility services and products, from both the airline and automotive sectors, has found that unlike developments within the automotive sector—airlines passenger services seem to learn more towards a mechanistic system. Therefore, AIAS is to be developed in a future study as an intervention model to wreak the benefits of SD logic. The process of AI integration within the aircraft cabin is not predicted to develop as quickly as its automotive counterpart. Nevertheless, this highlights a significant opportunity where a reconciliation between the two sectors can multiply SD logic in which service competency and centricity is integrated in both sectors as one MaaS ecosystem. What this means is that AIAS functionalities can now extend beyond personalising the aircraft cabin experience by allowing the passenger to plan and book a continuous journey from door to door on a global level, using one digital platform accessible within any cabin airborne or on ground. This can be regarded as a milestone towards SD logical implementation because it expands the service ecosystem, giving the user a less interrupted journey. An initial step towards this can commemorate when airlines, airport ground mobility services and MaaS technology place more efforts towards data integration within these multiple sectors. Meanwhile, more research can be allocated towards passenger transportation needs before and after the airline journey.

The cabin journey is a unique learning experience with potential of life changing decisions and transcendent values. Passengers' demand for enhanced solutions at the time of need makes it necessary to utilise the capacity of today's technology, processing more various data resources that offer quicker personalised solutions. It is inevitable that digital services and reliable cabin connectivity will continue to develop, opening doors to more possibilities than what is traditionally at offer, but such innovation must be met with service centricity for the user to make sense of the values technology may bring.

## References

ACI (Airports Council International). (2021). *Airport digital transformation. World airport digital transformation survey report*. Airport Council International (ACI). https://store.aci.aero/wp-content/uploads/2020/11/ACI_World_Airport_Digital_Transformation_Survey.pdf

AIX. (2022). Where cabin concepts take off. *Aircraft Interiors Expo | AIX. 6–8 June 2023 Hamburg Messe.*https://www.aircraftinteriorsexpo.com/en-gb.html

Akaka, M. A., Corsaro, D., Kelleher, C., Maglio, P. P., Seo, Y., Lusch, R. F., & Vargo, S. L. (2014). The role of symbols in value cocreation. *Marketing Theory, 14*(3), 311–326. https://doi.org/10.1177/1470593114534344

Akaka, M. A., & Vargo, S. L. (2014). Technology as an operant resource in service (eco)systems. *Information Systems and EBusiness Management, 12*(3), 367–384. https://doi.org/10.1007/s10257-013-0220-5

Akaka, M. A., Vargo, S. L., & Lusch, R. F. (2013). The complexity of context: A service ecosystems approach for international marketing. *Journal of International Marketing, 21*(4), 1–20. https://doi.org/10.1509/jim.13.0032

Almquist, E., Senior, J., & Bloch, N. (2016). The elements of value. *The Harvard Business Review, 2016*, 46–53.
Berggren, P. (2021). Exercises using simulators. Lecture notes and slides. 2021HT. *System engagements in service and design research*. Linköping University (LiU). 15-Nov-2021.
Bland, B. (2017). Sensor fusion: The only way to measure true emotion. *Medium*. https://medium.com/adventures-in-consumer-technology/sensor-fusion-the-only-way-to-measure-true-emotion-28af9b6fb0a4
Bland, B. (2018). How to build emotionally intelligent transportation. *Medium*. https://medium.com/hackernoon/how-to-build-emotionally-intelligent-transportation-e6c0e62b58ba
Braun, V., & Clarke, V. (2006). Using thematic analysis in psychology. *Qualitative Research in Psychology, 3*(2), 77–101. https://doi.org/10.1191/1478088706qp063oa
Burr, C., Taddeo, M., & Floridi, L. (2020). The ethics of digital well-being: A thematic review. *Science and Engineering Ethics, 26*(2), 1.
Chapman, J. (2004). *System failure: Why governments must learn to think differently*. Demos.
Clarke, A. (2022). *Situational analysis in practice: Mapping relationalities across disciplines*. Forthcoming 19-April-2022. Routledge & CRC Press. https://www.routledge.com/Situational-Analysis-in-Practice-Mapping-Relationalities-Across-Disciplines/Clarke-Washburn-Friese/p/book/9780367470999
Craven, L. (2021). Systems consciousness. Lecture notes and slides. 2021HT. *System engagements in service and design research*. Linköping University (LiU). 06-Dec-2021.
Davey, J., & Grönroos, C. (2019). Health service literacy: Complementary actor roles for transformative value co-creation. *Journal of Services Marketing, 33*(6), 687–701. https://doi.org/10.1108/JSM-09-2018-0272
Dodge, R., Daly, A. P., Huyton, J., & Sanders, L. D. (2012). The challenge of defining wellbeing. *International Journal of Wellbeing, 2*(3), 222.
Field, J. M., Fotheringham, D., Subramony, M., Gustafsson, A., Ostrom, A. L., Lemon, K. N., Huang, M.-H., & McColl-Kennedy, J. R. (2021). Service research priorities: Designing sustainable service ecosystems. *Journal of Service Research*. https://doi.org/10.1177/10946705211031302
Finne, Å., & Grönroos, C. (2017). Communication-in-use: Customer-integrated marketing communication. *European Journal of Marketing, 51*(3), 445–463. https://doi.org/10.1108/EJM-08-2015-0553
Flint, D. J., Lusch, R. F., & Vargo, S. L. (2014). The supply chain management of shopper marketing as viewed through a service ecosystem lens. *International Journal of Physical Distribution & Logistics Management, 44*(1/2), 23–38. https://doi.org/10.1108/IJPDLM-12-2012-0350
Frost, L. (2019, 04 Oct). Airline bankruptcies surge, leaving rivals vying for planes. *Reuters*. https://www.reuters.com/article/us-airlines-bankruptcy/airline-bankruptcies-surgeleaving-rivals-vying-for-planes-idUSKBN1WJ1JH
Gad, U. P., & Strandsbjerg, J. (Eds.). (2019). *The politics of sustainability in the Arctic: Reconfiguring identity, space, and time* (1st ed.). Routledge. https://doi.org/10.4324/9781351031981
Godsiff, P., Maull, R., & Davies, P. (2018). Systems behaviour and implications for service-dominant logic. In I. S. Vargo & R. Lusch (Eds.), *The SAGE handbook of service-dominant logic* (pp. 214–229). SAGE Publications Ltd. https://doi.org/10.4135/9781526470355.n13
Google. (2016). *How micro moments are reshaping the travel customer journey. Mobile, consumer insights, travel, micro-moments*. Available at: https://www.thinkwithgoogle.com/marketing-resources/micro-moments/micro-moments-travel-customer-journey/
Greer, C. R., Lusch, R. F., & Vargo, S. L. (2016). A service perspective: Key managerial insights from service-dominant (S-D) logic. *Organizational Dynamics, 45*(1), 28–38. https://doi.org/10.1016/j.orgdyn.2015.12.004
Grönroos, C. (1984). A service quality model and its marketing implications. *European Journal of Marketing, 18*(4), 36–44.
Grönroos, C. (1990). *Service management and marketing—Managing moments of truth in service competition*. Lexington Books.

Grönroos, C. (2007). *Service management and marketing: Customer management in service competition* (3rd ed.). Wiley-Blackwell.

Grönroos, C. (2008). Service logic revisited: Who creates value? And who co-creates? *European Business Review, 20*(4), 298–314.

Grönroos, C., & Gummerus, J. (2014). The service revolution and its marketing implications: Service logic vs service-dominant logic. *Managing Service Quality: An International Journal, 24*(3), 206–229. https://doi.org/10.1108/MSQ-03-2014-0042

Gummesson, E., & Grönroos, C. (2012). The emergence of the new service marketing: Nordic School perspectives. *Journal of Service Management*. https://doi.org/10.1108/09564231211260387

Holmqvist, J., Visconti, L. M., Grönroos, C., Guais, B., & Kessous, A. (2020). Understanding the value process: Value creation in a luxury service context. *Journal of Business Research, 120*, 114–126. https://doi.org/10.1016/j.jbusres.2020.07.002

Hughes, T., & Vafeas, M. (2018). Service-dominant logic as a framework for exploring research utilization. *Marketing Theory, 18*(4), 451–472. https://doi.org/10.1177/1470593118764019

Huileng, T. (2022, 31 March), *Passenger planes are being converted to cargo planes at record levels*. Businessinsider.com.https://www.businessinsider.com/passenger-planes-converted-cargo-planes-record-high-levels-inflation-2022-3?r=US&IR=Tssenger Planes Are Being Converted to Cargo Planes at Record Levels (businessinsider.com)

IATA History. (2022). *The founding of IATA*. Iata.org.http://www.iata.org/about/Pages/history.aspx

ICAO Structure. (2022). *The International Civil Aviation Organization (ICAO) - policy, Assembly, Council, Secretary General and Secretariat*. Nationsencyclopedia.com.http://www.nationsencyclopedia.com/United-Nations-Related-Agencies/The-International-Civil-Aviation-Organization-ICAO-STRUCTURE.html

Jones, P. H. (2014). Systemic design principles for complex social systems. In G. Metcalf (Ed.), *Social systems and design. Translational systems sciences* (Vol. 1). Springer. https://doi.org/10.1007/978-4-431-54478-4_4

Jones, P. H., et al. (2021). Seven steps to tackle complex issues. *Systemic Design Toolkit*. https://www.systemicdesigntoolkit.org/methodology

Jylkäs, T. (2020). *SHARED PATH service design and artificial intelligence in designing human-centred digital services* (unpublished doctoral dissertation). University of Lapland, Rovaniemi, Finland. https://luc.finna.fi/ulapland/Record/lauda.10024_64433

Jylkäs, T., Äijälä, M., Vuorikari, T., & Rajab, A. K. A., & Rjsé, V. (2018). *AI Assistants as non-human actors in service design*. Proceedings of the 21st DMI: Academic design management conference, Next Wave, pp. 1436–1444.

Kelly, P., Lawlor, J., & Mulvey, M. (2017). Customer roles in self-service technology encounters in a tourism context. *Journal of Travel & Tourism Marketing, 34*(2), 222–238. https://doi.org/10.1080/10548408.2016.1156612

Kujur, F., & Singh, S. (2019). Visual communication and consumer-brand relationship on social networking sites - uses & gratifications theory perspective. *Journal of Theoretical and Applied Electronic Commerce Research, 15*(1), 30–47.

Kökény, L., Kenesei, Z., & Neszveda, G. (2022). Impact of COVID-19 on different business models of European airlines. *Current Issues in Tourism, 25*(3), 458–474. https://doi-org.ezproxy.ulapland.fi/10.1080/13683500.2021.1960284

Lusch, R. F., Vargo, S. L., & Gustafsson, A. (2016). Fostering a trans-disciplinary perspectives of service ecosystems. *Journal of Business Research, 69*(8), 2957–2963. https://doi.org/10.1016/j.jbusres.2016.02.028

Layton, R. A. (2008). The search for a dominant logic: A macromarketing perspective. *Journal of Macromarketing, 28*(3), 215–227. https://doi-org.ezproxy.ulapland.fi/10.1177/0276146708320451

MaaS Alliance. (2022). *European Union's transparency register* (ID: 140593628433–52). https://maas-alliance.eu/homepage/what-is-maas/

Mele, C., & Russo-Spena, T. (2022). The architecture of the phygital customer journey: A dynamic interplay between systems of insights and systems of engagement. *European Journal of Marketing, 56*(1), 72–91. https://doi.org/10.1108/EJM-04-2019-0308

Miettinen, S., & Alhonsuo, M. (2019). Service designing a new hospital for Lapland hospital district. In M. A. Pfannstiel & C. Rasche (Eds.), *Service design and service thinking in healthcare and hospital management* (pp. 481–497). https://doi.org/10.1007/978-3-030-00749-2_27

Miettinen, S., & Koivisto, M. (2009). *Designing services with innovative methods*. Kuopio Academy of Design. Distributor: University of Art and Design.

Miettinen, S., Rytilahti, P., Vuontisjärvi, H.-R., Kuure, E., & Rontti, S. (2014). Experience design in digital services. *Research in Economics & Business: Central & Eastern Europe, 6*(1), 29–50.

Mills, A. J., Durepos, G., & Wiebe, E. (2010). Situational analysis. In *Encyclopedia of case study research* (Vol. 1, p. 871). Contributors: Adele E. Clarke. https://doi.org/10.4135/9781412957397.n321

OneWeb. (2022). Delivering LEO connectivity for new network solutions. *OneWeb.*https://oneweb.net/

Pihlajamaa, O., Eckhardt, J., Rajan, A., Kostiainen, J., & Kinnunen, T. K. (2018). *Matching mobility services with tourist traveller archetypes in rural destinations*. Paper presented at 25th ITS World Congress, Copenhagen, Denmark. EU-SP1617. Retrieved 16 May 2020, from https://cris.vtt.fi/en/publications/matching-mobility-services-with-tourist-traveller-archetypes-in-r

Rajab, A. K. A., & Rjsé, V. (2020). *DesİQual: Destination in motion. Emotional engagement as a determinant of service quality*. Service design for a personalised travelling experience and well-being (Master's Thesis). University of Lapland, Rovaniemi, Finland. https://lauda.ulapland.fi/handle/10024/64385

Rijn, M., & van der Burgt, R. (2021). *Explore The Big Picture - forces shaping the future of humanity* (1st ed.). Pearson Benelux. (23 December 2021. Macro, Meso, Micro environment explained. https://ivto.org/macro-meso-micro-environment-explained-foresightcards/

Rjsé, V., Jylkäs, T., & Miettinen, S. (2022). *AI enabled Airline Cabin Services: AI Augmented Services for emotional values. Service design for high-touch solutions and service quality*. Design Management as a Strategic Asset dmi: Academic Design Management Conference. OCAD University Toronto, Canada. https://www.dmi.org/page/ADMC2022proceedings. Direct PDF link: https://drive.google.com/file/d/1VDJdeOKH771XGCrlvcmdBB3Wanqv71nR/view?usp=share_link

Saldaña, J. (2014). Coding and analysis strategies. In P. Leavy (Ed.), *The Oxford handbook of qualitative research* (pp. 580–598). Oxford University Press. https://doi.org/10.1093/oxfordhb/9780199811755.013.001

Saunders, A. (2016). What is psychographic segmentation and how is it different from segmenting by customer demographics?. *Keep Business Success*. https://keap.com/business-success-blog/marketing/email-marketing/why-psychographic-segmentation-is-important

Siering, M., Deokar, A., & Janze, C. (2018). Disentangling consumer recommendations: Explaining and predicting airline recommendations based on online reviews. *Decision Support Systems, 107*. https://doi.org/10.1016/j.dss.2018.01.002

Stickdorn, M. (2018). *This is a service design: Doing a job in the real world: A practitioner's handbook*. O'Reilly Media.

Stickdorn, M., & Schneider, J. (2012). *This is service design thinking: Basics, tools, cases*. BIS Publishers.

Street, F. (2022, 16 June). 'Floating' airplane cabin could be the future of travel. *CNN Travel.*https://edition.cnn.com/travel/article/crystal-cabin-awards-2022-winners/index.html

Suoheimo, M., & Toni, L. (2020). *Process for mapping challenges of cross-border mobility in the Barents region*. Proceedings of the sixth international conference on design creativity (ICDC 2020), 168–175.

Sweeney, L., MacNamara, Á., & Horan, D. (2022). The Irish football player pathway: Examining stakeholder coherence throughout and across the player development system. *Frontiers in Sports and Active Living, 4*, 834633. https://doi.org/10.3389/fspor.2022.834633

Sweeney, L. (2017). All systems go! Developing a generation of "systems-smart" kids. In *EarthEd*. https://doi.org/10.5822/978-1-61091-843-5_12

Sweeney, L., & Sterman, J. D. (2007). Thinking about systems: Student and teacher conceptions of natural and social systems. *System Dynamics Review, 23*(2–3), 285–312. https://doi.org/10.1002/sdr.366

Tysk, J. (2021). Designing for the in-between ... Wellbeing. Lecture notes and slides. 2021HT. *System engagements in service and design research*. Linköping University (LiU). 13-Dec-2021.

Van Manen, M. (2014). *Phenomenology of practice: Meaning-giving methods in phenomenological research and writing*. Left Coast Press.

Vargo, S. L., & Akaka, M. A. (2009). Service-dominant logic as a foundation for service science: Clarifications. *Service Science, 1*(1), 32–41. https://doi.org/10.1287/serv.1.1.32

Vargo, S. L., & Lusch, R. F. (2004). Evolving to a new dominant logic for marketing. *Journal of Marketing, 68*(1), 1–17.

Vargo, S. L., & Lusch, R. F. (2010). From repeat patronage to value co-creation in service ecosystems: A transcending conceptualization of relationship. *Journal of Business Market Management, 4*(4), 169–179. https://doi.org/10.1007/s12087-010-0046-0

Vargo, S. L., & Lusch, R. F. (2011). It's all B2B ... and beyond: Toward a systems perspective of the market. *Industrial Marketing Management - IND MARKET MANAG, 40*, 181–187. https://doi.org/10.1016/j.indmarman.2010.06.026

Vargo, S. L., & Lusch, R. F. (2016). Institutions and axioms: An extension and update of service-dominant logic. *Journal of the Academy of Marketing Science, 44*(1), 5–23. https://doi.org/10.1007/s11747-015-0456-3

Vargo, S. L., & Lusch, R. F. (2017). Service-dominant logic 2025. *International Journal of Research in Marketing, 34*(1), 46–67. https://doi.org/10.1016/j.ijresmar.2016.11.001

Vázquez-Casielles, R., del Río-Lanza, A. B., & Díaz-Martín, A. M. (2007). Quality of past performance: Impact on consumers' responses to service failure. *Marketing Letters, 18*(4), 249–264. http://dx.doi.org.ezproxy.ulapland.fi/10.1007/s11002-007-9018-x

Weller, A. J. (2019). Design thinking for a user-centred approach to artificial intelligence. *She Ji: The Journal of Design, Economics, and Innovation, 5*(4), 394–396. https://doi.org/10.1016/j.sheji.2019.11.015

Xu, X., Liu, W., & Gursoy, D. (2019). The impacts of service failure and recovery efforts on airline customers' emotions and satisfaction. *Journal of Travel Research, 58*(6), 1034–1051. https://doi.org/10.1177/0047287518789285

Zins, A. H. (1998). Leisure traveler choice models of theme hotels using psychographics. *Journal of Travel Research, 36*(4), 3–15. https://doi.org/10.1177/004728759803600401

# Chapter 9
# Service Design for Older Adults Using Smart Digital Appliances: Person-Centred Service Design 4.0

**Lisa-Dionne Morris and Annelie Jordaan**

## Introduction

Insights show that older adults searched services for errors and usability from the operational interface on smart digital appliances, and gather information on the appliance from the associated source material, a manual and digital services (Morris, 2016). Morris et al. (2014) argues that emerging technology could respond better to the changing needs of older adults by responding to design priorities. Findings identify inconsistencies in the product, system and service design for older adults using smart digital appliances in two areas: information continuum and information categorisation (Morris, 2016).

Morris's (2016) study on operational interface design for an ageing population shows that older adults responded similarly when faced with appliance operational interface issues, which are linked to product, system and service design challenges. A more detailed understanding of older adults' service design needs has developed. Analysis of older adults' information-processing activities and interactions shows differences in older adults' processing, sequential operations and interaction points, suggesting opportunities for contributing to the knowledge of service design for emerging technologies. Morris (2016) details an understanding of older adults'

L.-D. Morris (✉)
School of Mechanical Engineering, Member of Institute of Design, Robotics and Optimisation, University of Leeds, Leeds, UK
e-mail: l.d.morris@leeds.ac.uk

A. Jordaan
Cape Peninsula University of Technology, Cape Town, Western Cape, South Africa
e-mail: jordaanan@cput.ac.za

 
U. Z. A. Hamid, M. Suoheimo (eds.), *Service Design for Emerging Technologies Product Development*, Springer Series in Design and Innovation 29,
https://doi.org/10.1007/978-3-031-29306-1_9

interactions and information processing. The findings have established and visualised 12-person-centred evaluative performance criteria to enable the design, development and assessment of alternative emerging service design and emerging technology applications for older adults using smart digital appliances.

## *Person-Centred Design and Service Design*

The focus of person-centred design is to explore solutions that give attention or consideration to users' capabilities. We interpret user capability to include both the characteristics of individual people, groups and communities. This research aims to improve the quality of life for older adults globally by creating new products, services and systems, which demonstrably support marginalised communities, older people and people with disabilities to remain active, productive, independent and socially connected, enabling them to complete everyday activities and reducing their dependence on others.

Key person-centred service capability themes: the dream, external factors, internal factors, assistive technology vision, strategic design priorities and gaps, and assistive actions have been proposed by Morris et al. (2022) and Morris (2011), laying the foundation for person-centred service design for marginalised communities. A situational perspective using a service design strategy to establish person-centred information needs, for transformed situations to equip key populations, has been advocated by Nyatuka and De la Harpe (2019). The goal of this chapter is to invoke the *conceptualise*, *collaborate* and *co-create*, of experiences and interactions of situational person-centred information needs intervention towards building resilient service design and systems for older adults using smart digital appliances. To achieve the above in a wellbeing context, Vink and Koskela-Huotari (2021) and Koskela-Huotari et al. (2021) have proposed four conceptual building blocks, namely, purpose, materials, processes and actors for developing a multilevel process model of service ecosystem design to advance the service design theory.

The 12-person-centred evaluative performance criteria presented in this chapter motivates the application of both proactive and reactive capability-sensitive design activities to determine the nature of the desired situation of person-centred service design. This is to be achieved through engaging purposive transdisciplinary service design research with a focus on person-centred information needs based on people's lived experiences and capabilities. Thus, the authors identified additional and relevant components to highlight the situational perspectives towards building person-centred service design for older adults. Person-centred service design components have been viewed through high-level person-centred service design and stakeholder involvement, based on Mitchell et al.'s (2015) situation, knowledge and learning.

## *Towards Person-Centred Design 4.0 and Person-Centred Service Design 4.0*

Product, service and systems are the focus of the sensitised centre of *interest* (marginalised communities such as older adults and people with a disability) and *activity* (product, system and service design). The merging of the information-processing acquisition behaviours, digital proficiency levels and reorganisation of service design ecosystems focus on the needs of the 'affected person'. There is an acceptance of the individualised product, system and service design solutions based on end-user information processing, experience and interactions and the distribution of resources and source material. The understanding is that appliances designed for older adults might better use smart digital technology that follows human needs. Capetillo et al. (2021) provide a list of 12 technologies (Industry 4.0) (the Fourth Industrial Revolution or 4IR) recognised as strategic for interactive design and manufacturing and include additive manufacturing, Internet of Things (IoT), big data, cyber security, software development, artificial intelligence, blockchain, cloud computing, robotics, simulations, advanced materials and virtual and augmented reality.

Capetillo et al. (2021) argue that when a focus has been placed on Industry 4.0 innovation, design ecosystems and smart industry, important discoveries for development technologically merge the boundaries between the physical and digital world. This is significant when considering the ways to interpret person-centred service design in three areas: sustainability, resilience and human activity. Castillo-López et al. (2019) confirm that traditional industrial patterns of business behaviour and value chains, which do not consider Industry 4.0, with high-level service design disenable the creation of novel products and services, changing business models. This chapter introduces new concepts of the use of Industry 4.0 applications to person-centred design and person-centred service design. A vision is presented where service design systems are capable of achieving the maximum efficiency of human interaction and information processing using conceptualised change to the design and implementation of emerging technologies, service and industry processes, responding to societal issues and patterns due to increasing interconnectivity of product, system and service design. The notion of person-centred design (12-person-centred evaluative performance criteria), person-centred service design and Industry 4.0 merges to create two new concepts proposed in this chapter on *Person-Centred Design 4.0 and Person-Centred Service Design 4.0*.

Neither person-centred design 4.0 nor person-centred Service design 4.0 are not yet mainstream design approaches. The aim of person-centred design 4.0 in the field of *design for user capability* is to, through the use of the 12-person-centred evaluative performance criteria and (Capetillo et al., 2021) list of 12 technologies, strengthen the cooperation and the characteristics of individual people, groups and communities including individuals' skills, abilities and aspirations. The aim of person-centred service design 4.0 is to improve the quality of life for people globally by creating new service design ecosystems and service design interface touchpoint tools, which demonstrably support service designers using person-centred design

4.0 to respond to the needs of marginalised communities, older people and people with disabilities to remain active, productive, independent and socially connected, enabling them to complete everyday activities, reducing their dependence on others, tackling social isolation and increasing independence, opportunities, employment, leisure, mobility and wellbeing. The people, organisations, institutes, businesses and government representatives we work with include people with disabilities, older adults and carers in the UK and developing countries. We focus on challenges of the future emerging with technologies as part of information and communication technologies for development (ICT4D) in the fields of socioeconomic development.

McKinsey's 'digital compass' wheel presented by Baur and Wee (2015) used in project, solution and change management illustrates various value drivers such as *customer experience and digitisation transformation*. Baur and Wee (2015) describe industry functions to release global opportunities for value creation in the area of service design. The authors argue that the emergence of their concept definition of person-centred design 4.0 and person-centred service design 4.0 merging with design for user emotions is where robust and independent processes to compare and contrast business opportunities with clear person-centred prioritisation based on the problem situation are needed for Industry 5.0. The associated Industry 5.0 levers are a useful tool for starting discussions and explorations of key improvement opportunities for human, culture and technology (HCT) for the extension to the conceptualisation of needs which is *person-centred design 5.0* and *person-centred service design 5.0*—especially, in designing services to address the touchpoints that create a user-product-interaction journey using the service design business interface. The introduction of a service design interface touchpoint tool is the focus of this chapter.

## Aims

This chapter aims to present an advanced assessment of a service design business interface touchpoint tool for older adults using smart digital appliances for design interventions using novel scalable person-centred performance criteria to use in simulating the more realistic experience and interactions of service design for emerging technologies. This chapter responds to the research question:

*What is most likely to constrain service research design and development of smart digital appliances for older adults?*

The goal is to show how each relevant stakeholder within the emerging technologies business architecture might use the 12-performance evaluation criteria in high-level service design and service product solutions development to improve service design interface touchpoint breakdowns and focus-shifts, common for relevant stakeholders. In addition, the goal is to provide a consistent method to enable relevant stakeholders to respond to priorities, emerging technologies and older adult issues associated with person-centred design 4.0 and person-centred service design 4.0.

## Literature Review

### *A Comprehensive Assessment of Older Adults' Common Issues, Problems and Challenges Using Smart Digital Appliances*

Personal compromise and loss of independence are key challenges for older adults. This provides a significant challenge in what is required of smart digital appliances, often referred to as white goods, task-specific appliances or domestic information-processing appliances (Morris, 2016). Focusing on examples of smart digital appliances in the home include washing machines, dishwashers and microwave ovens. Morris et al. (2014) point to a poor correlation between visual instructions and directions on associated cross-referenced source materials and the user interfaces of smart digital appliances. Older consumers refer to people 55 and older; their dissatisfaction is often the result of multiple errors in use caused by complex arrangements of visual instructions and directions on such appliance user interfaces which cause overlapping and intersection (Morris, 2016) and cause users' mental or emotional strain or tension resulting from adverse or demanding circumstance (Tams et al., 2018; Fraundorf et al., 2019).

Within the older adult context, the goal of this service design research was to enable older adults to have self-supported and fulfilled lives for longer through the design of improved operational user interfaces for smart digital appliances. In the future, this changing norm within society, from serving younger to older markets, will demand a change in the focus of products designed for older adults: from functional and technical operating services and systems to operational service interfaces designed for people with reduced capabilities. Table 9.1 combines activities for daily living from the Nottingham Extended Activities of Daily Living Scale (Nouri & Lincoln, 1987) with older adults' capabilities from the Cambridge Inclusive Design Cube (Keates & Clarkson, 2004, adapted by Morris, 2016). The matrix is used to frame the service design research study and the overlapping and

**Table 9.1** Overlapping and intersections of topic areas in the context of the activities of daily living adapted from Keates and Clarkson (2004)

| **Capabilities required** (Cambridge Inclusive Design Cube) **Activities of daily living** (Nottingham Extended Activities of Daily Living Scale) | **Motion capabilities** | | | **Sensory capabilities** | | **Cognitive capabilities** | |
|---|---|---|---|---|---|---|---|
| | **Locomotion** | **Reach and stretch** | **Dexterity** | **Vision** | **Hearing** | **Communication** | **Intellectual functioning** |
| | **Domestic tasks** | | | | | | |
| Complete full-clothes wash machine appliance | X | X | X | X | X | X | X |
| Heating a ready meal in a microwave oven appliance | X | X | X | X | X | X | X |

Source: Morris (2016)

intersections of topic areas and comprehensive assessment of older adults' common issues, problems and challenges using smart digital appliances.

## Presentation of Simulator and Scenario Builders for the Evaluation of the Deployment of Emerging Technologies

Evaluating the effects of emerging technologies or new replacement technologies for user-centred design has utilised standard protocols in Tables 9.2 and 9.3. However, in keeping with severe criticisms of sources used for the review of standards and user-centred design guidelines, these standards only provide a benign assessment of human activity and person-centred service design (Holmlid & Björndal, 2016). Moreover, they do not reflect more adverse conditions, which accelerate the resolve of the 'affected person' negative experience and inaccessibility. There are limited scenario formulators for high-level service design and service product solutions development in human activity and person-centred design. Simulators are relevant to the stakeholder and end-user priorities which improve service delivery protocols (Kim, 2021).

Being able to evaluate, prioritise and discuss emerging technologies for the most important attitudes, principles and skills of older adults' independence, which

**Table 9.2** Sources used for the review of standards and user-centred design guidelines

| Tools | British standards and European legal frameworks | Brief description of the use |
|---|---|---|
| A tool for understanding customer diversity to design better mainstream products | The Inclusive Design Toolkit Clark http://www.inclusivedesigntoolkit.com/betterdesign2/ | Understanding accessibility issues |
| Ten general principles for interaction design | Heuristics of Usability Nielsen http://www.nngroup.com/articles/ten-usability-heuristics/ | Understanding accessibility issues |
| A tool for measuring the usability of domestic information-processing appliances | System Usability Scale http://www.usability.gov/how-to-and-tools/methods/system-usability-scale.html | Describing end-user requirements and defining desirable properties of products |
| Ecological approach to designing | Activity-Centred Design: An Ecological Approach to Designing Smart Tools and Usable Systems (Acting with Technology) | Describing end-user requirements and defining desirable properties of products |
| Product usability guidelines | The Design of Everyday Things Norman | Describing end-user requirements and defining desirable properties of products |

**Table 9.3** Sources used for the review of standards and guidelines

| Source | British standards and European legal frameworks | Brief description of the use |
|---|---|---|
| A guide with information about understandability in Clause 6.2.6 | ISO-IEC Guide 71:2014 Guide for Addressing Accessibility in Standards | Understanding accessibility issues |
| A guide with information about people in Clause 7 and terminology in Clause B | ISO-IEC Guide 71:2014 Guide for Addressing Accessibility in Standards | Describing end-user requirements and defining desirable properties of products |
| Human-centred design for interactive systems—the international standard for designing human-centred systems | ISO 9241-210:2010 Ergonomics of human-system interaction: Part 210 | Re-engineering existing designs by redeveloping or adapting elements |
| Recommendations for how to develop a dementia-friendly community | PAS 1365:2015 Codes of practice for the recognition of dementia-friendly communities in England | Describing end-user requirements and defining desirable properties of products |
| Specification for the process of assessment of human-system issues | DD ISO/PAS 18152:2003 Ergonomics of human-system interaction | Understanding system issues |
| Usability methods supporting human-centred design | ISO/TR 16982:2002 Ed 1 Ergonomics of human-system interaction | Engineering or designing new systems of product parts by redeveloping or adapting interaction techniques |
| An overview of Mandate M 376 and Mandate M 473—describes the requirements for software accessibility defined in EN 301 549 and includes 'design for all' in relevant standardisation initiatives according to a feature-based approach | BS EN 301549:2015 Accessibility requirements suitable for public procurement of ICT products and services in Europe | Describing end-user requirements and defining desirable properties of products |
| Anatomical configurations, human body, size, ergonomics, and anthropometric characteristics | PP 7310:1990 Anthropometrics: an introduction | Re-engineering existing designs by redeveloping or adapting elements |
| Guidelines on balanced design selection criteria | The 7 Principles and 29 Guidelines for Universal Design http://universaldesign.ie/What-is-Universal-Design/The-7-Principles/ | Describing end-user requirements and defining desirable properties of products |
| A tool for rating a product on a Likert scale (Universal Design in reverse order) | Universal Design Product Evaluation Countdown http://www.ncsu.edu/ncsu/design/cud/pubs_p/docs/UDPEC.pdf | Describing end-user requirements and defining desirable properties of products |

is to carry out activities of daily living, is an important aspect of understanding independent living in the service design development process. Scenario builders for the evaluation of the deployment of emerging technologies can be used to learn effective ways to improve service design (Vink & Koskela-Huotari, 2021), specifically to reduce two impediments to service design product development proposed in this chapter, where there is an act or process of failing to function or continue in the movement of workflow in the service design process, which results in a stall in communication and optic (*service design business interface touchpoint focus-shift*), or a situation where there is an abrupt failure of relevant stakeholders within the emerging technologies business architecture to positively affect and restart the workflow (*service design business interface touchpoint breakdown*).

## *Verification and Validation of Evaluative Performance Criteria for High-Level Service Design and Service Product Solutions Development*

Assessment of the use of emerging technologies in product development has determined the need for a scalable solution for the management, verification and validation of evaluative performance criteria (Ojasalo & Ojasalo, 2015). The emerging technologies business architecture provides service solution components, identifying the main service design business interface. High-level service design and service solutions use interdisciplinary team members and stakeholders with nontechnical and technical emerging technologies business understandings. Service product solutions development for older adults should be understandable to all relevant stakeholders including the administrators, micro and macro external design consultancy, service designers, marketers, finance, user experience (UX) designers, customer experience (CX) designers, big data analysts and software and systems engineers. In this way, it provides what Ojasalo and Ojasalo (2018) define as adapting business model thinking to service logic.

Relevant stakeholders rely on consistent exact directions, information and instructions for high-level service design and service product solutions. However, emerging technologies business architects' priorities are not arranging the service design process to reflect the sequential process of the activities of service solution product development, nor are they making the information continuum and information categorisation materials consistent or accessible across relevant stakeholder groups. This means that emerging technologies business architects will respond differently when faced with the same service interface and product solution challenges, issues and problems. Differences in emerging technologies business architects' opinions and relevant stakeholder methods, processes, sequential operation and undefined service design interaction touchpoints are found to cause issues in service design for emerging technologies' product development. A more detailed understanding of older adults' needs and relevant stakeholders' high-level service design is presented in this chapter.

### *Application of Evaluative Performance Criteria to Adverse Scenarios and Emerging Technology Conditions in a Novel Service Design Intervention*

Emergent technology signifies the Fourth Industrial Revolution and involves the introduction of sustainable combinations of smart devices, applications, smart self-adaption systems, sensory technologies and technologies to enable connected intelligence and appliance autonomy. Industry 4.0 dramatically changes the landscape of service design innovation, research design and the starting of innovation in what Simanjuntak (2021) refers to as the designing of service-dominant logic, which explores how emerging technology concepts and technologies influence the way service design is conducted and how techniques, approaches, methods and tools for supporting service design for older adults advancements can take advantage of sustainable change and transformation suggested by (Sudbury-Riley et al., 2020).

## Methodology for 12-Person-Centred Service Design Evaluative Performance Criteria

### *Research Design, Data Collection and Data Analysis*

The research brought together descriptive and experimental research methods and surveys. The study research method and results are reported in Morris et al. (2014) and Morris (2016). Naturalistic inquiry methods described by Lincoln and Guba (1985) were used to identify operational interface design goals, through observations of older adults using newly developed and current operational interfaces on smart digital appliances. Nine adults were asked to carry out an activity of daily living: wash laundry using a washing machine and heat a ready meal using a microwave oven. Older adults were asked to interact and talk aloud. Data was video-recorded and transcribed using methods described by Ericsson and Simon (1998), Bødker (1996) and Suchman and Trigg (1991). Older adults' characteristics of interactions using the microwave oven and the washing machine were identified. Table 9.4 provides a summary of the combination of participants, case studies and research instruments used in the study of appliances. Evaluative performance criteria were derived and are shown in Table 9.5.

Case study findings were used to derive 12-person-centred service design evaluative performance criteria for the washing machine and the microwave oven and subsequently constrained to smart digital appliances. The remainder of this section provides details of the survey in the form of interviews, workshops and thematic analysis inquiry with international researchers and engineering innovation practitioners with results reported in section "Results and Findings".

**Table 9.4** Combination of participants, case studies and research instruments

| | Stage 1 | |
|---|---|---|
| **Combination of participants, appliances and interface used** | **Research instruments** | |
| **Participants** | **Operational user interface** | **Instruction of materials** |
| (GP) Female aged 55 | Washing machine interface | Manual and clothing labels |
| (MW) Female aged 68 | Microwave over interface | Manual and ready meal packaging |
| 14 responses to online feedback questionnaire | Microwave oven interface | Manual and online manufacture consumer services |
| | **Stage 2** | |
| **Combination of participants, appliances and interface used** | **Research Instruments** | |
| **Participants** | **Operational User Interface** | **Instruction of materials** |
| (GP) Female aged 55 | Washing machine interface | Manual and clothing labels |
| (MW) Female aged 68 | Microwave over interface | Manual and ready meal packaging |
| | **Stage 3** | |
| **Combination of participants, appliances and interface used** | **Research Instruments** | |
| **Participants** | **Operational User Interface** | **Instruction of materials** |
| (GP) Female aged 55 | Microwave open interface | Manual and ready meal packaging |
| (VA) Female aged 61 | Microwave oven interface | Manual and ready meal packaging |
| (FG) Female aged 68 | Microwave over interface | Manual and ready meal packaging |
| (TC) Male aged 58 | Microwave oven interface | Manual and ready meal packaging |
| | **Stage 4** | |
| **Combination of participants, appliances and interface used** | **Research Instruments** | |
| **Participants** | **Operational User Interface** | **Instruction of materials** |
| (DH) Aged 67 | Current microwave oven interface | Manual and ready meal packaging |
| (AD) Aged 68 | Current microwave oven interface | Manual and ready meal packaging |
| (JB) Aged 72 | Current microwave oven interface | Manual and ready meal packaging |
| (ST) Aged 74 | Current microwave oven interface | Manual and ready meal packaging |
| (SL) Aged 65 | New microwave oven interface on an IPad | New instructions and ready meal packaging |
| (CMK) Aged 69 | New microwave oven interface on an IPad | New instructions and ready meal packaging |
| (JH) Aged 70 | New microwave oven interface on an IPad | New instructions and ready meal packaging |
| (JH) Aged 74 | New microwave oven interface on an IPad | New instructions and ready meal packaging |

Source: Morris (2016)

**Table 9.5** 12-person-centred service design evaluative performance criteria to support service design business interface touchpoint breakdowns and focus-shifts for smart digital appliance design (Morris, 2016)

| Evaluative performance criteria | Emerging technologies responses |
|---|---|
| Purposeful (1) | Meaningful connections between the information-processing tasks, activities, operational goals and information used |
| Appropriate for use (2) | Ensure that participants can read the operational interface features from a minimum distance of 500 mm. Ensure that participants can reach and touch the operational interface features from a normal operating position. Position operational interface features along the horizontal line of sight between zero and 30 degrees. |
| Standardised elements (3) | Use consistent instructions and information on operational interfaces |
| Responsiveness to errors (4) | Use visual and auditory instructions on operational interfaces |
| Functional grouping (5) | Place operational interface features in a linear operational series constrain the series of actions if necessary |
| Sequential grouping (6) | User functional level language based on operational series and operational interfaces |
| Learnability (7) | Use intuitive words, phrases, symbols and imagery used in early learned experience of the activity or task |
| Flexible in use (8) | Allow users to tailor actions, operational series and activities |
| Perceptibility (9) | Use recognisable and understood words, phrases, symbols and imagery |
| Intelligibility (10) | Use interpretable words, phrases, symbols, and imagery |
| Level of cognitive ev ffort(11) | Use sequential series and auditory repetition of words, phrases, symbols and imagery to minimise cognitive overload |
| Actionability (12) | Use operational series and discriminate between operational interface features |

# Methodology for Thematic Analysis

## *Research Design, Data Collection and Data Analysis*

Qualitative research has been adopted as a methodology for this study. Given the qualitative nature of the data, the interpretive paradigm with a subjective stance has been selected, as the research focused on the ways grouped international researchers and engineering innovation practitioners attempt to make sense of the service design process and the service design world around them (Saunders et al., 2019, 141). The inductive approach, in alignment with the multi-method quantitative choice of the researcher, has been implemented using two strategies, namely, (i) a survey exemplified by interviews and observations with older adults, and (ii) thematic inquiry workshops, where the research has attempted to clarify and illustrate the significance and value of the international researchers and engineering innovation practitioners understanding of the topic (Saunders et al., 2019).

The researchers start with observations about the problem situation and service design for older adults using the video recordings obtained in section "Research Design, Data Collection and Data Analysis". The creation of protocols for methodological trustworthiness in qualitative data suggested by Amankwaa (2016) and implementing a triangulation protocol used by Farmer et al. (2006) were used to verify and validate themes. These themes were used to move from those particular experiences to a more general proposition about service design for older adults using smart digital appliances. Expressed differently, they move from data to theory, or from the specific to the general application for emerging technology (DeCarlo, 2018). The research strategy adopted was face-to-face workshops in groups using the video recordings obtained in section "Research Design, Data Collection and Data Analysis" and the 12-person-centred service design evaluative performance criteria presented in section "Service Design Performance Evaluated Criteria for Service Design for Emerging Technologies". For this study, the ethical principles of beneficence (do good) and non-malfeasance (not harm) have been adhered to. Thus, the risk of harm to the focus group participants has been minimised, the focus group participants' confidentiality and anonymity have been maintained, and the researchers did not use misleading practices (Lund Research, 2012).

## Results and Findings

The purpose of the workshops and the reasoning for how the practitioner and research experts met are understanding emerging technology and service design situational perspectives; informing multi-actor and service providers of 'affected person' intentional change; applying proactive and reactive capability-sensitive design activities and understanding information needs for transformed situations (Morris, 2011), which in this situation better equip older adults using smart digital appliances.

### *Service Design Performance Evaluated Criteria for Service Design for Emerging Technologies*

In this section, person-centred evaluative performance criteria have been introduced to resolve issues of high-level service design and service product solution development. The findings and results present design performance evaluation criteria in Table 9.5, and a process to explore relevant stakeholders' opinions, views, attitudes, experiences and propositions to employ emerging technologies solutions for older

**Fig. 9.1** Evaluation performance criteria responses to scenarios of emerging technologies given by the grouped international researchers and engineering innovation practitioners based on the video recordings

| | |
|---|---|
| *Group A* | 2 1 9 5 8 10 4 3 6 7 11 12 |
| *Group B* | 2 7 3 9 10 4 11 6 1 8 5 12 |
| *Group C* | 7 11 10 3 5 4 2 9 1 6 12 8 |
| *Group D* | 1 2 3 9 7 4 11 6 12 10 5 8 |
| *Group E* | 1 7 9 8 5 6 2 3 10 4 12 11 |
| *Group F* | 9 7 10 2 4 11 3 8 5 12 6 1 |
| *Group G* | 3 2 7 6 5 1 9 10 4 11 8 12 |

adults using smart digital appliances in Fig. 9.1. A service design business interface touchpoint tool to framework and assess the inter-relationship between older adults' needs and service design for emerging technologies delivery.

In this section, the 12-person-centred service design evaluative performance criteria in Table 9.5 have been introduced to respond to older adult needs and design priorities for smart digital appliances. Figure 9.1 shows the evaluation performance criteria responses to scenarios of emerging technologies given by the grouped international researchers and engineering innovation practitioners based on the video recordings obtained in section "Research Design, Data Collection and Data Analysis" and ordering of the 12-person-centred service design evaluative performance criteria.

## Analysis and Discussion

Nyatuka and De la Harpe (2019) argue that many authorities are invested in service design. For older adults, such authorities include user-centred design, engineering, domestic service adaptation, technology infrastructure, data analytics, public health and health economics. A systems design and design systems approach might provide an opportunity for diverse views on person-centred service design to be shared. The strength of the 12-person-centred evaluative performance criteria for improvements to service design interface business touchpoints is that it supports relevant stakeholders transact and arrange priorities with technical complexity and ill-defined or wicked problems, with the involvement of relevant stakeholders responding to priorities, emerging technologies and older adults' issues (Sudbury-Riley et al., 2020; Touloum et al., 2018; Roto et al., 2016; Yu, 2020). In this context,

**Fig. 9.2** Group G performance evaluation criteria responses to scenarios of emerging technologies

Langley et al. (2018) and Morris and Connolly (2010) suggest that there is seldom one single correct or best solution to emerging technologies service delivery. There are often interdisciplinary alternative solutions that appear because of separate teams acting on an ad hoc basis, commonly impassive or unaltered by the lived experience of older adults using smart digital appliances (Ghajargar & Bardzell, 2019a, b). Eliciting a reaction response to suggestions made by Dalsgaard (2014), Maedche et al. (2019) and Feine et al. (2019), 12-person-centred service design evaluative performance criteria for service design business interface touchpoint breakdowns and focus-shifts might be considered a service design business interface touchpoint situational tool with outcomes in interdisciplinary team interpretations of a specific situational problem experience shown in Fig. 9.2. A shared understanding of how these older adult competencies and understandings can support emerging technologies development and improve end-user service interactions (Yu, 2020) suggests it might also be an integrative service design framework for future service design agendas.

For Group G from Table 9.1, an example of person-centred service design evaluation performance criteria representation, priorities and sequential order of resolve is shown in Fig. 9.2. Figure 9.2 shows how the criteria might provide a high-level service design policy roadmap for service design delivery for the nine older adults using the smart digital appliance in the case study.

Figures 9.3, 9.4 and 9.5 show a process that significantly affects emerging technologies' success by having business architects work together to set expectations, review results and reward performance, process, appraise and explain the purpose of emerging technologies in older adult person-centred design solutions. The 12-person-centred evaluative performance criteria clarify expectations and set goals for improvement of emerging technologies weaknesses, rewards, accomplishments and overall performance. Explaining the steps involved during and after the high-level service design and service product solutions uses descriptions of the end-user interactions that should be understandable to the emerging technologies business architects.

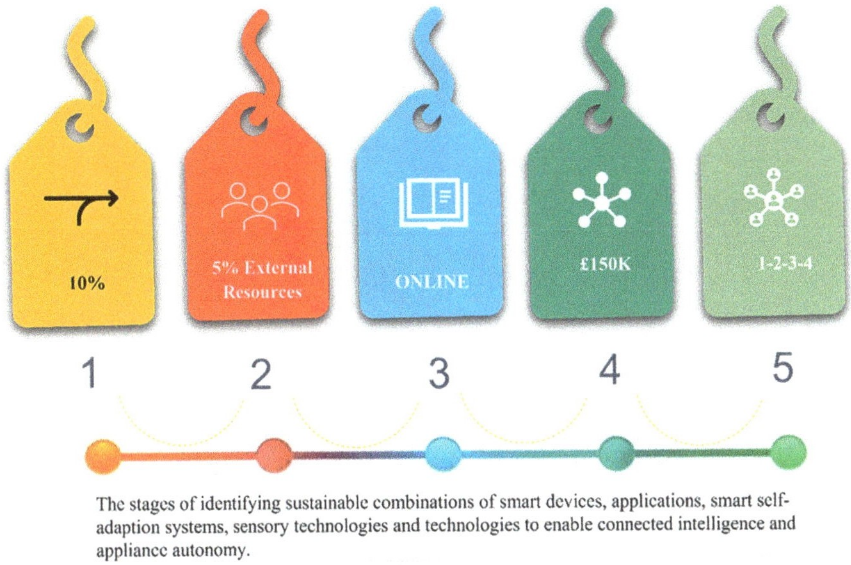


**Fig. 9.3** Group G performance evaluation criteria responses to scenarios of emerging technologies (extended)

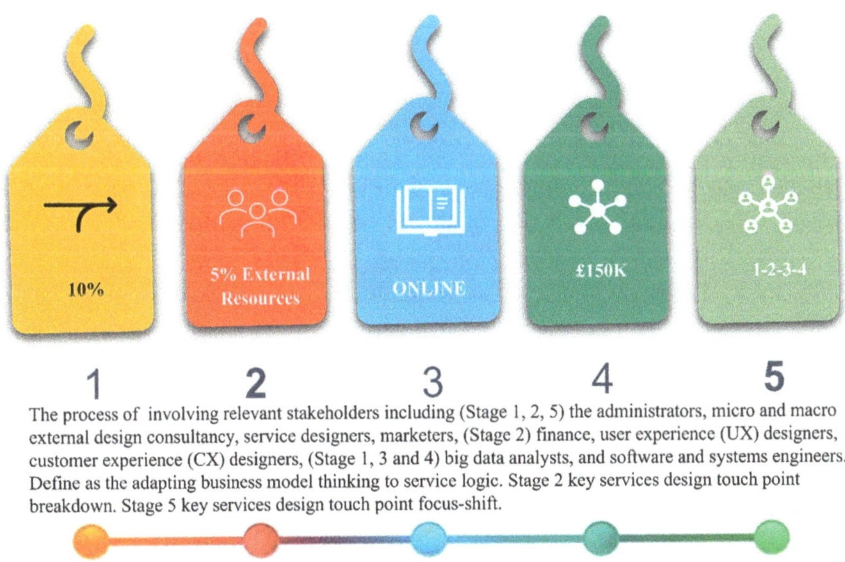


**Fig. 9.4** Group G performance evaluation criteria responses to scenarios of emerging technologies (contextualised)

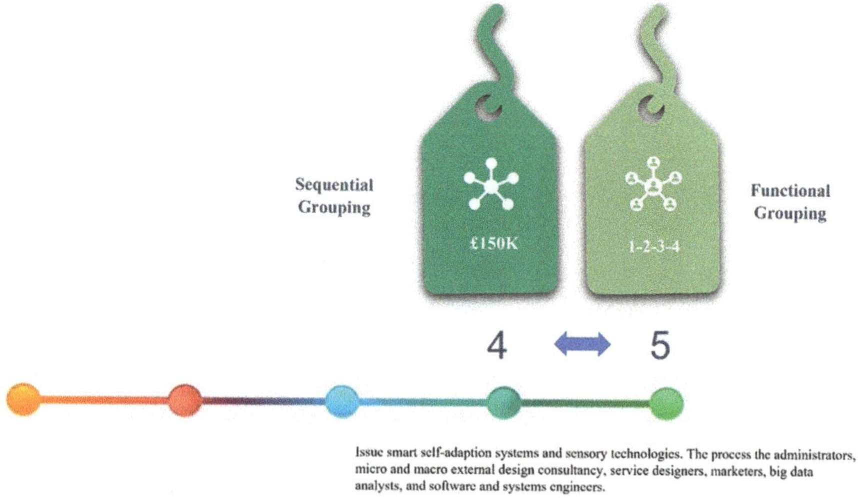


**Fig. 9.5** Group G performance evaluation criteria responses to scenarios of emerging technologies (prioritised)

Using person-centred evaluative performance criteria provides start, intermediate and end assessment points for service design development to set developmental objectives focused on finding emerging technologies' strengths and weaknesses and developing secondary relations. Overall, the 12-person-centred evaluative performance criteria achieve the following objectives during service design business interface touchpoint breakdowns and focus-shifts.

- *Fostering relevant stakeholder areas of competencies.*
- *Provisional confirmation of aims, objectives, strategy and a shared vision for emerging technologies implementation.*
- *Identifying person-centred technology development.*
- *Regulating and monitoring process for service design business interface touchpoint breakdown and focus-shift.*
- *Providing constructive criticism and guidance to the emerging technologies development implementation strategy.*
- *Improving communication for dialogue between the service design emerging technologies business architects, underscoring areas of concern. This can also have the effect of increasing the trust between relevant stakeholders within the business* architecture.
- *Providing a tool for academic and industry research areas such as emerging technologies selection, emerging technologies training, emerging technologies change and transformation and emerging technologies' effective impact.*

## Conclusion

The chapter outcomes suggest that the use of the 12-person-centred service design evaluative performance criteria might positively affect the inconsistent information on the operator interface of the smart digital appliance. In the past, Industry 4.0 has not resolved issues that lead to operator interface problems. The chapter's findings help to propose a theory that *person-centred design 4.0* and *person-centred service design 4.0* integrating the 12-person-centred service design evaluative performance criteria might provide the technology, industry partners and societal patterns and service processes to achieve this. The interesting divergence between this chapter and another is that the findings suggest that researchers and practitioners using the person-centred service design evaluative performance criteria show a readiness to set design priorities when faced with a better understanding of older operator difficulties, thus beginning to deliver *person-centred design 4.0*. Analysis of older adults' information-processing activities and interactions shows that these might become application areas for *person-centred service design 4.0*, through service design models and tools that respond to differences in older adults' experience, processing, sequential operations and interactions using smart digital appliances.

The benefit of this chapter on service design for emerging technologies product development is that it provides a method for service designers to explore approaches for supporting independent living through improved service design for older adults using smart digital appliances, thus answering the research question. The chapter focuses on evaluations and explorations of different smart digital appliances using the 12-person-centred evaluative performance criteria. Engaging service designers of emerging technologies to connect and empathise with diversity, equity and inclusion in older adults could uncover divergent and convergent reports of experiences and pathways for new and emerging technologies applications.

Recommendations made here are that service design practice settings have made the idea of designing globally relevant inclusive smart services, systems and product solutions a pertinent and achievable requirement. The 12-person-centred evaluative performance criteria support the development of emerging technologies integration, business architecture and the entire service solutions components to identify the principal components developed through service design solutions for *person-centred design 4.0* and *person-centred service design 4.0*. Emerging technologies business architects might prioritise and then arrange the service design process to reflect the sequential process of the activities of service solution product development using the criteria, making the information continuum and information categorisation consistent, thus, providing a formal design policy that is accessible across relevant stakeholder groups for improved service design business interface touchpoints. Areas of emerging complexity in service design might benefit from service designers who enact the recommendations and processes presented in this chapter.

This means that emerging technologies business architects will respond in an interconnected and cohesive way when faced with the same service interface product solution challenges, issues and problems. The chapter findings identify, uncover and define service design business interface touchpoints found to cause issues in service design for emerging technologies development and definitions of *person-centred design 4.0* and *person-centred service design 4.0*. As outlined in this chapter, the target population is older adults using smart digital appliances; therefore, a series of key service design business interface touchpoint breakdown and focus-shift common indicators will be explored to extend the research to help develop and verify service design for emerging technologies product development for older adults.

## References

Amankwaa, L. (2016). Creating protocols for trustworthiness in qualitative research. *Journal of Cultural Diversity, 23*(3), 121–127.

Baur, C., & Wee, D. (2015). *Manufacturing's next act*. McKinsey & Company. http://www.mckinsey.com/insights/manufacturing/manufacturings_next_act

Bødker, S. (1996). Applying activity theory to video analysis: How to make sense of video data. Human-computer interaction. In B. A. Nardi (Ed.), *Context and consciousness-activity theory and human-computer interface* (pp. 147–174). The MIT Press.

Capetillo, A., Abraham Tijerina, A., Ramirez, R., & Galvan, J. A. (2021). Evolution from triple helix into penta helix: The case of Nuevo Leon 4.0 and the push for industry 4.0. *International Journal on Interactive Design and Manufacturing (IJIDeM), 15*(4), 597–612.

Castillo-López, S. E., Flores Guirao, S. K., Roldán-Reyes, E., Maldonado Macias, A. A., & Duque, D. M. (2019). A service design process based on the business model CANVAS and the CK theory. In *Managing innovation in highly restrictive environments* (pp. 279–301). Springer.

Dalsgaard, P. (2014). Pragmatism and design thinking. *International Journal of Design, 8*(1), 143–155.

DeCarlo, M. (2018). Scientific inquiry in social work. *Open Social Work Education*. https://creativecommons.org/licenses/by-nc-sa/4.0/

Ericsson, K. A., & Simon, H. A. (1998). How to study thinking in everyday life: Contrasting think-aloud protocols with descriptions and explanations of thinking. *Mind, Culture, and Activity, 5*(3), 178–186.

Farmer, T., Robinson, K., Elliott, S. J., & Eyles, J. (2006). Developing and implementing a triangulation protocol for qualitative health research. *Qualitative Health Research, 16*(3), 377–394. https://doi.org/10.1177/1049732305285708

Feine, J., Gnewuch, U., Morana, S., & Maedche, A. (2019). A taxonomy of social cues for conversational agents. *International Journal of Human-Computer Studies, 133*, 138–161. https://doi.org/10.1016/j.ijhcs.2019.07.009

Fraundorf, S. H., Hourihan, K. L., Peters, R. A., & Benjamin, A. S. (2019). Aging and recognition memory: A meta-analysis. *Psychological Bulletin, 145*(4), 339–371. https://doi.org/10.1037/bul0000185

Ghajargar, M., & Bardzell, J. (2019a). What design education tells us about design theory: A pedagogical genealogy. *Digital Creativity, 30*(4), 277–299.

Ghajargar, M., & Bardzell, J. (2019b). Synthesizing opposites: Technical rationality and pragmatism in design. *The Design Journal, 22*(sup1), 2031–2044.

Holmlid, S., & Björndal, P. (2016). Mapping what actors know when integrating resources: Towards a service information canvas. In *Service design geographies. Proceedings of the ServDes. 2016 conference* (Vol. 125, pp. 544–550). Linköping University Electronic Press.

Keates, S., & Clarkson, P. J. (2004). *Countering design exclusion: An introduction to inclusive design*. Springer-Verlag.

Kim, M. (2021). A study of dignity as a principle of service design. *International Journal of Design, 15*(3), 87–100.

Koskela-Huotari, K., Patrício, L., Zhang, J., Karpen, I. O., Sangiorgi, D., Anderson, L., & Bogicevic, V. (2021). Service system transformation through service design: Linking analytical dimensions and service design approaches. *Journal of Business Research, 136*, 343–355. https://doi.org/10.1016/j.jbusres.2021.07.034

Langley, J., Wolstenholme, D., & Cooke, J. (2018). 'Collective making' as knowledge mobilisation: The contribution of participatory design in the co-creation of knowledge in healthcare. *BMC Health Services Research, 18*(1), 1–10. https://doi.org/10.1186/s12913-018-3397-y

Lincoln, Y. S., & Guba, E. (1985). *Naturalistic inquiry*. Sage.

Lund Research. (2012). *Principles of research ethics*. Laerd. https://dissertation.laerd.com/principles-of-research-ethics.php

Maedche, A., Legner, C., Benlian, A., Berger, B., Gimpel, H., Hess, T., Hinz, O., Morana, S., & Söllner, M. (2019). AI-based digital assistants: Opportunities, threats, and research perspectives. *Business & Information Systems Engineering., 61*(4), 535–544. https://doi.org/10.1007/s12599-019-00600-8

Mitchell, C., Cordell, D., & Fam, D. (2015). Beginning at the end: The outcome spaces framework to guide purposive transdisciplinary research. *Futures, 65*, 86–96.

Morris, L. D. (2011). Helping your lecturers creatively infuse ubiquitous computing technology into teaching. In T. T. Kidd & I. Chen (Eds.), *Ubiquitous learning: Strategies for pedagogy, course design and technology* (pp. 207–226). IAP.

Morris, L. D. (2016). *The design of operational interfaces for older adults*. Doctoral dissertation. University of Leeds.

Morris, L. D., & Connolly, A. (2010). Involving students in the development and evaluation of a ubiquitous learning application for a design practice setting. *Ubiquitous Learning: An International Journal, 2*(4), 21–38. https://doi.org/10.18848/1835-9795/CGP/v02i04/40480

Morris, L. D., McKay, A., & Cassidy, T. (2014). A coding scheme for use in the design of electronic consumer products for older adults. In D. Marjanovic, M. Storga, N. Pavkovic, & N. Bojcetic (Eds.), *DS 77: Proceedings of the DESIGN 2014 13th international design conference* (pp. 2115–2124). https://collaborative.designsociety.org/publication/35352/A+CODING+SCHEME+FOR+USE+IN+THE+DESIGN+OF+ELECTRONIC+CONSUMER+PRODUCTS+FOR+OLDER+ADULTS

Morris, L. D., Abu Alghaib, O., & Northridge, J. (2022). Capability-sensitive principles for assistive technology to support young graduates with disabilities in Bangladesh and Kenya into employment. *Journal of International Development, 34*(5), 964–987. https://doi.org/10.1002/jid.3691

Nouri, F. M., & Lincoln, N. B. (1987). An extended activities of daily living scale for stroke patients. *Clinical Rehabilitation, 1*, 301–305. https://doi.org/10.1177/026921558700100409

Nyatuka, D. R., & De la Harpe, R. (2019). Evaluating mHealth interventions in an underserved context using service design strategy: A case of Kenya. In *Proceedings of the third international conference on medical and health informatics 2019* (pp. 153–160). https://doi.org/10.1145/3340037.3340060

Ojasalo, K., & Ojasalo, J. (2015). Adapting business model thinking to service logic: An empirical study on developing a service design tool. In J. Gummerus & C. Von Koskull (Eds.), *The Nordic school-service marketing and management for the future* (pp. 309–333). Hanken School of Economics.

Ojasalo, J., & Ojasalo, K. (2018). Service logic business model canvas for lean development of SMEs and start-ups. *Journal of Research in Marketing and Entrepreneurship, 20*(1), 70–98. https://doi.org/10.1108/JRME-06-2016-0015

Roto, V., Väätäjä, H., Law, E., & Powers, R. (2016). Experience design for multiple customer touchpoints. In *NordiCHI'16: Proceedings of the 9th Nordic conference on human-computer interaction* (pp. 1–3). https://doi.org/10.1145/2971485.2987685

Saunders, M. N. K., Lewis, P., & Thornhill, A. (2019). *Research methods for business students* (8th ed.). Pearson.

Simanjuntak, M. (2021). Designing of service dominant logic and business model canvas: Narrative study of village tourism. *Golden Ratio of Marketing and Applied Psychology of Business, 1*(2), 73–80. https://doi.org/10.52970/grmapb.v1i2.60

Suchman, L., & Trigg, R. H. (1991). Understanding practice: Video as a medium for reflection and design source. In J. Greenbaum & M. Kyng (Eds.), *Design at work: Cooperative design of computer systems*. CRC Press.

Sudbury-Riley, L., Hunter-Jones, P., Al-Abdin, A., Lewin, D., & Naraine, M. V. (2020). The trajectory touchpoint technique: A deep dive methodology for service innovation. *Journal of Service Research, 23*(2), 229–251. https://doi.org/10.1177/1094670519894642

Tams, S., Thatcher, J. B., & Grover, V. (2018). Concentration, competence, confidence, and capture: An experimental study of age, interruption-based technostress, and task performance. *Journal of the Association for Information Systems, 19*(9), 857–908. https://doi.org/10.17705/1JAIS.00511

Touloum, K., Idoughi, D., & Seffah, A. (2018). Leveraging user experience and touchpoints analysis for services design: Case of crisis management. *The Journal of Interaction Science, 6*(1), 25–25.

Vink, J., & Koskela-Huotari, K. (2021). Social structures as service design materials. *International Journal of Design, 15*(3), 29–23.

Yu, E. (2020). Toward an integrative service design framework and future agendas. *Design Issues, 36*(2), 41–57. https://doi.org/10.1162/desi_a_00589

# Part III
# Organizational Transformations and Management for Practicing Service Design

# Chapter 10
# Organizational Transformation Through In-House Service Design: A Case Study of a Multinational Manufacturing Corporation

**Krista Korpikoski and Satu Miettinen**

## Introduction

Traditionally, product development, especially in the area of manufacturing industry, has been viewed as a phase where 'the value and role of product development to the organization is minimised since its activities are merely something that the organization needs to contend with temporarily' (Junginger, 2008, p. 28). Hence, product development actions are usually focused on the object-related functions of usability and form, and are not viewed as processes of change, nor design in itself as an active intervention that creates change within an organization (Meurer, 2001). However, an organization's internal operations, such as research and development (R&D) processes and activities, may be closely linked to its overall customer experience and service experiences (Junginger, 2008; Yu & Sangiorgi, 2018). In addition, they affect employee experiences in the phases of service development, implementation, and delivery of newly developed services and end-to-end processes. Yet, transforming into a customer-centric organization might be a tough call for engineering-based corporate cultures in the context of the manufacturing industry. This can be due to strong existing routines, rites, and heroes (Borja de Mozota, 1998). Such tendencies in decision-making attitudes, regarding early closures on problem-solving, contrast with design attitudes based on higher-order human-centric approaches (Boland & Collopy, 2004). The latter approach is defined as allowing time for openness and closure to find the best possible answer (Boland & Collopy, 2004). Therefore, this chapter concentrates on discovering how service design as a new in-house methodology and practice

K. Korpikoski (✉) · S. Miettinen
Faculty of Art and Design, University of Lapland, Rovaniemi, Finland
e-mail: krista.korpikoski@ulapland.fi



U. Z. A. Hamid, M. Suoheimo (eds.), *Service Design for Emerging Technologies Product Development*, Springer Series in Design and Innovation 29,
https://doi.org/10.1007/978-3-031-29306-1_10

supports the transformation from an expert-driven organization to a more human- and customer-centric one.

In academic literature, organizational transformation (e.g. By, 2005; Bustinza et al., 2017; Francis et al., 2003) and especially change management (e.g. Kotter, 2007; Lauer, 2010; Paton & McCalman, 2008) are well-covered areas. They are mainly treated in the area of organization and management studies. Since in-house service design is becoming more commonly used, it inevitably starts changing service systems and organizations by bringing in human- and customer-centric working cultures. Hence, more focus should be put on combining organizational transformation and change management studies with design research. Such understanding would increase organizations' understanding of the meaning and benefits of in-house service design as well as how to manage, lead, and support efficient use of service design. In addition, such understanding would increase knowledge of the required change aspects that organizations must put focus on at individual and organizational levels when moving towards human- and customer-centric working cultures. Therefore, this study discovers the under-researched area of service design as an in-house development methodology by combining it with organizational transformation studies. The study asks:

How has in-house service design supported an organization's working culture to transform towards human- and customer-centricity?

The results of this research are based on a qualitative case study of a multinational manufacturing corporation, which has used service design as an in-house practice as a part of the R&D department since the autumn of 2014. Thus, the organization is seen as a valuable information provider. Twenty-six semi-structured interviews including 36 interviewees provide the data for this study. Interviews were conducted in 2018 at all organizational levels except for the chief executive officer (CEO) and shareholders. By then the in-house service design team had been within the organization for four years. The first author has a working background as an in-house service designer in the organization under study.

The data analysis follows coding methods according to the rules of Saldaña (2016) in addition to thematic analysis following the rules of Braun and Clarke (2022). As a result, we present that in-house service design has supported an organization's working culture to transform towards human- and customer-centricity through the change aspects of individuals and an organization. The individual change aspects consist of changes in individuals' attitudes, mindsets, and beliefs, which are achieved through experiential learning through service design. These form the preconditions for further organizational changes in terms of organizational paradigm transformation through strategy renewal affected by human- and customer-centric values.

### *Description of the Company Under Investigation and the Role of Service Design*

The participating organization of this study employs approximately 60,000 persons globally and it is a matrix organization with over a hundred years of history in engineering and manufacturing. The role of services is based on sales, maintenance, and consulting services. Along with digitalization and the Internet of Things (IoT), the role of services has become increasingly important for the organization in the last decade. Hence, these changes within societies and increased customer demands have put services into a more central role. Before the use of in-house service design, product development activities followed Cooper's (2001) Stage-Gate process model as a project management technique and end-users of the products had mainly been the information sources for designers. Since 2014, in-house service design expertise and consultants' help have introduced human- and customer-centric development processes and activities. Customer in this context means business-to-business customers.

The role of in-house service design regarding service development is strategic in the organization under study. Service designers are located in the company headquarters to work side-by-side with key business stakeholders, management, and other departments such as marketing and information technology (IT). In addition, some country-level service designers have been hired. In-house service designers have an active role in the early phases of service development regarding customer research and the ideation and creation of new service concepts. Design actions such as co-creation and prototyping methods are used together with customers, frontline employees, and other internal stakeholders and experts. In-house service designers hold a strong communicative and facilitative role within the projects. They guide discussions with other project and business stakeholders and engineers regarding the current front- and back-end processes versus future visions. In addition, they provide strong support for management regarding decision-making due to understanding customer value in depth.

## Theoretical Background

Service design is a networked bottom-up activity (Meurer, 2001, p. 52), which can contribute to front- and back-end business processes (Yu & Sangiorgi, 2018, p. 103) with the help of human-centric methods and skills (Junginger, 2008). As a human- and customer-centric holistic methodology, service design uses outside-in strategies in problem-solving activities (Andreassen et al., 2016; Junginger, 2008) and it leans on experiential learning and knowledge-building (Buhring & Liedtka, 2018; Kolb, 1984; Stock et al., 2018). Within organizations, service designers participate actively in creating and making changes happen through iterative development processes together with customers, users, organization's partners, and internal stakeholders from different levels, units, and departments (Junginger, 2008; Miettinen, 2009). In this study, service design is seen as a human- and customer-centric development methodology and practice, which holds a strategic (outside-in) development process – a process where analytical and creative reflections intellectually take turns. People (customers and employees), front- and back-end of services, products, digital systems and tools, processes, and practices are naturally the subjects of service design.

## *Transformation Design Embedded with Service Design*

As a social practice, service design can be naturally seen to fit with the fields of social change and organizational studies (Sangiorgi, 2011). During the last two decades, design research has increasingly studied design's transformative role within organizations (Bate & Glenn, 2007; Buchanan, 2004; Junginger, 2008; Junginger & Sangiorgi, 2009; Kurtmollaiev et al., 2018; Sangiorgi, 2011; Trullen & Bartunek, 2007; Van Aken, 2007; Yu & Sangiorgi, 2018). Service design approaches that support shaping behaviours, processes, and organizations are conceptualized as transformation design (Burns et al., 2006; Sangiorgi, 2011). The focus of transformation design is not only on developing and shaping the final solutions but also on reinforcing abilities to brace sustainable innovation (Bailey, 2012; Terrey, 2013).

Burns et al. (2006) define six characteristics of transformation design: (1) defining and redefining the brief since designers participate in defining the problems to create the brief; (2) collaborating between disciplines since designers facilitate collaboration to solve complex issues; (3) employing participatory design techniques since bottom-up innovation strategies with the help of frontline personnel and users support problem-solving; (4) building capacity, not dependency, since transformation design projects should leave organizational capacities and skills to answer ongoing change; (5) designing beyond traditional solutions since designers shape the behaviour of people, systems, and organizations, not just form, and they consider issues more holistically regarding high levels of systems thinking; (6) creating fundamental change since transformation design projects can initiate cultural changes of human-centeredness within organizations. According to Burns et al. (2006), transformation design is not a change management process, but participating in the design processes might help to move towards the desired outcomes. Such processes offer participating stakeholders' ownership of the vision, tools, and capacity to adapt and innovate in addition to 'initial steps towards changing the culture, aligning thinking and focusing around the end user' (Burns et al., 2006, p. 22). Pinheiro et al. (2012) support this notion by arguing that if service designers involve internal stakeholders and units in customer-centred conversations, the object of change may expand to organizational cultures and norms.

According to Yu and Sangiorgi (2018), service design impacts transforming both service systems and organizations. When user- and customer-centred activities, such as service design, are integrated into the organization, it can result in different qualities and impacts on service development and operations; it can impact the entire innovation process; it integrates multidisciplinary functions; it can affect stakeholders' perspectives and behaviours. Hence, service design can catalyse organizational transformation (Yu & Sangiorgi, 2018). This is also due to the skillset and toolsets that the designers hold (Yu & Sangiorgi, 2018). Bailey (2012) and Lin et al. (2011) support these arguments by bringing forward how service design can trigger changes in organizational contexts by incorporating human-centred cultures with design tools and knowledge. Andreassen et al. (2016) argue that service design can reform organizations since customer-driven service development practices require changes at the organizational levels.

Junginger and Sangiorgi (2009) elaborate on three levels where service design can have impacts and outcomes in organizations: (1) artefacts and behaviours regarding service interaction design – the impact will remain small or temporarily if improvements touch upon new or improved artefacts and hence, organizational norms and values are not questioned behind them; (2) norms and values regarding service design interventions – changes might not be radical unless the new service concept affects deeper fundamental assumptions such as norms and values of the organization, and service designers demonstrate the value of change by engaging the organization and rethinking the organizational elements around the new service; (3) fundamental assumptions regarding organizational transformation – a long-term collaboration and strong commitment from the organization is required since service concepts require deep transformations, which touch the fundamental assumptions of the organization and hence, design is used to unveil deep assumptions regarding the current situation versus an agreed future vision. In addition to long-term commitment, achieving sustainable and effective transformation within organizations requires genuine interest, change of cultures and attitudes through trust-building, ongoing dialogues, and co-created vision (Junginger & Sangiorgi, 2009; Sangiorgi, 2011). Hence, service designers must be able to generate transformative insights regarding the fundamental assumptions, norms, values, and behaviours of the organization (Junginger & Sangiorgi, 2009).

When design confronts organizational and behavioural change, pilot projects can work as seeds and vehicles for change since they offer a fundamental role in opening the way to transformative changes and knowledge exchange within longer transformation processes (Junginger & Sangiorgi, 2009; Junginger, 2008). Kurtmollaiev et al. (2018) support these notions based on the study they carried out on the telecommunication company Telenor. Service design causes changes in organizational mindset and practices (Kurtmollaiev et al., 2018). In addition, their study shows that service design 'becomes the new service development process itself and grows into a powerful transformative force that is capable of changing institutions' (Kurtmollaiev et al., 2018, p. 70). Managers who start implementing service design 'should prepare for organization-wide transformation that includes changes in employees' mindsets and routines' (Kurtmollaiev et al., 2018, p. 71).

Sanders (2009) claims that the most critical component in getting companies to transition from designing for customers to designing with customers is the mindset and attitudes held by individuals. Before co-creation can happen, such people who think it makes sense to design with customers and people are needed. Once the mindset is there, there is the ability to change processes and the culture of the organization (Sanders, 2009). In addition, Liedtka et al. (2013) and Rousseau (1995) argue that the integration of service design regarding transformation requires challenging fundamental assumptions, beliefs, norms, and values that individuals and staff members hold in addition to mental models in organizations brought forward by Vink et al. (2019). Borja de Mozota (1998, p. 250) argues that 'the integration of design is not likely to happen unless the company is going through a crisis where past beliefs and mechanisms have proved inefficient and there is a willingness to be receptive to new kinds of information'.

## *Experiential Learning with Help of Service Design*

The authors argue that before any transformation in organizations can happen, learning of individuals must occur first. Therefore, it must be considered as well. Research and development processes used in service design are similar to Von Hippel's (2005) iterative learning cycle in product development and Kolb's (1984) model of experiential learning. Experiential learning has proven to be one of the successful strategies when embedding design thinking in the organization (Stock et al., 2018). It facilitates and fosters strong peer-to-peer learning where the benefits of user and stakeholder inclusion, design process, and methods are experienced through practical case studies in company contexts. This may create strong buy-in among stakeholders since the co-design approach used in service design is one of the methods of engaging not only stakeholders but also the leadership in evaluating possible solutions, learning about them, and making decisions.

According to Sangiorgi (2011, p. 34), 'within organizational development studies, a strong emphasis is given to participatory research and learning processes within organizations seen as drivers for transformational change'. From the organization's perspective, outside-in strategies along with human-centred activities of customer research, participatory design, and iterative processes provide the organization with an avenue to learn about customers and themselves (Andreassen et al., 2016; Junginger, 2008). Along this process of making and creating new solutions, designers bring people, structures, and resources of the organization into alignment, and learning is put into action (Junginger, 2008). The engagement of internal organization stakeholders as project participants empowers them to be co-creators of new solutions (Sangiorgi, 2011). This is required since transformational changes cannot happen without a deep involvement of psychological engagement among stakeholders in the systems (Chapman, 2002). These go in line with Adcroft et al. (2008, p. 44), who claim that people, managers in particular, 'learn best when they are active learners and reflect their own experiences'.

According to Buhring and Liedtka (2018) design's emphasis on learning in action offers a powerful contribution to enhancing strategic planning processes in conjunction with foresight where assumptions of future scenarios can be tested through experiments. When ecosystem players at different levels are engaged in the design and execution of experiments, learning becomes an ongoing process and future scenarios can be adjusted as real-world feedback informs the process (Buhring & Liedtka, 2018). Hence, 'instead of regarding design and management as different entities, there are grounds for focusing on the similarities between the two and to examine the learning that could occur between design and management' (Borja de Mozota, 1998, p. 257). When the design is valued as a process, design can lead to sustained innovation, higher customer value, and improved competitive advantage (Borja de Mozota, 1998). This is achieved due to interdisciplinary conversations, which designers facilitate to enhance stakeholders' ability to align, learn, and change together (Liedtka, 2017).

## Research Design

This is a qualitative case study, which follows an inductive research approach. The concentration of the research is to find out the *what* and *how* (Yin, 2009). The results follow Yin's (2009) and Stake's (2005) definitions of context dependency since the topic under investigation is seen as socially constructed within its context, place, and time. The interviewees' statements have been interpreted as each person's individual and unique experiences of service design within the organization under study after four years of service design usage. Hence, interviewees' opinions do not represent official statements of the organization. This was brought up with each informant before data gathering. Hence, following Stake (2005) and Merriam and Tisdell (2015), the aim is to produce a better understanding of the phenomenon through participants' experiential knowledge, which is intrinsically bounded with the case in its real-life context, in this case, the particular organization, the multinational manufacturing corporation.

### *Data Collection and Participants*

The organization under study was chosen due to the use of in-house service design expertise as a part of R&D projects since the autumn of 2014. Altogether, 33 semi-structured individual, pair, and group interviews were conducted in 2018 including 45 participants from all organizational levels except for the CEO and shareholders. Interview consent was collected from each informant before interviewing them.

In this chapter, the results are based on data from 26 semi-structured interviews including 36 interviewees. Eighteen individual interviews, two group interviews, and six pair interviews were conducted. Hence, 18 interviewees participated in pair and group interviews. Twelve of the 36 interviewees come from the R&D department. Twenty-four interviewees represent other departments of the organization. All

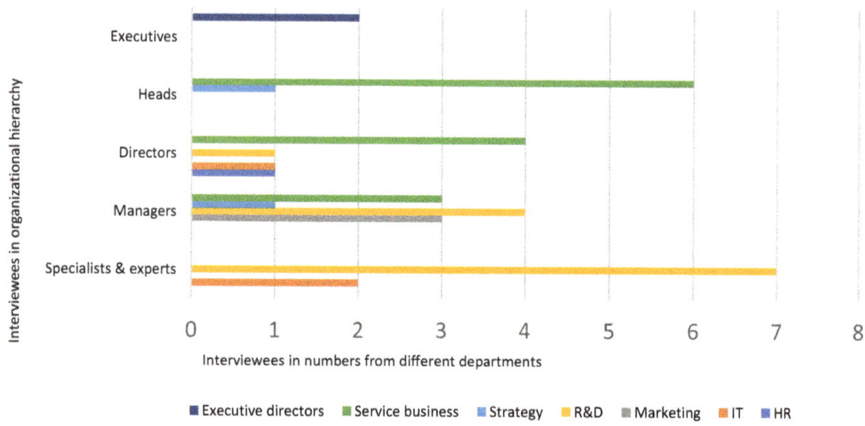


**Chart 10.1** Description of the interview participants

of the interviewees were chosen based on their experience in service design projects. Both service designers and non-service designers were interviewed. The purpose was to get a holistic perspective and understanding of service design's impacts and evolvement within the organization in addition to how it is perceived by stakeholders from different levels and departments of the organization (Chart 10.1).

The interviewees had an average of ten years of work experience in the organization by August 2018. Deviation from the average of ten years varied from four to 20 years of experience. Only a few of the participants had working experience from a few months to three years and one of the participants for 30 years. Demographics in terms of the interviewees' age were not gathered as their age was not seen as relevant information since service design as a methodology was new or fairly new for all research participants besides service designers. Half of the interviewees were female and half were male.

## *Data Analysis*

Data analysis has been performed in three phases (Table 10.1). During the first phase, all 33 interviews were coded by the use of Descriptive Coding, Concept Coding, and Sub-coding methods by following the coding rules of Saldaña (2016). As a result, 1205 descriptive codes emerged, which were then categorized into 25 concept code groups. In the second analysis phase, the code group *Service Design Value* was chosen for further analysis. The content of the code group was categorized into ten sub-themes with the help of thematic analysis. In the third analysis phase, one of the sub-themes, *transforming business and working culture*, was further thematized. Then subheadings within the data were created to organize interview

**Table 10.1** Journey of the research analysis process

| Coding | Thematic analysis | Further thematization | Results |
|---|---|---|---|
| **The 1st phase of analysing data** | **The 2nd phase of analysing data** | **The 3rd phase of analysing data** | **Interpretation of the content of a sub-theme** |
| **Used methods**<br>Descriptive coding<br>Concept coding<br>Sub-coding | **Used methods**<br>Thematization of the interview content within a concept code group *Service design value* | **Used methods**<br>Further thematization of a sub-theme: 1) Transforming business & working culture | Three subheadings including the interview content<br>Synthesizing |
| **Deliverable**<br>1205 descriptive codes arranged into 25 concept code groups | **Deliverable**<br>10 sub-themes / text files | **Deliverable**<br>Subheadings of the sub-theme | **Deliverable**<br>Results written |

quotations based on their meanings. Next, a synthesis of the topics was made to provide results.

## Deriving Research Results

As described above, all of the conducted 33 interviews were coded into 1205 codes, which were then categorized into 25 themed concept code groups. All of the themes were derived from the codes. These formed the key interpretations for further data analysis. The data analysis was conducted in three phases: (1) coding, (2) thematic analysis, and (3) further thematization, which are presented and discussed next.

### *The First Phase of Analysing Data: Coding*

During the first analysis phase, two coding rounds of the 33 interview transcriptions were done with the help of Atlas.ti – a qualitative data analysis program. Descriptive Coding, Concept Coding, and Sub-coding methods were used. As a result of the initial coding phase, 1205 empirical codes were created based on the Descriptive Coding rules of Saldaña (2016). The method offered a straightforward way to create descriptive nouns based on the topics discussed. Saldaña (2016, p. 105) states that nouns alone 'may not enable more complex and theoretical analysis as the study progresses'. Due to this, the Concept Coding method was also used. Concept Coding provides an analytical task where so-called lumping can be done to create larger units of data (Saldaña, 2016). The method in this study helped to form macro-level concepts of micro-level nouns.

Since coding is an iterative and cyclical process, it is natural that the codes evolve as the researcher revisits the data. Hence, codes may become more focused (Gibson & Brown, 2009). This is what happened in this research. Sub-codes were generated to provide 'further nuances within a given code structure' (Gibson & Brown, 2009,

**Table 10.2** 25 Concept code groups

| | | | | |
|---|---|---|---|---|
| *Innovation Capability* 8 codes | *Communication* 5 codes | *Key Performance Indicators (KPIs)* 87 codes | *The Changing Role of Marketing* 18 codes | *Organizational Processes* 41 codes |
| *Organizational Readiness for Service Design* 118 codes | *Competence Management* 17 codes | *Service Design Versus Business* 21 codes | *Service Design Challenges* 130 codes | *Service Design Value* 208 codes |
| *Service Design Value/Challenges* 14 codes | *Managing Service Design* 13 codes | *Development of Service Design* 49 codes | *The Possibilities of Service Design* 24 codes | *Service Design Methods* 15 codes |
| *The Role of Service Design in the Organization* 83 codes | *The Role of Service Design in the Projects* 28 codes | *Service Design Processes* 76 codes | *The Need for Service Design in Internal Organizational Development* 24 codes | *Service Design Versus Marketing* 7 codes |
| *Project Management* 31 codes | *Stakeholder Group Work* 45 codes | *Strategy* 40 codes | *Performance Pyramid* 83 codes | *Product Development Processes* 20 codes |

p. 142). As a result, third- and/or fourth-ordered tags, which work as siblings for the Descriptive Codes, were assigned to the codes to highlight more specific contexts of each code. This made further analysation and thematization of the data easier. Overall, as a result of the first analysis phase, 25 concept code groups were formed (Table 10.2).

## *The Second Phase of Analysing Data: Thematic Analysis*

Thematic analysis was used as the analysis method in this phase. Following Braun and Clarke (2022), the concept code group *Service Design Value* of 208 codes formed the analytical entity and the conceptualized building block for thematization. The content of this code group was reviewed based on the descriptive codes within the code group. Commonalities in addition to differences and relationships were examined by following the rules of Gibson and Brown (2009). Revisits to interview transcriptions had to be naturally done to ensure further code grouping. Codes which included associated themes and meanings were grouped. Then headlines were created for each group to identify implicit and explicit ideas of the data as guided by Guest et al. (2012). The formed headlines worked as sub-themes of the *Service Design Value* concept code group.

1. Transforming business and working culture
2. Knowledge-building and organizational learning
3. Service design's implications for the work of internal stakeholders
4. Impacting strategy and implementing strategy
5. Impacts on innovation capabilities
6. Impacts on competitiveness and differentiation
7. Impacts on brand value
8. Short- and long-term impacts on business
9. Support for managerial decision-making (role-based view)
10. Facilitative role within the organization (role-based view)

Next, all quotations of the named groups (sub-themes) were copied and pasted into ten separate text files. The headlines of the text files were as described above.

## *The Third Phase of Analysing Data: Further Thematization*

Finally, after a few reading rounds, a text file (a sub-theme), *Transforming business and working culture*, was chosen to provide data for this chapter. This is due to the meanings, which answer the research question discussed in this chapter. All quotations were read through carefully. As a result, three categories were created to highlight the topics based on the meanings of the quotations.

1. Experiential learning through service design supports changing individuals' attitudes
2. Understanding the meaning and benefits of service design support changing individuals' mindsets and creating belief
3. Towards a new organizational paradigm through strategy renewal

Next, a synthesis of the content was made to provide results. These are presented and discussed next.

# Results

## *Experiential Learning Through Service Design Supports Changing Individuals' Attitudes*

Here, the first topic, (1) experiential learning through service design supports changing individuals' attitudes, is discussed related to aspects of change for individuals. The data set of this study showed that the process of changing attitudes from a technology-oriented engineering organization towards customer-centricity has not been straightforward. It became clear from the interviews that learning at the level of individuals, and thus, building knowledge and understanding, is required first to change technology-oriented expert-driven attitudes towards human- and customer-centricity. Such learning experiences have been strongest by being part of human- and customer-centric projects where service design as a methodology has been in use.

> Those who have been involved in service design projects have gone through a eureka moment and realisation to understand what it means altogether. (n27)
>
> Leaving the expertise, the fact that you know things, should dare to be left out altogether. Instead, listen. It took me three months before I realised how it happens. (n26)
>
> I've had to put my hat on my hands and be humbler, so there has happened some change in my head to the direction of listening to customers even more. You start thinking differently. You're not just thinking of things from your perspective, but taking the opposite side and you start thinking of the problem first instead of the possible solution. (n42)

These statements show how people have started to understand through experiential learning what human- and customer-centric development means in terms of service design and how it is done. As a result of this, the attitudes of individuals have started to change, which has led to the realization of the transforming effect of service design. This becomes clear also in the interview comments below.

> . . . it's also that attitudinal change ( . . . ) I think already by (name of the first official service design project) ( . . . ) people were starting to ask, and pull, and now, I think even more so ( . . . ). I think it's a very natural pull ( . . . ), it has more demand now than supply. (n13)
>
> Making decision-makers understand the value we got from the field by having open discussions and asking questions, that we got there to see customers' daily lives and their needs, it has changed attitudes, or at least in that project. (n28)

Thus, based on the comments above, experiential learning of service design through service design projects has supported stakeholders to learn and understand what it means, how customer-centric work is done, and why it makes sense to develop in such a way. This has resulted in changing the attitudes of individuals, and hence, buy-in and 'pull' of service design within the organization have gotten created. Hence, service design has supported changing attitudes from technology-oriented engineering-based working cultures towards human- and customer-centricity.

## *Understanding the Meaning and Benefits of Service Design Support Changing Individuals' Mindsets and Creating Belief*

Here, the second topic, (2) understanding the meaning and benefits of service design support changing individuals' mindsets and creating belief, is discussed related to aspects of change for individuals. The interviews showed that experiencing and understanding how to do human- and customer-centric development in addition to learning from customers about the customers and productivity issues of the organization have caused shifts in stakeholders' mindsets. As a result of this, changes in thinking from technology-oriented mindsets towards human- and customer-centricity have started appearing.

> What I see is the greatest value that has changed, people's thinking and understanding have changed. People want to do things differently and they find it natural to ask the customer. Then everything else (...) follows. (n34)

The data also showed that project rehearsals and pilots have been in a key role in creating belief in service design as a methodology. Based on the interviews, also the first official service design project has caused transforming experiences among the top management.

> (...) we wanted to first create belief in this thing (service design) through project rehearsal and piloting. (...) Of course, it helped that there was such a success story. It created a belief that this could work more widely, and service design is a way we can win together with the customer. (n27)

> We started with this listening and observation exercise, the service design exercise, then that was powerful and this was transforming our business. (n44)

> (Name of a director) turned to the dark side and he became an advocate for service design. When he understood and internalised it (service design), saw and heard things, and how it has great power in all this, he started thinking it must be used as an exporter of the message in a company like this that only believes in the facts and the customer. (n12)

Hence, in addition to changing attitudes, the results showed that understanding the meaning and benefits of service design has caused changes in individuals' mindsets and created belief. Thus, from the perspectives of individuals, first, reserved attitudes towards human- and customer-centricity have started to turn into cooperative ways of behaving and acting through experiential learning of service design. Second, the mindsets have started to evolve from technology-oriented thinking towards human- and customer-centricity through understanding the meaning of service design. Third, understanding the benefits of service design has supported creating belief in service design as a human- and customer-centric development methodology and practice. These presented individual change aspects form preconditions for organizational change aspects since the acts of an organization follow individuals' thinking and acting. Thus, the third and last topic is discussed next in terms of the organizational change aspects.

### *Towards a New Organizational Paradigm Through Strategy Renewal*

The data showed that once the attitudes, mindsets, and beliefs of individuals started altering, service design became a part of the organization's strategy. Hence, a transition towards a human- and customer-centric organizational paradigm started to happen, and hence, behaviours and decision-making started to change.

> (...) in the past we were in a world where headquarters made demands, product development developed, then they were launched, and then we hoped that the country-level

> organizations would manage to sell. This has now, in a way, gotten reversed. We now start with the customers. (. . . ) it then became part of one aspect of our strategy. There we then found that for us to remain a winner and innovate in this changing world, we need to work with customers, with partners, and also within the organization much more broadly. (n27)

> It's about the ability to go deeper in understanding the customer needs, and the ability to help the organization to frame this feedback and what it means, in terms of, potential offering, service innovations, process innovations. (n30)

In addition, the interviews showed that the human- and customer-centric ways of thinking and behaving have increased courage in the organization. If before customers were met in the latter phases of product development, they are now involved regularly in the development of services since the beginning of the projects. Continuous inclusion of business-to-business customers in service development has supported the implementation of an outside-in strategy. However, shifting the focus towards human- and customer-centricity has challenged the organization's existing norms and practices.

> In a new way, this (service design) challenges what the organization does, how things are done, and how to be customer-centric. We boldly just go and try and see, and test new ideas. It has brought new perspectives and increased courage. (n5)

> It has brought a whole new culture. (n27)

> Outside-in and co-creation are strongly connected. (n25)

According to the interviews, the organization is now willing to apply service design across the full chain from service development to service delivery and operations as well as utilizing service design in the traditional product business. However, the interviews brought forward that despite the benefits of service design and the strong commitment from the top management, in-house service design has still not achieved a stable base culturally within the organization after four years.

> We have not generated a new culture yet, but we have achieved such a situation with the help of the strong commitment of the company's management and key executives. If we left service design out now, we would go back to the old days and ways of developing with incredible speed. (n12)

Hence, in terms of the organizational change aspects, the results show that in-house service design has supported transforming the organizational working culture by becoming part of the organization's strategy. Thus, the organizational paradigm transformation has begun. When human- and customer-centric values were built to be a part of the organization's strategy, the organization started increasingly to develop and innovate in such ways.

## Discussion

Our results show that in-house service design has supported an organization's working culture to transform towards human- and customer-centricity through change aspects for individuals and an organization. These mean changes in individuals regarding attitudes, mindsets, and beliefs, which are preconditions for further organizational changes. This is where experiential learning of individuals is essential in understanding the meaning and benefits of service design as a human- and customer-centric methodology and practice. When understanding of its meaning and benefits exists, changes in an organization can happen in terms of organizational paradigm transformation through strategy renewal affected by human- and customer-centric values.

Let us first discuss the individual aspects of change. Bartunek and Louis (1988) claims that before any organization can change, changes in attitudes, beliefs, and cultures must happen to support transformational changes. This study claims that individuals' attitudes, mindsets, and beliefs must change first to trigger changes in organizations and their working cultures. The results of the study show that changing attitudes, mindsets, and beliefs of individuals happens most efficiently through experiential learning by being part of the development projects where service design is used. Here pilot projects offer a fundamental role since they open the way to transformative changes and knowledge exchange within longer transformation processes (Junginger & Sangiorgi, 2009; Junginger, 2008; Sanders, 2009). This is due to learning in action and experiential learning, which are powerful contributors since learning happens most efficiently through the inclusion of internal stakeholders and units (Buhring & Liedtka, 2018; Liedtka, 2017; Kolb, 1984; Pinheiro et al., 2012; Stock et al., 2018). This creates commitment and ownership in addition to trust-building (Junginger & Sangiorgi, 2009; Sangiorgi, 2011) among in-house service designers and stakeholders. This, in turn, clears the path for socially sustainable and effective transformation within organizations in terms of human- and customer-centricity.

The results of this study prove that learning to understand the meaning and benefits of service design as a human- and customer-centric methodology supports changing the decision-makers' decision attitudes towards design attitudes (Boland & Collopy, 2004). In addition, the results show that understanding the meaning and benefits of service design affects changing mindsets as well from technology-oriented thinking towards human- and customer-centricity. Hence, this study shows, in line with Chapman (2002), that deep involvement of psychological engagement supports changing attitudes and mindsets. In terms of changing beliefs, Borja de Mozota (1998) argues that the company must go through a crisis of past beliefs and inefficient mechanisms to be able to integrate design. Unlike Borja de Mozota (1998), this study claims that changing past beliefs does not require a crisis. Instead, the results of this study show that beliefs can be changed through changing attitudes and mindsets by being part of human- and customer-centric projects where the path is cleared for socially sustainable ways of working when developing services and organizations. Such participatory processes support learning and thus, changing beliefs towards human- and customer-centric ways of working. Hence, these results support the views of Andreassen et al. (2016), Junginger (2008), Meurer (2001), and Sangiorgi (2011) who argue that service design is a networked bottom-up activity, which uses outside-in strategies along with human- and customer-centric activities to provide the organization stakeholders an avenue to learn about their customers and themselves.

In terms of the organizational aspects of change, Junginger and Sangiorgi (2009) and Sangiorgi (2011) claim that service design impacts in transforming organizations require a long-term commitment and genuine interest of an organization regarding transformative changes and transformation processes. The results of this study show that once the transforming experiences among the top management occur related to attitudes, mindsets, and beliefs, the next step is to consider the organizational paradigm through the organization's strategy and shared values. Thus, a transition towards a human- and customer-centric organizational paradigm is enabled.

Hence, based on the discussion, the findings of this study are in line with Kurtmollaiev et al. (2018) who argue that service design can become more than a practice, and it can grow into a powerful transformative force, which starts changing institutions. From the perspective of transformation design, service design can create fundamental change since co-creative/participatory projects can initiate cultural changes of 'human-centeredness' within organizations, and thus, the object of change may expand to organizational cultures (Burns et al., 2006; Pinheiro et al., 2012). However, turning the mindset of a technology-oriented organization to human- and customer-centricity is not straightforward in engineering-based corporate cultures (Borja de Mozota, 1998; Boland & Collopy, 2004). We argue that in addition to changes in individuals and organizational paradigms through strategic renewals, norms and practices must also be developed accordingly to get human- and customer-centric working cultures to flourish. When norms at the level of people, processes, and systems are in place as human- and customer-centric standards of behaviour, human- and customer-centric practices and decision-making can start to flourish. Hence, a more complete transition in working cultures can happen.

Overall, the novelty of this study lies in the dimensions related to the individual and organizational change aspects, which in-house service design has supported to get the organization's working culture to transform towards human- and customer-centricity:

1. In-house service design supports changing individuals' attitudes from technology-oriented working cultures towards human- and customer-centricity. Here experiential learning through service design is the key.
2. Understanding the meaning of service design supports changing individuals' mindsets and understanding the benefits of it supports creating belief in service design. Once the attitudes, mindsets, and beliefs of individuals support service design as a human- and customer-centric methodology and practice, changes in organizations can start more widely.
3. Through affecting individuals, in-house service design can initiate the transformation of an organization by becoming part of the organization's strategy, which starts the organizational paradigm transformation. Hence, the organization

can start implementing human- and customer-centric values through outside-in strategies.

## *Limitations*

The following limitations of this study should be acknowledged. The data analysis was performed only by the first author who has a working background within the organization under study. There were no meetings to discuss the codes and findings of the data between the authors. Thus, research bias natural in qualitative research must be considered (Mehra, 2002). In addition, quantification of the content brought forward by the interviewees under the sub-theme of *Transforming business and working culture* was not done. This was not seen as relevant because all interviewees did not have transparency on all topics discussed. Instead, similarities and differences were looked for in terms of what interview content was repeated. In addition, by whom and with what emphasis the content was brought forward were seen as valuable. Thematization was done based on the similarities and differences between the topics. The results are based on the most repetitive topics and the topics that were brought forward with strong emphasis. The quotations brought forward in the results have been chosen in such a manner that they represent the content in the best possible way. In addition, the age of the interview participants was not gathered, which might affect the interpretation of the results. Despite the limitations, we believe that the data and data analysis provide a representative overview of the topics discussed in this study.

## *Future Research*

More research would be needed regarding organizational transformation and service design management in terms of support structures, norms, and practices. This understanding would increase organizational know-how on how to enable frictionless human- and customer-centric working cultures, and thus, how to improve the internal efficiency and quality of an organization overall. Such understanding would be important not only for the manufacturing industry but also for other industries such as the automotive industry and transportation, which are going through significant changes in the Fourth Industrial Revolution, for example, regarding robotics and artificial intelligence. Technology must not become an absolute value. The boundaries between the physical and digital worlds are increasingly getting blurred. Hence, understanding the needed organizational norms, practices, and support structures to support human- and customer-centricity is highly needed in emerging technologies, and organizations are in key positions in doing this.

## Conclusion

In-house service design has supported an organization's working culture to transform towards human- and customer-centricity through the change aspects of individuals and an organization. Experiential learning of stakeholders is the key. Once understanding of the meaning and benefits of service design occurs, changes in individuals in terms of attitudes, mindsets, and beliefs can happen. These form the preconditions for change aspects of an organization, which are related to an organizational paradigm transformation through strategy renewal affected by human- and customer-centric values. Hence, according to the results of this study, in-house service design has affected the organization's working culture by initiating changes in individuals who then have initiated changes in an organization. However, we argue that human- and customer-centric organizational paradigms must also be supported by norms and practices accordingly to get human- and customer-centric working cultures to thrive. Once the norms and practices produce human- and customer-centric organization paradigms systematically and accordingly, a more holistic organizational transformation can take place. If this does not happen, the transformation of an organization in terms of human- and customer-centricity will remain incomplete.

**Acknowledgements** This research has been supported by the European Regional Development Fund as part of the *Design in Smart Mobility Business Services* project (2018–2019) in addition to the Ph.D. grant received from the Finnish Cultural Foundation (2022–2023).

## References

Adcroft, A., Willis, R., & Hurst, J. (2008). A new model for managing change: The holistic view. *The Journal of Business Strategy, 29*(1), 40–45.

Andreassen, T. W., Kristensson, P., Lervik-Olsen, L., Parasuraman, A., McColl-Kennedy, J. R., Edvardsson, B., & Colurcio, M. (2016). Linking service design to value creation and service research. *Journal of Service Management, 27*(1), 21–29.

Bailey, S. (2012). Embedding service design: The long and the short of it. In *Proceedings of the 3rd ServDes. Conference on service design and service innovation* (pp. 31–41). Helsinki, Finland.

Bartunek, J., & Louis, M. R. (1988). The interplay of organization development and organizational transformation. *Research in Organizational Change and Development, 2*, 97–134.

Bate, P., & Glenn, R. (2007). Toward more user-centric OD: Lessons from the field of experience-based design and a case study. *The Journal of Applied Behavioral Science, 43*(1), 41–66.

Borja de Mozota, B. (1998). Chapter 11: Challenge of design relationships: The converging paradigm. In M. Bruce & B. Jevnaker (Eds.), *Management of design alliances: Sustaining competitive advantage* (pp. 243–259). Wiley.

Boland, R. J., & Collopy, F. (2004). Design Matters for Management. In R. J. Boland & F. Collopy (Eds.), *Managing as Designing*, (pp. 3-18). Stanford, CA: Stanford Business Books.

Braun, V., & Clarke, V. (2022). *Thematic analysis: A practical guide*. SAGE Publications.

Buchanan, R. (2004). Chapter 4: Management and design. Interaction pathways in organizational life. In R. J. Boland & F. Collopy (Eds.), *Managing as designing*. Stanford Business Books.

Buhring, J., & Liedtka, J. (2018). Embracing systematic futures thinking at the intersection of Strategic Planning, Foresight and Design. *Journal of Innovation Management, 6*, 134–152.
Burns, C., Cottam, H., Vanstone, C., & Winhall, J. (2006). *RED Paper 02: Transformation design*. Design Council.
Bustinza, O. F., Vendrell-Herrero, F., & Baines, T. (2017). Service implementation in manufacturing: An organisational transformation perspective. *International Journal of Production Economics, 192*, 1–8.
By, R. T. (2005). Organisational change management: A critical review. *Journal of Change Management, 5*(4), 369–380.
Chapman, J. A. (2002). A framework for transformational change in organisations. *Leadership & Organization Development Journal, 23*(1), 16–25.
Cooper, R. G. (2001). *Product leadership: Creating and launching superior new products*. Perseus Books.
Costa, N., Patrício, L., Morelli, N., & Magee, C. L. (2018). Bringing service design to manufacturing companies: Integrating PSS and service design approaches. *Design Studies, 55*, 112–145.
Francis, D., Bessant, J., & Hobday, M. (2003). Managing radical organisational transformation. *Management Decision, 41*, 18–31.
Ghauri, P. N., & Grønhaug, K. (2005). *Research methods in business studies: A practical guide* (3rd ed.). Financial Times Prentice Hall.
Gibson, W. J., & Brown, A. (2009). *Working with qualitative data*. Sage.
Guest, G., MacQueen, K. M., & Namey, E. E. (2012). *Applied thematic analysis*. Sage Publications.
Junginger, S. (2008). Product development as a vehicle for organizational change. *Design Issues, 24*(1), 26–35.
Junginger, S., & Sangiorgi, D. (2009). Service design and organizational change. Bridging the gap between rigour and relevance. In *Proceedings of the 3rd IASDR conference on design research* (pp. 4339–4348). Seoul, South Korea: Korean Society of Design Science.
Kolb, D. (1984). *Experiential learning: Experience as the source of learning and development*. Prentice Hall.
Kotter, J. P. (2007). Leading change: Why transformation efforts fail. In *Museum management and marketing* (pp. 20–29). Routledge.
Kurtmollaiev, S., Fjuk, A., Pedersen, P. E., Clatworthy, S., & Kvale, K. (2018). Organizational transformation through service design: The institutional logics perspective. *Journal of Service Research, 21*(1), 59–74.
Lauer, T. (2010). *Change management*. Springer Berlin Heidelberg.
Liedtka, J. (2017). Beyond better solutions: Design thinking as a social technology. In E. Bohemia, C. de Bont, & L. S. Holm (Eds.), *Conference proceedings of the design management academy* (Vol. 1, pp. 1–18). London: Design Management Academy.
Liedtka, J., King, A., & Bennett, K. B. (2013). *Solving problems with design thinking: Ten stories of what works*. Columbia University Press.
Lin, M. C., Hughes, B. L., Katica, M. K., Dining-Zuber, C., & Pisek, P. E. (2011). Service design and change of systems: Human-centered approaches to implementing and spreading service design. *International Journal of Design, 5*(2), 73–86.
Mehra, B. (2002). Bias in qualitative research: Voices from an online classroom. *The Qualitative Report, 7*(1), 1–19.
Merriam, S. B., & Tisdell, E. J. (2015). *Qualitative research: A guide to design and implementation*. Wiley, Incorporated.
Meurer, B. (2001). The transformation of design. *Design Issues, 17*(1), 44–53.
Miettinen, S. (2009). Designing with innovative methods. In S. Miettinen & M. Koivisto (Eds.), *Designing services with innovative methods* (pp. 10–25). Publication series of the University of Art and Design Helsinki B 93. Kuopio Academy of Design, Taitemia Publication Series 33. Otava.
Paton, R. A., & McCalman, J. (2008). *Change management: A guide to effective implementation*. Sage Publications.

Pinheiro, T., Alt, L., & Mello, J. (2012). Service design creates breakthrough cultural change in the Brazilian financial industry. *Touchpoint: Journal of Service Design, 3*, 18–23.
Rousseau, D. M. (1995). *Psychological contracts in organizations: Understanding written and unwritten agreements*. Sage Publications.
Saldaña, J. (2016). *The coding manual for qualitative researchers* (3rd ed.). Sage.
Sanders, E. (2009). Exploring co-creation on a large scale: Designing for new healthcare environments. In P. J. Stappers (Ed.), *Designing for, with and from user experiences. Symposium*. Delft, The Netherlands: Faculty of Industrial Design Engineering, TU/Delft.
Sangiorgi, D. (2011). Transformative services and transformation design. *International Journal of Design, 5*(2), 29–40.
Stake, R. E. (2005). Qualitative Case Studies. In N. K. Denzin & Y. S. Lincoln (Eds.), *The SAGE handbook of qualitative research* (3rd ed., pp. 443–466). Sage Publications.
Stock, K. L., Bucar, B., & Vokoun, J. (2018). Walking in another's shoes: Enhancing experiential learning through design thinking. *Management Teaching Review, 3*(3), 221–228.
Terrey, N. (2013). Managing by design: Enacted through situated networks. *Design Management Journal, 8*(1), 52–61.
Trullen, J., & Bartunek, J. M. (2007). What a design approach offers to organization development. *The Journal of Applied Behavioral Science, 43*(1), 23–40.
Van Aken, J. (2007). Design science and organization development interventions. Aligning business and humanistic values. *The Journal of Applied Behavioral Science., 43*, 67–88.
Vink, J., Edvardsson, B., Wetter-Edman, K., & Tronvoll, B. (2019). Reshaping mental models – Enabling innovation through service design. *Journal of Service Management, 30*(1), 75–104.
Von Hippel, E. (2005). *Democratizing Innovation*. The MIT Press.
Yin, R. K. (2009). *Case study research: Design and methods* (4th ed.). Sage.
Yu, E., & Sangiorgi, D. (2018). Service design as an approach to implement the value cocreation perspective in new service development. *Journal of Service Research, 21*(1), 40–58.

# Chapter 11
# The Challenges of In-House Service Design in Organizational Transformation: A Case Study of a Multinational Manufacturing Corporation

**Krista Korpikoski**

## Introduction

During the last two decades, design has increasingly been implemented as an in-house development activity within the manufacturing industry and service organizations, with regard to strategic renewal (Kolko, 2015; Kurtmollaiev et al., 2018; Ravasi & Lojacono, 2005; Yoo & Kim, 2015). Despite this, design is seen only as a support function in many organizations (Oakland et al., 2021) and not as a dynamic, ongoing activity. Many "organizations still exist at the delivery end of the thinking life cycle, not at the discovery end" (Golsby-Smith, 2007, p. 22). Hence, design is not seen as a strategic toolkit shaping new futures. Predominantly customers in this kind of "non-design" organization come last, at the end of the delivery life cycle, as objects to whom the developed services and products are sold and delivered. In opposition, design offers a new approach to strategy immersed in a social process of argumentation and debate, with the customers and users at the discovery end (Golsby-Smith, 2007). However, embedding design expertise might be challenging in organizational contexts, which strongly emphasize operational management in terms of the delivery end, to defend the status quo (Golsby-Smith, 2007). Hence, design is not always effectively managed and might not be integrated into business processes (Borja de Mozota, 1998).

The strategic business impact of design, economic design value, and return on investment of design have been elaborated in many studies (Candi et al., 2010; Cheng et al., 2012; Danish Design Centre, 2003; Den Ouden, 2012; Design Council, 2007; Grzecznowska & Mostowicz, 2010; Hertenstein et al., 2010; Rae, 2014, 2015, 2016; Whicher et al., 2011; Zec & Jacob, 2010). These studies bring forward design

K. Korpikoski (✉)
Faculty of Art and Design, University of Lapland, Rovaniemi, Finland
e-mail: krista.korpikoski@ulapland.fi

 
U. Z. A. Hamid, M. Suoheimo (eds.), *Service Design for Emerging Technologies Product Development*, Springer Series in Design and Innovation 29,
https://doi.org/10.1007/978-3-031-29306-1_11

impacts in terms of improved economic output/profitability and better competitive position of the organizations that use design. However, such knowledge alone is not enough. More understanding is needed in terms of the input, in terms of what is required from the organizations that start applying service design as a bottom-up activity (Meurer, 2001). Such understanding can offer knowledge on which areas to improve and develop within organizations, to enable more efficient management and use of human- and customer-centric working cultures in practice. Therefore, this chapter asks the following question: What are the challenges from the perspectives of different organizational stakeholders that occur when in-house service design is used in service development?

The results of this study are based on a qualitative case study of a multinational manufacturing corporation conducted by the author. The author has a working background as an in-house service designer within the organization under research. By the summer of 2018, when the research data was collected between May and August, in-house service design had been in use in the organization approximately for four years. The data of this study consists of 17 semi-structured interviews including 19 interviewees and a group discussion of a service design value workshop of four in-house service design experts and a service design manager. Experiences of stakeholders from all organizational levels are brought forward except for the chief executive officer (CEO) and shareholders. The data has been analysed using the conventional content analysis method. The results discuss the qualitative internal organizational challenges by presenting five challenge categories: (1) lack of organizational understanding of service design, (2) conflicting development processes, (3) tight scheduling and resources management, (4) colliding decision-making cultures, and (5) absence of organizational performance measurement alignment.

## Description of the Organization Under Investigation and the Role of Service Design

The organization under study is a matrix organization, a multinational manufacturing corporation with over a hundred years long history in engineering and manufacturing. The organization structure is similar to that of other manufacturing businesses. The organization employs approximately 60,000 employees globally and it produces equipment, maintenance, and consultation services. The role of services has become more crucial for the organization due to the transition from manufacturing economies to service economies, a revolution of Information Technology (IT) in the latter part of the twentieth century (Shah et al., 2006), Internet of Things (IoT), and increased customer demands. Hence, the organization is required to understand customer value in depth, to understand customers' processes and contexts that affect their experiences (Andreassen et al., 2016). Therefore, service design as a practice has become a useful and valuable approach for the organization regarding service development.

Prior to service design, product development activities followed more or less Cooper's (2001) Stage-Gate process model within the organization under investigation. End-users had been the main source of information for product and user interface designers. Business-to-business customers had not been regularly involved in the development projects of R&D. In addition, the development of services had mostly been in the hands of business stakeholders. As a follow-up of in-house service design, human- and customer-centric work practices have become increasingly central. Hence, the role of service design has become strategic in the organization. The service design team is located in the company headquarters to work together with other business stakeholders and different departments. In addition, service designers have been hired elsewhere to support area- and country-level development. They perform an active role, especially in the early phases of service development. Co-creation activities such as customer research, ideating, prototyping, testing, and piloting together with customers and internal stakeholders are a part of service designers' work. They hold a strong communicative and facilitative role within the development projects. Service designers handle projects holistically when examining detailed processes, and solving complex issues and challenges. Thus, they provide strong support for management regarding decision-making.

## Theoretical Background

Service design is a development approach that covers the entire process of service development (Goldstein et al., 2002; Holopainen, 2010). It has extended its focus also to service systems and organizations instead of merely concentrating on developing services (Polaine et al., 2013; Yu & Sangiorgi, 2014). It is a methodology that builds heavily on design thinking, which "allows the firm to see through the eyes of the customer" (Andreassen et al., 2016, p. 23). As a creative, holistic, and customer-centric bottom-up activity, it differs from the conventional views of practicing business (Kurtmollaiev et al., 2018; Meurer, 2001). In this study, service design is seen as a human- and customer-centric development methodology and practice, which holds a strategic (outside-in) development process – a process where analytical and creative reflections intellectually take turns. People (customers and employees), front- and back-end of services, products, digital systems and tools, processes, and practices are naturally the subjects of service design.

## Challenges of Design Practitioners in the Area of Transformation Design

Today, design is used for solving complex problems in different contexts by enabling a wide range of stakeholders and disciplines to collaborate in the design process to develop practical and desirable solutions (Burns et al., 2006). Hence, design is becoming transformational due to its role in affecting services, organizations, and societies. This is called transformation design, originally set forth by Burns et al. (2006). Design practitioners face both philosophical and practical challenges in the area of transformation design (Burns et al., 2006). The philosophical challenges designers face are: (1) the loss of personal creative authorship by collaboratively facilitating others' ideas; (2) shaping behaviour of systems, interactive platforms, and people's roles and responsibilities rather than form; (3) transformation design is never done since emergent systems keep changing; (4) creativity happens in run time in short design cycles of inflow and in situ of the market, not just in design-time in studios as safe zones; (5) diversity over the quality of accepting what's good enough; and (6) design becomes a Pro-Am community of shared practice including professional educated designers and amateurs, or the non-design trained (Burns et al., 2006, p. 26).

The practical business difficulties and barriers, which designers face, relate to questions about (a) leadership and value; (b) the development of new business models, tools, and processes; and (c) the encouragement of new skills and orientation among designers (Burns et al., 2006, p. 27). Hence, regarding the practical challenges that designers face, "an appropriate shared language and evidence base" should be built up to support "communicating the value and impact of a transformation design process" among internal stakeholders (Burns et al., 2006, p. 27). In addition, better tools and techniques must be developed to be shared in multidisciplinary collaboration among stakeholders, and non-designers within organizations must be induced into design practices (Burns et al., 2006; Junginger, 2008; Pinheiro et al., 2012; Sangiorgi, 2011).

## Service Design Is Transformational Since It Facilitates Change

Service design "grows into a powerful transformative force that is capable of changing institutions", and hence, the organizations, which start implementing it, should prepare for organization-wide transformation regarding changes in the mindsets and routines of employees (Kurtmollaiev et al., 2018, p. 70). Sangiorgi (2011) argues that service design is becoming transformational due to its capabilities in facilitating change within organizations since it provides the tools and capacities for human-centred innovation. Thus, designers are required to (a) engage in strategic processes of business problem definitions, not just respond to given briefs as problem statements; (b) think systematically; (c) apply design in broader social, economic, and political contexts; (d) collaborate fruitfully with other disciplines; and (e) champion a human-centred design approach at the highest levels (Burns et al., 2006, pp. 27–28).

According to Andreassen et al. (2016, p. 24) service design, especially service system design, can "enhance an organization's process, structure, and culture in creating value for customers". They argue that service design might facilitate organizational change by affecting organizational performance and customer experience through external touchpoints. Internal organization change happens by affecting the touchpoints that affect internal IT systems and administrative procedures, personnel satisfaction by frontline and support personnel in addition to virtual and physical service delivery (Andreassen et al., 2016). Hence, the "organizational structures, processes and culture will inevitably be impacted by implementing an outside-in perspective in service design" (Andreassen et al., 2016, p. 24).

Junginger and Sangiorgi (2009) argue that service design can facilitate organizational changes on three levels, which are (1) artefacts and behaviours; (2) norms and values; and (3) fundamental assumptions of an organization. Pinheiro et al. (2012) state that the object of change may extend to organizational norms and cultures from artefacts and behaviours if service design practitioners involve organizations in customer-centric conversations. Shah et al. (2006) propose how an organization can transform from product-centric to customer-centric through changes in organizational culture, structure, processes, and financial metrics. However, for any transformation to be successful, it requires strong leadership commitment concerning the organizational realignment of horizontal and vertical structures through process and systems support, revised metrics, learning, and continuous improvement (Shah et al., 2006).

Due to the transformational force of service design, its use may contradict with organization's existing innovation routines, processes, and practices (Kurtmollaiev et al., 2018; Yu & Sangiorgi, 2018). Hence, the author of this study argues that organizational challenges may occur since service design facilitates change through iterative customer-centric outside-in working principles that may be contradictory to the existing linear innovation processes and fact-based decision-making cultures. Therefore, in this study, transformation design is seen through the perspectives of service design.

## Organizational Challenges That Arise from the Use of Service Design

Service design requires open and experimenting cultures to support its ways of working. Hence, the success of embedding service design depends on the level of understanding of the value of service design principles and tools (Kurtmollaiev et al., 2018). In "non-design", organizations imagining customer emotions and guessing their needs is common, and hence, managers might not see value in interacting with customers (Kurtmollaiev et al., 2018). Thus, in this kind of institution, the right mindset of using service design does not exist (Kurtmollaiev et al., 2018). The experimental nature of service design might be seen as too messy, fuzzy, and risky for the organizations – "nothing like the conventional formal stage-gate process" (Kurtmollaiev et al., 2018, p. 68). Therefore, in institutions where performance orientation and formal and linear processes are favoured, embedding service design may be hindered (Kurtmollaiev et al., 2018). Service design work might be tried to be squeezed into the existing models, which is not fruitful ground for service design (Kurtmollaiev et al., 2018). This causes high frustration among service designers while resistance to real changes in innovation and delivery processes keep existing within the organizations (Kurtmollaiev et al., 2018). In addition, service design might be seen as trendy, and as a consequence of this, it might be seen as an instrument to be used only with a symbolic meaning of using it (Kurtmollaiev et al., 2018).

The existing practices, meeting business key performance indicators (KPIs) and personal goals to sustain the status quo, might inhibit the use and implementation of service design on a wider scale (Kurtmollaiev et al., 2018). Issues also arise due to tight resources. Several projects run at the same time. Hence, they suffer from a lack of time, funding, and human capital. In addition, managers might be unwilling to "release their best resources to work on interdisciplinary projects within cross-functional teams" since this may imply "the loss of project ownership and direct control" (Kurtmollaiev et al., 2018, p. 68).

Top management should encourage the creation of service design-based common vocabulary to enable shared corporate language by realigning KPIs with service design principles and objectives and provide room for experimentation (Bailey, 2012; Kurtmollaiev et al., 2018). Encouragement and rewarding methods should be in place (Bailey, 2012) such as performance-appraisal systems (Kotter, 1995). Design should be integrated into management styles to encourage the use of design amongst staff (Bailey, 2012). In addition, specialized teaching and training in service design should be offered to managers and employees to familiarize them with it in parallel with business practices throughout the organization to initiate behavioural change (Bailey, 2012). This is due to employees' personal experiences with service design, which "can further ensure a smooth transition from a rigid shareholder-value-focused firm to a more flexible customer-centric and design-driven organization" (Kurtmollaiev et al., 2018, p. 71). Hence, all of these are crucial elements when building in-house service design capacities and capabilities to support "delivering input to the business strategy in the long-term, while also delivering value to the business through projects in the short-term" (Bailey, 2012, p. 2). However, innovating change and new practices might take time for large organizations, and hence, support must come from all management levels if a change in the culture is to be achieved and sustained (Bailey, 2012).

The organizational challenges and development needs brought forward by Kurtmollaiev et al. (2018) and Bailey (2012) are in line with Deserti and Rizzo (2014), Holmlid et al. (2017), Junginger and Bailey (2017), Lin et al. (2011), Sangiorgi et al. (2012, 2017), and Sangiorgi and Prendiville (2017). They all bring forward the need to develop and do changes in terms of structures, processes, systems, and practices to support design work since they work as crucial elements in shaping internal organizational behaviour regarding innovation practices. When these are viewed as existing support structures, they can either enhance or hinder participatory and co-creative micro-scale innovation in the development of future macro-scale visionary solutions to enhance organization strategies (Deserti & Rizzo, 2014).

## Research Design

This research is a qualitative case study, which follows an inductive research approach. Perspectives provided by Stake (2005), Merriam and Tisdell (2015), and Ghauri and Grønhaug (2005) are applied. Hence, a case study is not seen as a method (Stake, 2005). Instead, the organization under study, a multinational manufacturing corporation provides the context to discover and learn about the phenomenon: the challenges that are experienced from the perspectives of different organizational stakeholders through the use of in-house service design. Hence, the phenomenon under investigation is seen as intrinsically linked with the case (Merriam & Tisdell, 2015). Thus, it is difficult to quantify and prove the phenomenon outside its natural setting and context (Ghauri & Grønhaug, 2005).

This research follows inductive reasoning, which is natural for qualitative research. In inductive reasoning, the results are achieved through theorizing by making sense of the data (Gillham, 2000; Ghauri & Grønhaug, 2005). Thus, the results provided in this study are based on interpretations of the data to provide more abstract generalizations and ideas. Everything that comes up from the research data is interpreted as each person's individual and unique experiences with in-house service design within the organization. Hence, informants' opinions do not represent official statements of the organization under investigation. This was brought up orally and in writing in interview and workshop consents, which were collected from each informant before interviewing them.

## Participants and Data Collection

The data set of this study consists of 15 individual and two pair interviews, and a service design value workshop. The 17 semi-structured interviews include 19 interviewees from all five organizational levels except for the CEO and shareholders. The service design value workshop includes five participants: four service design experts and a service design manager. The total number of informants is 21 regarding the data used in this study including the workshop participants who did not join the interviews. Eight of the informants are females at the levels of specialists, managers, and heads. The rest are male. The research data were collected from May

**Table 11.1** The data set of this study

| DATA SET 1 | DATA SET 2 |
| --- | --- |
| 17 Semi-structured Interviews | A Service Design Value Workshop |
|  |  |
| 19 Interviewees | 5 Workshop Participants |

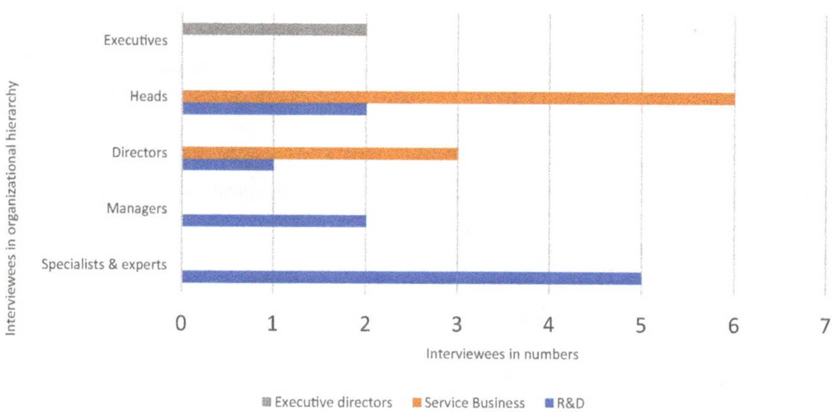


**Chart 11.1** Interview and workshop participants

to August 2018 when in-house service design had been in use approximately for four years within the organization (Table 11.1, Chart 11.1).

The semi-structured interviews provide perspectives on management's experiences from the areas of service business and R&D related to the organizational challenges in terms of the use of in-house service design. Two of the individual R&D interviews provide also an in-depth understanding of the existing product development processes and project management systems. In addition to these, two semi-structured pair interviews provide perspectives on the challenges experienced by in-house service designers: (1) two service design specialists, and (2) a service design manager and a service design specialist. Below are two examples of the interview questions, which were presented related to the challenges of in-house service design:

1. Have you faced any challenges with service design or with service designers? What are the challenges?
2. What are the drivers and barriers when integrating service design in the organization? Are there any experienced obstacles? What? When? How? Why?

The meaning of the service design value workshop was to understand what in-house service designers experience as the positive impacts of service design utilization within the organization in addition to the challenges that they face. Two posters were hung on the wall, which included the following headlines: (1) positive values of service design, and (2) progressing values of service design. The latter topic regarding the challenges (progressing values) constructs the data regarding the topic of this study since it provides answers to the research question. The workshop was chosen to be analysed to provide more in-depth perspectives on the experiences of in-house service designers in comparison to the considerable amount of management interviews. The workshop lasted for 2.5 h. The topics were considered first individually by writing thoughts down on post-it notes and attaching them to the posters. Then the individuals freely presented, discussed, and reflected upon their thoughts.

The data is limited to the perspectives of executive directors, service business stakeholders, and R&D personnel since the purpose of this study was to understand service design challenges from the perspectives of business and service design. In terms of the R&D department, the views of user experience and industrial design specialists were left out since their views are a matter of another research publication. In addition, views of marketing, IT, human relations (HR), strategy, and customer experience were left out since the concentration of this study is to compare the perspectives between service design and management from the areas of service business, R&D, and executive directors.

## Results

Conventional content analysis was used in this qualitative case study. According to Hsieh and Shannon (2005), it is an appropriate method when no previous theory exists or the research literature on a phenomenon is insufficient. In conventional content analysis, the categories are not predefined, but they are formed based on the data, for example, basis on the perspectives provided by the interviewees (Hsieh & Shannon, 2005). Content analysis aims to develop new concepts, models, or phenomena, but not new theories (Hsieh & Shannon, 2005).

In this study, the data analysis began by importing direct quotes from interview and workshop transcripts into an Excel file. To provide answers to the research question, the quotes naturally were chosen related to the challenges of in-house service design. They were organized into five columns according to the roles of the interviewees at five levels. Then the quotes were subdivided into meaning units based on the phrases related to the same meaning, as instructed by Graneheim and Lundman (2004). Next, the meaning units were labelled with a few words to highlight the meanings. Altogether 247 labels emerged. The labels offered reduced expressions of the meanings, which supported the second phase, clustering. The data were clustered by grouping the labels according to content similarities into five main categories, which were then named corresponding to their content:

1. The lack of understanding of service design causes inefficiencies within projects and frustration among stakeholders.
2. Existing linear organizational processes challenge the use of in-house service design.
3. Tight schedules and small resources inhibit continuous development, scalability of the developed solutions, and service design know-how.
4. Value-based decision-making collides with fact-based and product-oriented decision-making cultures.
5. Organizational performance measures are not aligned with service design at the project level.

The formed categories thus represent the most important information of the study (Tuomi & Sarajärvi, 2013), meaning the five categories of challenges of in-house service design. Then, a more simplified Excel table was created since the categories were studied through the roles at five levels. The horizontal axis of the Excel table consisted of the five main categories. The vertical axis consisted of the interviewees' roles at five levels. Condensed descriptions of the meanings of each main category were formed at each interviewee level followed by quote examples. This method provided an easier comparison of the meanings within categories between the interviewees in different roles. Table 11.2 shows a simplified visual representation of the Excel table including quote examples of each main category at each organizational level.

Next, the synthesis of each five main categories is discussed below through the roles at five levels.

1. *The lack of understanding of service design causes inefficiencies within projects and frustration among stakeholders*

**Executives** The executive directors expressed that it is not clear to all stakeholders what service design means, and therefore, involvement in human- and customer-centric projects is a learning experience for many. They stated that it is not straightforward nor evident for people who are not used to dealing with service designers about what and how they are going to contribute. In addition, they argued that the role of service design has not been clearly defined in the organization, which might cause ambiguity related to it.

**Heads of service business and R&D** Heads of service business argued that there is no clear understanding of service design and how other stakeholders can support service designers' work. They expressed that the lack of understanding of service design causes challenges and inefficiencies in the level of project management since service design roles, deliverables, tasks, responsibilities, and milestones are not

**Table 11.2** The five challenge categories of in-house service design including the stakeholders' quote examples

| The roles at five levels | Category 1 | Category 2 | Category 3 | Category 4 | Category 5 |
|---|---|---|---|---|---|
| Executives | *It's not always, straightforward and evident from the very beginning for people who are not used to, dealing with service designers* (about) *what, how they are going to contribute.* (n30) | *The more we see that as our offering is evolving towards (more) (–) service or a combination of products and services, we're going to probably need to reinvent ourselves in this innovation process (...). And today for us it's still to be done.* (n30) | *... all resources are today a bit (...) scarce, (...), compared to what we would need I would say, with all our needs.* (n30) *When enthusiasm is strong, launches happen early, but there is still a lot to develop to get things done to be able to scale the solutions successfully.* (n27) | *... we are challenged as a corporation (...), agility to use customer input and to iterate constantly. (...) It requires leadership and clarity about the direction where we want to go with it.* (n30) | *In product and IT development projects we have the general metrics, the KPIs that we use, and process measures as complete on time and cycle time. So, about service design, they* (the measures) *are not up-to-date with it yet.* (n27) |
| Heads of service business and R&D | *... these different sections are called swim lanes, but do you see service design here anywhere? No.* (n40) *... you have to describe who owns the task and who contributes, (...) which milestone needs an additional bullet, what's the specific content* (of service design). (n21) | *... product development process descriptions, (...), they don't sufficiently take into account service design needs, (...) someone could do an additional project that complements the old deliverable document lists and milestones, and says what is the role of service design in all of them.* (n21) | *... the needs change over time.* (n45) *The challenge is how to manage further development of services.* (n34) | *The challenge is that (...) we have the burden of proof.* (n21) *... as a development organization, we should be more agile, (...), we should create an operating model that works differently from the traditional waterfall models. It needs to be iterative.* (n31) | *The challenge is the length of the lead time. The benefits of service design projects come after many years. Incentive models could be built, but it is difficult, (...) so how to measure the success of service design and how to give bonuses become a challenge.* (n34) |
| Directors | *I think it still has a bit of a narrow role, so is the service design understanding at a good enough level? Maybe it comes back to how well we as an organization understand service design, what it means, and* what are its possibilities (n20). | *... there are many handovers and along the way, something gets lost (...) it probably starts from how we define the function and role of service design, tasks, and participation in our service development process or ways of working.* (n20) | *It* (service design) *requires an awful lot of money and resources, and I think we should have even more of them on our own payroll. We don't have a service designer for every project, or what happens that they are there, to begin with. (...) then they disappear.* (n42) | *It (service design) is difficult for middle management because the order of importance of things and the decision-making logic change.* (n12) | *The two most important things are monitored in projects, the budget and schedule. Once they are locked, they are strictly followed.* (n42) *Qualitative measures are difficult because they are considered as subjective.* (n12) |

(continued)

**Table 11.2** (continued)

| The roles at five levels | Category 1 | Category 2 | Category 3 | Category 4 | Category 5 |
|---|---|---|---|---|---|
| R&D managers | *The value is not in the end result, but in the fact that* (all) *those things have been thought through and understood that there are such and such things also here, that this* (development of a service and service design) *is not such a simple thing.* (n41) | *. . . it* (the process) *is product oriented. It is not ( . . . ) fitting to the* (development of) *services, even at a generic level. ( . . . ) we push too big entities forward in gate-like ways, which is a very straightforward way to do development.* (n25) | *. . . the problem is that ( . . . ) learning stops when something is pushed out, ( . . . ) so how to enable ( . . . ) constant development, constant learning becomes the matter.* (n25) | *. . . we have measurement and fact-based decision-making, ( . . . ) but how can experimentation and other qualitative values support decision-making, and how does that lead to an efficient organization that makes good decisions by aligning decisions with the right values? That's the challenge in my opinion.* (n25) | *. . . service design aims are difficult to connect with business KPIs, but we must create our own KPIs for service design projects to give direction and focus.* (n25) |
| In-house service design specialists | *If this person totally understands what you are doing and understands your value, things go, but then, if this person doesn't, it's a pain, ( . . . ), you still need to kind of prove your value and that takes a lot of energy and a lot of time.* (n46) | *Agile development doesn't happen if we follow the existing product development process.* (n26) | *. . . seems like we put a lot of money, we develop, and then we just multiply, everywhere.* (n47)<br>*. . . we don't have enough people. If we would be like, ( . . . ) 20 people, I'm sure the impact would be much more.* (n46) | *Ability to listen is difficult because it is drowned under the consistency and scalability discussions and the culture of harmonisation.* (n26) | *. . . there is a problem with the* bonus (system) *and the mindsets won't change and* they (other stakeholders) *won't collaborate if you're kind of giving bonus the way it is now.* (n47) |

marked in project management tools at a system level. They stated that this causes the dilemma where the heads of service business must go to service designers to ask and make sure that all the required and additional work relating to service design is done within projects.

**Directors** It became clear from the interview data that understanding service design and its methods among personnel at this level varies. Directors of service business presented that there is a lack of understanding of what must be done in terms of service design and how long things take. In addition, service design tasks, deliverables, responsibilities, and milestones are not part of project management systems either to guide stakeholders' work. An R&D director highlighted that the lack of transparent service design responsibilities might cause in-house service design to have a narrow role within the projects, and business directors get a false idea that makes it look like the projects are on target.

**R&D managers** Interviews with R&D managers brought forward that service design roles, responsibilities, tasks, and level of involvement in different project phases are not clear to all stakeholders. This is because deliverables, milestones, tasks, and dedicated persons responsible for service design are not fed into the project management tools to externally support in guiding work.

**In-house service design specialists** In-house service design specialists expressed in the workshop discussions that doing and delivering service design work may be hindered or even prevented in development projects where service design is not understood by project owners. As a result of this, in-house service designers experienced being left aside in such projects. They stated that this causes service design to have a narrow role, which leads to strong frustration among them. In-house service designers also brought forward that they experience having to fight for their position to prove the meaning and value of their work. This is very demanding since it takes a lot of energy, time, and effort.

2. *Existing linear organizational processes challenge the use of in-house service design*

**Executives** The executive directors argued that the existing linear development processes do not yet support service design and service development since they have been built from the perspectives of product development and IT. They expressed that now the organization is challenged to develop the existing linear innovation processes in a more agile direction.

**Heads of service business and R&D** Heads of service business presented that service design is not industrialized as a process within the organization, and the existing linear processes do not sufficiently consider the deliverables, milestones, and role of service design. Interviewees also brought forward that the existing linear processes are not flexible, iterative, and agile enough to allow the flexibility of service design usage in different phases of the project's lifecycle.

**Directors** Directors at this level expressed that service design's role is too short in the development projects, which creates a lot of knowledge gaps in terms of project handovers along the processes. They argued that a continuous presence of service design in all the process phases and along the projects' lifecycle is a challenge since the function and role of in-house service design within the organization are not clearly defined.

**R&D managers** R&D managers brought forward that the challenges appear when service design and iterative ways of working are tried to be squeezed and fit into the existing linear development processes. The managers stated that the organization is strongly product-oriented, and hence, it focuses a lot on operations, efficiency, and pushing solutions out in gate-like ways, which does not fit with the development of services, which is agile in nature.

**In-house service design specialists** In-house service designers expressed that the existing linear process models are product-oriented and they do not sufficiently support service design in terms of ideation and agile experimentation. In addition, they brought forward that they must leave projects once concepts are developed, which is the reason that they are no longer intensively involved when the developed solutions proceed into productization. They argued that this causes knowledge gaps along the process in terms of handovers and a lack of ownership of the responsibilities of the developed solutions.

3. *Tight schedules and small resources inhibit continuous development, scalability of the developed solutions, and service design know-how*

**Executives** The executive directors expressed that service design resources are scarce in comparison to what the organization would need since the organization has a strong enthusiasm for developing solutions, and therefore, the developed solutions are launched early. However, the projects have long-term development needs that must be taken care of. The executives concluded that the presented reasons create the challenge that there is still a lot to develop to get things to the finish line to be able to scale the developed solutions successfully.

**Heads of service business and R&D** Heads of service business brought forward the challenges related to timetables and costs. A danger is that the solutions are developed fast and pushed early into the market, which end up not being scalable. In addition, heads of service business argued that very often the actions of service designers may require additional money and time. R&D personnel at this level expressed that due to the high demand for service design combined with small resources, issues start emerging in the phase of productization since that is where the constraints and limitations of the organization appear, but in-house service designers might no longer be involved.

**Directors** Directors expressed that service design requires an awful lot of money and resources, and due to tight schedules, small resources, and many projects running at the same time, service design ends up having a narrow role. There is no service designer for every project, or service designers are in projects, to begin with, then they must leave for new projects. The directors argued that a follow-up is missing of the developed solutions to provide the continuity of service design work. In addition, an R&D director stated that tight schedules combined with business priorities pull service design to happen in business centres, which leads service design to happen in a silo, and this does not support the scalability of service design on a wider scale.

**R&D managers** R&D managers argued that when the developed service concepts proceed into the productization, organizational constraints appear, but then service designers are already working on other projects due to a small number of them. This creates the challenge of how to enable constant learning and continuous development within the organization of the minimum viable services to achieve fully developed service concepts.

**In-house service design specialists** In-house service designers expressed that they must exit the projects before the developed concepts have been productized, and the new project personnel are lacking tacit knowledge in terms of the project background and in-depth customer understanding. Due to this, in addition to tight schedules and constraints of the organization, they argued that there is a danger that crucial features that create the core customer value will be wiped out. They also expressed that budgets are decided well ahead, which might not contain the costs of continuous development. Due to all these reasons, in-house service designers brought forward that minimum viable services enter the market early, which then lack ownership and control of continuous development to reach the fully designed service concepts. Instead, scaling up, standardizing, and multiplying the developed solutions are seen as easier and faster from the organization's perspective. In addition, in-house service designers expressed that the number of them is small in comparison to the rest of the organization, and service design's role has not been clearly defined within the organization. This is also the reason why the role of service design is limited to the development, and further, continuous development of the solutions in addition to scaling up service design know-how within the organization is challenging.

4. *Value-based decision-making collides with fact-based and product-oriented decision-making cultures*

**Executives** The executive directors argued that the organization is challenged in terms of the agility to use customer input and to iterate constantly in addition to the speed of development. They stated that in comparison to developing products, which are expected to be at the level of perfection before going to market, services might not be fully developed when they are released. This is a challenge in terms of continuous development and reliable solutions from the perspective of the frontline. The executive directors concluded that developing internal productivity and efficiency through service design is challenging since changing innovation processes and operative models require leadership and clarity about the direction where the organization wants to go with service design.

**Heads of service business and R&D** Heads of service business expressed that the challenge of the organization is the burden of proof. R&D personnel brought forward that it is challenging to operate in iterative and agile ways since they are very different from the existing linear decision-making models of the organization.

**Directors** An R&D director argued that service design is a big change and challenge for middle management since priorities, the order of importance of things, and the decision-making logics change. Business directors expressed the fear of losing project ownership and direct control due to the use of service design, which led a business director to ask "do we act as a company for the common good, or do we try to optimize our responsibilities and goals" (n20)? According to the argument of an R&D director, this has led to contradictions between management and design, which challenges the continuation of the projects and the quality of the solutions.

**R&D managers** R&D managers brought forward the challenge of fact-based decision-making cultures of knowledge. When this is combined with fear of failure, learning and experimentation are corrupted, and hence, value-based decision-making is inhibited. This leads to the following question related to decision-making mechanisms, "how can experimentation and other qualitative values support decision-making, and how does that lead to an efficient organization that makes good decisions by aligning decisions with the right values" (n25)?

**In-house service design specialists** In-house service designers highlighted the challenge of the ability to listen, which is difficult due to scalability discussions. They argued that the organization is used to a culture of standardization and harmonization, and hence, the consistency roller works centrally from the headquarters. In addition, in-house service designers concluded that in such an organization, which comes financially along well, there is no sense of urgency to do things differently than what the organization is used to, which does not support enhancing value-based decision-making.

5. *Organizational performance measures are not aligned with service design at the project level*

**Executives** The executive directors stated a challenge in terms of the existing quantitative business measures, market research metrics, and performance measures. The measures are not aligned nor up-to-date with service design. In addition, they expressed that no qualitative measures are in place to support service design. They concluded that service design is challenging to distinguish and measure since there are always other stakeholders involved in the development, such as marketing, IT, sales, and so on. However, the executive directors argued that validating concepts with customers is the first way to measure the success of service design, which gives a signal if the customer expectations have been captured properly and translated correctly into the features of the offering.

**Heads of service business and R&D** Heads of service business argued that the challenge is the length of lead times since the benefits of service design projects come after many years. They presented that incentive models could be built, but it is difficult since many of the employees might change roles and/or organizations, and the organization itself change in a meantime. This is why measuring the success of service design and how to give bonuses become challenging. However, heads of service business expressed that the use of qualitative metrics has been discussed in terms of leading indicators and how to learn quickly from customers through iterations and new proposals, and if the methods that are used in service design provide high quality.

**Directors** Directors of service business stated that the budget and schedule are strictly followed in each project after they have been locked. Qualitative measures were expressed as difficult since they are interpreted as being subjective. In addition, directors at this level argued that KPIs and other quantitative measures are project-specific and they might seem abstract and as being far away for many employees besides the management.

**R&D managers** R&D managers expressed that having qualitative metrics to support human- and customer-centric development in an engineering-based organization is challenging since there is a burden of evidencing, measuring, and showing results quantitatively. The managers argued that what adds to the challenge is that there are two different things to measure: (1) how service design influences the solutions during the development (input), and (2) what comes out as a result of the customer interface (output). According to the managers, aligning business KPIs with service design qualitatively at the project level is in the hands of service designers only, but it must be done since they give direction and focus. However, they argued that in terms of service design, KPIs are more of a guiding element than a measuring element.

**In-house service design specialists** In-house service designers argued that challenges appear when the measurement and bonus systems are directed differently from what service designers are trying to achieve. First, they stated that it is challenging to measure service design quantitatively in terms of business metrics. Lead times are long, which means that the time frame for validating the output of the realized projects financially is challenging in comparison to the input of service design and development. Second, they brought forward that the existing bonus systems do not support the other departments to collaborate with service design in the early phases of service development. Due to this, project features and elements might go out of scope since stakeholders are not willing to invest time in things for which they do not receive bonuses.

## Discussion

The results of this study show from the perspectives of organization stakeholders at five levels that when in-house service design is used in service development, challenges occur in the areas of (1) organizational understanding of service design, (2) development processes, (3) scheduling and resources management, (4) decision-making cultures, and (5) organizational performance measurement. In terms of transformation design, the results show, in line with Burns et al. (2006), that in-house service design is transformational due to its role in affecting organizations. However, unlike Burns et al. (2006), I argue that the difficulties and challenges that emerge are not only faced by designers, but also by other stakeholders and management since they hold the power to further develop organizational norms and practices to better support in-house service design and agile service development.

The results present that the level of understanding of in-house service design and its role, responsibilities, and deliverables within innovation and development is not yet sufficient among the stakeholders. These findings are in line with Bailey (2012), Kurtmollaiev et al. (2018), and Burns et al. (2006). They argue that focus should be put especially on training service design to management and experts to familiarize them with it since the success of embedding service design depends on the level of understanding of the value of service design principles. In addition, service design training would decrease frustration among service designers, who face constant challenges in communicating the meaning and value of it among internal stakeholders (Burns et al., 2006). This study also shows that centralized toolkits, rules, and guidelines of service design should exist in a self-study format to increase internal understanding of where and how to use service design, what professional in-house service designers do, and what belongs to the rest of the project personnel. In addition, the results reveal that the lack of organization-wide understanding of service design causes challenges and inefficiencies in the level of project management. Hence, in-house service design ends up having a narrow role within the projects. To make project management easier, service design roles, deliverables, tasks, and responsibilities should be marked in the project management tools at the system level. In addition, how and by whom to support in-house service design should be prominent as well. These would offer support for stakeholders whose understanding of service design is not sufficient.

In terms of the development processes, this study shows that challenges emerge when linear development process models, which do not support iterative service development, must be followed. Hence, the study proves the views of Cooper (2001) and Kurtmollaiev et al. (2018). The nature of iterative service design work, which is based on experimentation, learning, and continuous development, gets challenging when big entities are pushed through the gates of stage-gate processes (Cooper, 2001). This is not a fruitful ground for service design when the work is trying to be squeezed into the existing linear development models (Kurtmollaiev et al., 2018). Based on the results of this study, redefining the process models towards a more flexible direction depends on where the organization is willing to go with in-house service design. Stakeholders bring forward the need of defining its role and function since it would support understanding its role in further phases of service development, such as productization. Now the role of in-house service design is experienced as too short since it is limited to the early stages of service development. However, the transformation towards flexible and agile innovation processes and operating models will take time. Due to this, heads of service business highlight the need to define the role of in-house service design in terms of the existing linear development processes.

Despite the lack of understanding of service design, the narrow role of it also happens due to tight schedules and small resources, and many projects running at the same time (Kurtmollaiev et al., 2018). Hence, issues arise from a lack of time, funding, and human capital (Kurtmollaiev et al., 2018). This study shows that the budgets are limited and the number of in-house service designers within the organization is too small. Hence, they are required to be in new development projects before the developed concepts have been productized. Therefore, handovers must happen, which leads to knowledge gaps and a lack of ownership and control of continuous development. The results of this study also suggest that tight schedules combined with business priorities pull service design to happen in business centres. This leads service design to happen in a silo, which does not support the scalability of service design on a wider scale within the organization. Also, for this reason, non-designers at all levels should be induced into design practices (Burns et al., 2006) and more widely trained (Bailey, 2012). In addition, the results of this study present a wider need for in-house service designers. Business stakeholders wish for regular in-house service design interventions since managing and maintaining the continuous development of the developed services is a matter. Hence, business-minded service designers would be needed in every team since stakeholders would involve them throughout the entire lifecycle of projects and also during the delivery of services. This would require the organization to further define the role of in-house service design.

According to the executive directors and middle management, in-house service design challenges the fact-based decision-making logic and cultures in terms of value-based decision-making. The organization is thus challenged to use customer input and to iterate constantly. The results of this study show that as a consequence of this, contradictions between management and design start to emerge since business priorities and aims do not necessarily match with the needs of the customers. Thus, giving control from management to designers and people working closer to the frontline might be challenging. According to Kurtmollaiev et al. (2018), this might be the case, especially in traditional non-design matrix organizations, which lean heavily on the cultures of knowing by imagining customer emotions and guessing their needs while fearing failure. This study suggests that giving control from management to designers and iterative ways of working requires trust, which in-house service design specialists yearn for to enable frictionless work. However, defining clarity about the direction of where the organization is willing to go with service design is required first. This would also enable further inspection of innovation processes and operative management models through in-house service design and value-based decision-making.

In terms of organizational performance measures, this study suggests, in line with Shah et al. (2006), that quantitative and financial metrics should be aligned with qualitative project metrics to support achieving service design goals. These findings support also the views of Kurtmollaiev et al. (2018) and Bailey (2012). The creation and use of service design-based corporate language and common vocabulary should be encouraged by middle management by realigning KPIs with service design principles and objectives, and by providing room for experimentation (Kurtmollaiev et al., 2018). However, this study shows that it is not straightforward to do when service design is still such a new thing within the organization. According to top management, it is hard to measure the success of service design quantitatively since it is hard to distinguish and lead times are long. In addition, translating quantitative business performance goals into qualitative aims of the projects is left in the hands of in-house service designers. The results of this study suggest that project owners and directors should take a stronger role in the creation of project-specific qualitative metrics by helping to translate KPIs to match the qualitative aims of service design. In this way, KPIs would not seem as being too far away or abstract for project personnel.

In addition, the results of the study are in line with Golsby-Smith (2007). He argues that the existing bonus systems reward operational efficiency by considering the delivery end. From the perspectives of in-house service designers in this study, rewarding operational efficiency by considering only the delivery end does not support stakeholders collaborating with service design, nor contribute to the early phases of service development at the discovery end. Hence, this study's findings also support the views of Kurtmollaiev et al. (2018) and Bailey (2012). The existing practices and meeting personal KPIs in sustaining the status quo and local goals inhibit the use and implementation of service design on a wider scale (Kurtmollaiev et al., 2018). To enhance the application of design principles, also rewarding methods must be in place (Bailey, 2012) such as performance-appraisal systems.

Overall, this study shows that in-house service design has a transformational force due to its role in affecting organizations while developing services. That is why organizations must understand the key issues that appear when starting to apply in-house service design as a human- and customer-centric practice. The novelty of this study has been in defining the organizational challenges that occur when in-house service design is brought to be a part of service development. These were inspected from the perspectives of stakeholders at five levels. The researched knowledge provided by this research offers knowledge on what areas to develop to get a culture shift to happen systematically when the organization is heading towards human- and customer-centric development cultures with help of in-house service design. This study argues that the challenges in the five areas are not faced only by the design practitioners, but also by the stakeholders at all levels within organizations. Since service design is becoming an organic part of service development, it should gradually be developed into a really strong part of what the organization does. That is why it is crucial to understand the challenges that emerge from the use of in-house service design and to define its role within the organizations since it determines the rest: (1) the level of training to provide a deeper organizational understanding of service design, (2) definition of continuous iterative and agile development processes, (3) scheduling and resources management, (4) enabling value-based decision-making cultures, and (5) qualitative organizational performance measurement supported by bonus systems in the early phases of service development.

## Limitations

The age of the informants was not gathered. This is because the interviewees' age was not seen as relevant information since service design as a methodology was new or fairly new for everyone besides service designers. However, this might affect the interpretation of the results. In addition, as the publication is written solely by the author, no research triangulation for the case study and analysis appears, which might affect the credibility and validity of the research findings. Also, as the author has been a part of the in-house service design team in question, a larger risk of research bias may appear. Despite the limitations, the data analysis performed and the results presented by the author are seen as a representative overview of the topics discussed.

## Future Research

This research shows the challenges of in-house service design in the organizational transformation from the perspective of the manufacturing industry. A more in-depth understanding of service design's role and organizational challenges would be needed from different contexts where emerging technologies play strong roles in product development. Such areas are, for example, automotive, shipping, cargo, and aviation industries. Providing knowledge of the organizational challenges would offer a more in-depth understanding of what areas to improve to enhance internal efficiency to support iterative, and human- and customer-centric development cultures. Such knowledge is crucial since providing an avenue to understand customers helps organizations learn the market needs. Hence, the iterative and agile human- and customer-centric ways of working offer the possibility to do decisions closer to customers, and thus, target decisions accordingly. In addition, research in different contexts could also reveal how service design as a profession should develop to better answer different organizational needs.

## Conclusion

To answer the research question of this chapter, what are the challenges from the perspectives of different organizational stakeholders, which occur when in-house service design is used in service development, organizational challenges occur in the areas of (1) organizational understanding of service design, (2) development processes, (3) scheduling and resources management, (4) decision-making cultures, and (5) organizational performance measurement. Organizations applying in-house service design must first and foremost define its role and clarify the direction where the organization aims to go with it. Special focus must be put on internal training, continuous iterative and agile development processes, scheduling and resources management, value-based decision-making cultures, and qualitative performance measures, to support more efficient use of in-house service design. Hence, in-house service design requires strong leadership and building teams and expertise that integrate it systematically with the rest of the organization.

**Acknowledgements** This research has been supported by the European Regional Development Fund as part of the *Design in Smart Mobility Business Services* project (2018–2019) in addition to the Ph.D. grant received from the Finnish Cultural Foundation (2022–2023).

## References

Andreassen, T. W., Kristensson, P., Lervik-Olsen, L., Parasuraman, A., McColl-Kennedy, J. R., Edvardsson, B., & Colurcio, M. (2016). Linking service design to value creation and service research. *Journal of Service Management, 27*(1), 21–29.
Bailey, S. (2012). Embedding service design: The long and the short of it. In *Proceedings of ServDes 2012* (pp. 31–41). Linköping University Electronic Press.
Borja de Mozota, B. (1998). Challenge of design relationships: The converging paradigm. In M. Bruce & B. Jevnaker (Eds.), *Management of design alliances: Sustaining competitive advantage* (pp. 243–259). Wiley.
Burns, C., Cottam, H., Vanstone, C., & Winhall, J. (2006). *RED paper 02: Transformation design*. Design Council.
Candi, M., Genser, G., & Van den Ende, J. (2010). *Design effectiveness industry report*. Rotterdam School of Management.
Cheng, H., Keinänen, K., & Salo, M. (2012). *DROI measurable design*.
Cooper, R. G. (2001). *Product leadership: Creating and launching superior new products*. Perseus Books.
Danish Design Centre. (2003). *The economic effects of design*.
Den Ouden, E. (2012). *Innovation design. Creating value for people, organizations and society*. Springer.
Deserti, A., & Rizzo, F. (2014). Design and organizational change in the public sector. *Design Management Journal, 9*(1), 85–97.
Design Council. (2007). *The value of design factfinder report*.
Ghauri, P. N., & Grønhaug, K. (2005). *Research methods in business studies: A practical guide* (3rd ed.). Financial Times Prentice Hall.
Gillham, B. (2000). *Case study research methods*. Continuum.

Goldstein, S., Johnston, R., Duffy, J., & Rao, J. (2002). The service concept: The missing link in service design research? *Journal of Operations Management, 20*(2), 121–134.
Golsby-Smith, T. (2007). The second road of thought: How design offers strategy a new toolkit. *Journal of Business Strategy, 28*(4), 22–29.
Graneheim, U. H., & Lundman, B. (2004). Qualitative content analysis in nursing research: Concepts, procedures and measures to achieve trustworthiness. *Nurse Education Today, 24*(2), 105–112.
Grzecznowska, A., & Mostowicz, E. (2010). Industrial design: A competitive strategy. *Design Management Review, 15*(4), 55–60.
Hertenstein, J., Platt, M., & Brown, D. (2010). Valuing design: Enhancing corporate performance through design effectiveness. *Design Management Review, 12*(3), 10–19.
Holmlid, S., Wetter-Edman, K., & Edvardsson, B. (2017). Breaking free from NSD: Design and service beyond new service development. In D. Sangiorgi & A. Prendiville (Eds.), *Designing for service: Key issues and new directions*. Bloomsbury Academic.
Holopainen, M. (2010). Exploring service design in the context of architecture. *Service Industries Journal, 30*(4), 597–608.
Hsieh, H.-F., & Shannon, S. (2005). Three approaches to qualitative content analysis. *Qualitative Health Research, 15*, 1277–1288.
Junginger, S. (2008). Product development as a vehicle for organizational change. *Design Issues, 24*(1), 26–35.
Junginger, S., & Bailey, S. (2017). Designing vs. designers: How organizational design narratives shift the focus from designers to designing. In D. Sangiorgi & A. Prendiville (Eds.), *Designing for service: Key issues and new directions*. Bloomsbury Academic.
Junginger, S., & Sangiorgi, D. (2009). Service design and organizational change. Bridging the gap between rigour and relevance. In *Proceedings of the 3rd IASDR conference on design research* (pp. 4339–4348). Korean Society of Design Science.
Kolko, J. (2015). Design thinking comes of age. *Harvard Business Review, 98*(9), 66–71.
Kotter, J. P. (1995). Leading change: Why transformation efforts fail. *Harvard Business Review, 73*, 59–67.
Kurtmollaiev, S., Fjuk, A., Pedersen, P. E., Clatworthy, S., & Kvale, K. (2018). Organizational transformation through service design: The institutional logics perspective. *Journal of Service Research, 21*(1), 59–74.
Lin, M., Hughes, B., Katica, M., Zuber, C., & Plsek, P. (2011). Service design and change of systems: Human centered approaches to implementing and spreading service design. *International Journal of Design, 5*(2), 73–86.
Meurer, B. (2001). The Transformation of Design. *Design Issues, 17*(1), 44–53.
Merriam, S. B., & Tisdell, E. J. (2015). *Qualitative research: A guide to design and implementation*. Wiley.
Oakland, J. S., Oakland, R. J., & Turner, M. A. (2021). *Total quality management and operational excellence: Text with cases* (5th ed.). Routledge.
Pinheiro, T., Alt, L., & Mello, J. (2012). Service design creates breakthrough cultural change in the Brazilian financial industry. *Touchpoint: Journal of Service Design, 3*, 18–23.
Polaine, A., Løvlie, L., & Reason, B. (2013). *Service design*. Rosenfeld.
Rae, J. (2014). What is the real value of design? *Design Management Review, 24*(4), 30–37.
Rae, J. (2015). Design value index. *Design Management Review, 26*(1), 4–8.
Rae, J. (2016). Design value index exemplars outperform the S&P 500 index (again) and a new crop of design leaders emerge. *Design Management Review, 27*(4), 4–11.
Ravasi, D., & Lojacono, G. (2005). Managing design and designers for strategic renewal. *Long Range Planning, 38*(1), 51–77.
Sangiorgi, D. (2011). Transformative services and transformation design. *International Journal of Design, 5*(2), 29–40.
Sangiorgi, D., & Prendiville, A. (2017). Designing for interdependence, participation and emergence in complex service systems. In D. Sangiorgi & A. Prendiville (Eds.), *Designing for service: Key issues and new directions*. Bloomsbury Academic.

Sangiorgi, D., Fogg, H., Johnson, S., Maguire, G., Caron, A., & Vijayakumar, L. (2012). Think services: Supporting manufacturing companies in their move toward services. In *Proceedings of ServDes 2012* (pp. 253–263). Linköping University Electronic Press.
Sangiorgi, D., Patrício, L., & Fisk, R. (2017). Introduction. In D. Sangiorgi & A. Prendiville (Eds.), *Designing for service: Key issues and new directions*. Bloomsbury Academic.
Shah, D., Rust, R., Parasuraman, A., Staelin, R., & Day, G. (2006). The path to customer centricity. *Journal of Service Research, 9*(2), 113–124.
Stake, R. E. (2005). Qualitative case studies. In N. K. Denzin & Y. S. Lincoln (Eds.), *The SAGE handbook of qualitative research* (3rd ed., pp. 443–466). Sage Publications.
Tuomi, J. K., & Sarajärvi, A. (2013). *Laadullinen tutkimus ja sisällönanalyysi* (10., uud. laitos.). Tammi.
Whicher, A., Raulik-Murphy, G., & Cawood, G. (2011). Evaluating design: Understanding the return on investment. *Design Management Review, 22*(2), 44–52.
Yoo, Y., & Kim, K. (2015). How Samsung became a design powerhouse. *Harvard Business Review, 93*(9), 73–78.
Yu, E., & Sangiorgi, D. (2014). Service design as an approach to new service development: Reflections and future studies. In *Proceedings of the 4th ServDes. Conference on service design and service innovation*.
Yu, E., & Sangiorgi, D. (2018). Service design as an approach to implement the value co-creation perspective in new service development. *Journal of Service Research, 21*(1), 40–58.
Zec, P., & Jacob, B. (2010). *Design value: A strategy for business success*. Red Dot Edition.

# Chapter 12
# Designing Human and Artificial Intelligence Interactions in Industry X

**Stefan Walter**

## Introduction

The pressure for competitiveness is forcing companies in the industry to improve their effectiveness and efficiency (Ketchen Jr. & Hult, 2007; Lii & Kuo, 2016; McDougall et al., 2022; Sakuramoto et al., 2019). Additional constraints arise from the need to adapt quickly to changing customer demands or other unforeseen events, volatility, for example, in customer and supply markets, and, of course, the aspects of sustainable development, especially regarding environmental and social needs (Del Vecchio & Paschalidis, 2006; Delpla et al., 2022; Hayat et al., 2021; Khan et al., 2019; Sarkar & Chung, 2019; Schmidt & Schwegler, 2008; Wang et al., 2017; Werner, 2020).

This requires technologies that generate knowledge dynamically and make it available to support decisions, but also management approaches that have continuous improvement as their basis. Artificial intelligence, as one of the most important megatrends on the way to the next industrial revolution, is a core element of these types of technologies (Gupta et al., 2022; Masood et al., 2021; Wan et al., 2021).

In the context of Industry X – signifying a continuous industrial evolution beyond Industry 4.0,[1] artificial intelligence enables an increase in quality, flexibility,

[1] In a typical classification, the fourth industrial revolution follows a series of previous ones, starting with the first industrial revolution, which resulted from mechanisation by means of water and steam power, the second industrial revolution, which is characterised by mass production, and the third revolution, which reflects the use of electronics and information technology. The fourth industrial revolution is above all inspired by comprehensive networking, including software,

S. Walter (✉)
VTT Technical Research Centre of Finland Ltd, Oulu, Finland
e-mail: stefan.walter@vtt.fi

 
U. Z. A. Hamid, M. Suoheimo (eds.), *Service Design for Emerging Technologies Product Development*, Springer Series in Design and Innovation 29,
https://doi.org/10.1007/978-3-031-29306-1_12

and agility in particular. With the help of integrated supply chains, industry can apply corresponding adaptive strategies, increase competitiveness in terms of adjustments and rapid reactions to changing customer requirements, and generally make processes more robust and resilient (Adobor, 2020; Maozhu et al., 2020).

In addition, we realise that the problems society faces can no longer be solved from the perspective of one discipline alone, such as engineering or computer science. Therefore, the approaches we work with need to broaden their focus. It is therefore not surprising that there are a number of visions and concepts that represent the processes of the future in industry in an interdisciplinary way (Schipper et al., 2021; Wesselink, 2008; Wohlgezogen et al., 2020).

For example, the concept of Industry 5.0 recognises that, in addition to responding to economic challenges, which every company must do to remain competitive and continue existing, industry should also have the responsibility to simultaneously respect ecological limits and consider human well-being (European Commission, 2019; European Commission et al., 2021). The concept includes considerations of co-creation for the benefit of greater sustainability and resilience and is in line with the policy of increasing technological innovation in Europe, which should be in harmony with people (European Commission, 2021).

A vision for the future industry called Sustainable Industry X also considers the importance of the digital industry and aims to combine its development with approaches from the fields of risk management, sustainability, and circular economy (Wahlström et al., 2020). While Industry 4.0 is still concerned with connecting the digital world with the physical world, the vision in the context of Industry 5.0 emphasises the topic of co-working of humans and intelligent technology. Co-working arises precisely because the different strengths of humans and machines can be considered through the combination. As a result, a paradigm shift in the interaction between humans and technology can be expected, leading to new forms of collaboration in manufacturing.

So, even though the use of artificial intelligence brings economic sustainability, problems still exist such as complexity of technology and application, inadequacies or faults in results, and shortage of skilled workers (Anderson et al., 2020). Furthermore, despite technological advances, many processes and steps rely on manual labour. On the one hand, this may be because human labour is in part cheaper, and on the other hand, human activity is advantageous due to the greater flexibility and motor and learning skills (Gorlach & Wessel, 2008).

However, classical management approaches still essentially focus on the improvement of processes in terms of their economic context, which thus refers to the reduction of costs. These approaches clearly show the inadequacy in dealing with human work and the interaction possibilities of technology and people (Glock

mechanical, and electronic parts, to so-called cyber-physical systems. All these revolutions were or are accompanied by far-reaching consequences for the economy and society (Rupp et al., 2021). The fifth industrial revolution can be seen as a correction of the negative consequences of past revolutions, with a strong focus on people and labour, sustainability, and resilience (European Commission et al., 2021).

et al., 2017; Sgarbossa et al., 2020). As a response to this challenge, the article can be seen in the context of the transition that is needed to emphasise the human dimension (Pyykkö et al., 2021).

According to Sgarbossa et al. (2020), embedding technology and new information-based tools in human work processes requires the development of a corresponding perspective that takes human-centred design and management into account. On the one hand, the digital transformation of companies can succeed, and on the other hand, demographic change can be mastered, the possibilities and efficiency of human-technology interaction can be decisively improved, and the basis for further, more sustainable processes in industry can be laid. Here, process designers can take things into their own hands and use their skills in a targeted way. Consequently, service design supports the transition from a more engineering-oriented approach as reflected in Industry 4.0 to a more human-centred and sustainability-oriented approach as reflected in Industry 5.0.

It is important to classify these requirements correctly. Due to the diversity of factors, the challenge is very complex.[2] Therefore, an approach is needed that can systematically produce an understanding of highly complex systems. The general systems approach, for example, emphasises the interaction between different elements and the effects of this interaction. There is also the requirement to include the whole, that is, the global understanding of how a system works. The focus for considering the above-mentioned requirements will here be on the concept of service ecosystems, which play the role of complex systems in the context of service design. Understanding service ecosystem design includes, for example, the role of human users as actors, their processes and organisational arrangements, which in turn leads to the emergence of co-created values and consequences (Vink et al., 2021).

To give a practical framework to what is described in this chapter, reference is made to the European knowlEdge project (www.knowledge-project.eu). In response to the above aspects, both the general developments and pressures of the market and the needs and requirements regarding the constraints and opportunities of human labour, the project *knowlEdge – Towards Artificial Intelligence powered manufacturing services, processes, and products in an edge-to-cloud-knowledge continuum for humans [in-the-loop]* has developed an architecture for an information technology platform that meets these requirements. Within this platform, artificial intelligence solutions are developed with features that support collaboration between human users and artificial intelligence, primarily to design better manufacturing processes and organisational advances.

[2] According to Luhmann (1987), complexity represents the totality of possibilities of experiencing and acting that are realised in a context of meaning. However, this also means that complexity is both an observer-dependent and an observation-dependent fact. Every choice of possibilities reduces complexity. This makes it clear why design exists at all. In an abstract sense, design reflects the necessity of selection, since there are always more possibilities of experience and action than can be realised.

Before describing the service ecosystem design approach, to provide a rationale and purpose for the design, the author first discusses the benefits of artificial intelligence in industry and the characteristics of adaptive supply chains and then describes some basics of information technology as the foundation of intelligent decision support and the logic of interactive information systems. Then, the service ecosystem design approach to understand the complex interactions of its building blocks and their application to a design example of the *knowlEdge* project is explained. Finally, the author concludes by discussing some of the design impacts of the emerging service ecosystem resulting from the human-AI interaction, especially in terms of corporate management and supply chain resilience.

## Artificial Intelligence for Adaptive Supply Chains

In giving purpose and a rationale for understanding the design effort described in this chapter, some explanations about the benefits of applying artificial intelligence will be made. This sheds light on human interactions and the reasons for its increased use in industry and especially in its supply chains.

According to Werner (2020), artificial intelligence is a branch of computer science. This subfield is about transferring the human ability to learn and think to the machine, the computer. Once this is done with the appropriate algorithms, IT systems can find solutions to problems based on collected data all by themselves. As the data changes over time, if data continues to be collected, the algorithms become dynamic models. Incidentally, the data basis does not have to correspond to the data collected by the system itself, but, and this is also a feature of modern information technology, a wide variety of data sources can be included. This can mean that factors unknown to humans lead to different conclusions than those humans would have come to on their own. Thus, the conclusions can be used to create a wide variety of behavioural patterns and probabilities of occurrence, depending on the need.

The benefits of using artificial intelligence are thus obvious. Since the management of companies, and even more so the management of global supply chains, is very information-intensive, networked decision-making processes are necessary that allow joint requirements and forecasts to be made between the partners of networked value creation. This not only improves the agility, that is, responsiveness, of the chains but also increases efficiency, which has an impact on energy and resource consumption. In addition, joint planning can also increase resilience, which is a positive factor in the context of risk management (Min, 2019).

Learning is of central importance when using artificial intelligence. On the one hand, the algorithm is constantly learning by improving the data basis (Iarovyi et al., 2015; Li et al., 2021; Mavrikios et al., 2013). On the other hand, it enables the employees of an organisation, be it a company or a supply chain organisation, to draw conclusions from the results of the automatic analyses and thus, for example, to change processes in a corporate workflow. This development is reflected in the concept of the learning organisation, in which technologies are used that perceive, recognise, improve, etc. In this sense, artificial intelligence is a very transformative technology and, depending on the business sector, can not only have enormous potential to improve operations but also determine long-term competitiveness (Zheng et al., 2021). The concept of cognitive manufacturing or, in the context of supply chains, the concept of the cognitive supply chain is comparable to the idea of the learning organisation (Werner, 2020).

Therefore, it can be assumed that if information exchange takes place across company boundaries and activities and their coordination in a supply chain increasingly merge, the classic rigid and hierarchical supply chain organisation will sooner or later change into a flexible and adaptable structure (Bak, 2021; Zhang et al., 2009). Here it is possible to make a conscious distinction between the characteristics of Industry 4.0 and the characteristics of Industry 5.0 as well as other future developments in relation to supply chains. While the former chains are characterised primarily by efficiency through technology and external planning, the latter are undergoing a shift towards an emphasis on human labour, the ability to learn, resilience and sustainability, and control in a decentralised way from within. Technology such as AI assists here (Frederico, 2021).

And as mentioned above, integration into a network can synchronise events and activities (Werner, 2020). This has, as one can easily imagine, extensive consequences for the planning of material flow, but also for the organisation of employees in companies of the entire network. For example, in an environment without a rigid hierarchy, employees can react independently to events and influence work processes (Snowdon & Saunders, 2021). This is a contribution to resilience.

The pervasive ability to plan and synchronise naturally leads, on the one hand, to an increase in flexibility and speed of response to change. This is an increasingly important aspect for responding quickly to changing customer needs, depending on the industry (Maozhu et al., 2020). On the other hand, employees can also react autonomously to unexpected events because recommendations for action are developed with the help of artificial intelligence. Employees and artificial intelligence can thus form a unit within the framework of a learning organisation and a dynamically evolving supply chain.

Nevertheless, what has been described is still a passive variant in the cooperation between humans and artificial intelligence. This interaction should not only function in this one direction, in which information technology makes suggestions for action to humans. The interaction should also work in the opposite direction, in which humans actively improve the data basis of the artificial intelligence model based on their own experience. Appropriate methods for capturing experience and domain knowledge should become a normal part of the design of human-artificial intelligence interactions in the future, providing a crucial quality improvement to the algorithms involved.

## Decision Support and Human-Computer Interactions

Now that the implications of using artificial intelligence in industry and supply chains have been explained, this chapter will turn more specifically to decision support systems. Decision support systems are important touchpoints for the design of human-artificial intelligence interactions. In a sense, they provide the infrastructure for the development and emergence of services.

From the intentions of corporate management, the role that advanced information and communication technology and decision support systems have is about ensuring effective utilisation of available resources. This is done regarding the goals of a process, for example, manufacturing a product based on a customer order or ensuring a certain service level in logistics planning. For logistics, it can be a matter of analysing transportation requirements, vehicle availability, and the corresponding transportation costs in such a way that direct connections are maximised, and unnecessary empty runs are minimised. This is, of course, while keeping overall freight costs as low as possible (Bowersox et al., 2016).

Clearly, such planning becomes extremely complex when it is no longer a single company within whose boundaries the planning is performed, but sometimes global supply chain. In addition to the large number of factors to be considered regarding aspects such as sourcing, production, warehousing, and market, the data intensity required to make decisions must be managed. In this regard, Bowersox et al. (2016) note that sophisticated modelling and analysis techniques must be used to cope with this complexity. The decision-making systems developed for this purpose are generally based on mathematical optimisation. This is usually done with a view to increasing a utility value, which also includes the avoidance of possible consequential problems. Thus, there is always an alternative when making decisions (Vahs & Schäfer-Kunz, 2015).

Heylighen (1991) states that decision support systems can be divided into two classes. First, there are so-called closed systems in which the user of the system is completely replaced except for the data entry function. According to Heylighen, the system here takes over the function of thinking, so to speak, that is, replacing the human as an expert, processing the input data and outputting a solution based on it. In comparison, the open system is merely a medium that allows the user to be creative. In fact, in the open system, the user is the expert whose work is merely supported by the system.

Another difference between open and closed systems is defined by the fact that the open system is much more flexible compared to the closed one. In an open system, where the user is the expert as already described, there are few constraints on the application. Heylighen (1991) gives as examples of limitations the type of problem to be solved and the format in which the data is to be entered and the resulting output. A closed system, in contrast, has a very specific scope of application or expertise. Because an open system is built on the user as the expert, one might assume that the open system is much less knowledgeable or intelligent than the closed system. One gains expertise, so to speak, with the closed system, but loses flexibility.

The solution is to combine both approaches so that their respective advantages come together. However, from the point of view of interactivity, none of the above systems would be satisfactory. Heylighen (1991) points out that for interaction the user in a closed system cannot intervene in the process of data processing and problem-solving but can only wait for the result. And in an open system, the system cannot direct or influence the work or problem the user is addressing. The goal should be that both parts, user and system, can interact. Both should work on the problem simultaneously. Thus, opportunities would have to be created so that data and results would be constantly exchanged. In such a case, the decision about the problem to be solved or the decision between alternatives to be chosen would be with the user, but both, system and user, would not have complete control over the process as a whole.

In a business environment, interactive approaches can be of great use, which, in addition to the cost calculation, also include a simulation that illuminates the consequences of certain alternatives. The alternatives in the decision process are specified by the decision maker or programmed into the decision support system. The decision maker may interactively evaluate alternatives based on various performance characteristics and then refine an action strategy. Bowersox et al. (2016) write that a clear disadvantage of such an approach is the dependency on the knowledge and skills of the user, that is, the decision maker, which becomes especially problematic when the complexity of the problem increases. As we can see here, this results in an obvious need for artificial intelligence.

## Designing for Complexity

As mentioned in the introduction, the use of technology and its embedding in evolving organisations leads to a high level of complexity. Where complexity is prevalent, it is important not to limit oneself to showing how to optimise a particular or narrow process, but to work with the whole. Vink et al. (2021) refer to this often-prevalent issue as reduction of service design, which will lead to problems in practice dealing with the results. The reason for this is that reduction would lead to superficial solutions, and thus probably cause numerous consequences that would be both unintended and negative in nature.

Considering that organisations and supply chains are made up of interactions of many different actors and technologies, as well as the rules of social practices and norms of work processes, we need a design approach here that can deal with the obvious complexity. Only in this way can the result contribute to the actual increase in value creation.

To understand a little more, the connection between the logic of service design and the method of human-centred design, which will be discussed here later in the context of the case study, the following explanations should still be made.

Service design has established itself as a practice that enables the delivery of services with a human-centred approach. This approach is thus able to place human needs at the centre of a process of change or development (Miettinen, 2017; Stickdorn & Schneider, 2011). Reflecting on human needs reveals their multifaceted character. Therefore, they cannot simply be defined without an application context, but a service design goes through several stages, including a definition of the problem to be solved, the exploration of a solution, the development of a corresponding concept as well as a prototype or demonstrator, and finally their implementation (Penin, 2018).

Thus, the practice of service design can be compared to the metaphor of a journey of exploration and discovery. According to this, service design and the passage through the aforementioned stages describe important user touchpoints that reveal human needs in dealing with the technology to be developed in the context of the work process. Furthermore, a concept for the service is developed at the same time, which includes the infrastructure and resources necessary for its delivery, such as workplace, equipment, and personnel.

Both Young (2012) and Kuure and Miettinen (2017) recognise that service design plays an important role in improving social situations. When users are involved in their respective contexts, that is, as a co-design process, service design can create a more equal position of different stakeholders such as employers, employees, and their technology. The motivation for this type of co-design is the transformation of the organisation in question, be it a company, a department, or a work group, that can be expected as a result, and its change for the better (Kurtmollaiev et al., 2018). This is particularly relevant in the context of human-technology interaction, which in many cases requires changes in the associated work processes.

Such co-evolution could be seen as a development process that involves the innovation capacity of stakeholders. Strategically, service design can also be seen as a process of empowerment and capacity building (Miettinen & Vuontisjärvi, 2016).

Thus, from an overall perspective, the outcome of the design process is a service ecosystem that, according to Vink et al. (2021), co-creates value through the interactions of multiple actors. In the context of this account, a service ecosystem can be defined as the intentional design of institutional arrangements and their physical implementation by actors. This is done through reflexivity and reformation to enable the emergence of desired forms of value creation. In this ecosystem, with the help of social institutions, such as the rules of a company or the roles of workers, the available resources are pooled and work results and products or services are made available and governed for exchange. Due to the high level of abstraction, a systemic understanding can thus be ensured.

An important aspect of these remarks is to understand that with the concept of design there is an intention, that is, a conscious design with the aim of bringing about a change. The perspective clarifies that the involved actors themselves influence changes, often of a long-term nature, in the ecosystem of which they are a part. To this end, it is first necessary to discuss the structure of a service ecosystem according to Vink et al. (2021).

| | Service design building blocks | | | |
|---|---|---|---|---|
| | Purpose | Material | Processes | Actors |
| Design of services | Developing new services | Touchpoints, interfaces | Phases in services development | Experts driving new services |
| Service Ecosystem Design | Faciliates emergence of new forms of value co-creation | Institutions and their physical representations | Feedback loops | Collective design |

**Fig. 12.1** Service design conceptualisation. (Adapted and simplified from Vink et al. (2021): 173)

Following their research, service design consists of four essential building blocks, namely the purpose of a design enterprise, materials, processes, and actors (Fig. 12.1). Based on these building blocks, a logical context was developed leading to the concept of service ecosystem. At this point, it should be noted that the path to collective design is intentionally described in a simplified way to emphasise the difference between the intention of the design and the spontaneity of the outcome. This context illuminates the collaborative nature of the ecosystem, which means that the system itself promotes and at the same time reflects the emergence of shared value creation.

The purpose thus asks for the essence of the reason for the design. Design without a reason or purpose is unthinkable. Design in this sense contains a normative orientation when we imagine that it is about how a product, an organisation, or a system should be. Therefore, the purpose of the service ecosystem can be seen in the co-creation of value. This is done through integration, transformation, and exchange of different means, such as resources and so on (Martelaro & Ju, 2018). This social process, in a sense, creates the value (Wetter-Edman, 2012).

Furthermore, it must be stated that a service ecosystem is the subject of spontaneous developments. This means, firstly, that the system does not simply exist as a static structure, but that the value created is aimed at change and the system must be considered dynamic. Secondly, spontaneity implies the presence of emergence.

According to Heylighen (1989), emergence can be explained in the context of self-organisation. Self-organisation can be defined as a spontaneous process of organisation of a system. This means that this organisation is not controlled by an external system, but that an organised whole, as in the case of the service ecosystem, emerges from the disorganised collection of parts interacting with each other. Furthermore, we can attribute to self-organising systems the ability to build internal complexity (Luhmann, 1984). Such systems are therefore more than the sum of their parts. Ultimately, this means that although design is a purposeful activity, the continuous transformation of corresponding self-organising systems is never fully predictable.

In the traditional sense, design is about reshaping or transformation of substances or materials in a given context. This is the case, for example, in classical product design. In the context of the service ecosystem, however, the meaning of materials necessary for the design process lies with institutional frameworks, such as rules, norms, and meanings, which are visible through material symbols and interactions and the like.

Material conditions are therefore still present. Touchpoints or interfaces can be mentioned here as examples, as they also occur in the analysis of the working environment and development of the interactions between humans and artificial intelligence (Vink et al., 2017). However, these materials should only be understood as physical manifestations of the more invisible or immaterial institutional and social arrangements (Vink et al., 2021).

This means that the shared value creation common to all service ecosystems is not so much determined by the materials, but is enabled, but also constrained, by the implicitly existing institutions and social practices.

As a third building block, we refer to the design processes themselves. Besides purpose and materials, it is now the processes that reject the static and emphasise continuity. In the context of the service ecosystem, process stands for ongoing reproduction of the institutions mentioned earlier. Design focuses on the reshaping of these institutions.

According to Vink et al. (2021), the actors concerned must first become aware of the institutional arrangements to change them. This phenomenon is called reflexivity. It is only through reflexivity that it is possible to recognise the changeability and design potential and to start reforming or redesigning existing institutional arrangements.

The concept of self-organisation has already been described. The concepts of reflexivity and reformation follow on from this. They represent feedback loops that can be used to incorporate complementary alternative perspectives that arise in the reproduction of institutions. In this way, the possibilities and forms of the creation of shared value can ultimately be influenced.

Actors as designers have already been mentioned. But here they are emphasised again as the fourth building block of service ecosystems. Everything we have described so far in connection with the systems in question is brought about by actors. Actors shape in the sense of continuously reproducing institutional arrangements. According to Vink et al. (2021), they are involved in the design of service ecosystems when this is done with an intention.

An essential element in understanding service ecosystems is that value creation takes place collectively. Collective design does not mean that all actors pull together, so to speak. It is possible and quite realistic that the diversity of actors leads to dynamic and multidirectional design processes. The ecosystem is thus the result of all actions and design processes. The diversity that will always be present also always leads to new uncertainty about the evolution of the system (Wetter-Edman, 2012).

## Designing the Interaction and Understanding the Building Blocks in Action

The building blocks of service design can be identified within the design approach of interaction or collaboration between humans and artificial intelligence. In the example, Human-Centred Design (HCD) was used, which is a problem-solving process for the development of interactive systems (Fig. 12.2). This primarily aims to consider the needs and requirements of the users involved (Boy & Narkevicius, 2014; Forbrig, 2016; Hashizume & Kurosu, 2013; Nguyen Ngoc et al., 2022). The approach is defined by the International Organization for Standardization (ISO) under ISO 9241-210. The basic concept is based on expert-led meetings with users, such as the staff of a workshop, and encourages and requires collaboration with these users. The users in turn provide input, suggestions, etc., for the further development of the use case and the technology to be used (International Organization for Standardization, 2019).

The approach encourages the participation of experts, who have unique domain knowledge and know the conditions of their working environment better from their experience than outsiders (Maguire, 2001). Secondly, human-centred design has an integrating effect because it tries to bring together different perspectives in a systemic, holistic way (van der Bijl-Brouwer & Dorst, 2017; Coulton & Lindley, 2019). This is an aspect that should not be underestimated, as many interrelationships in the world of work are becoming more complicated and sustainable solutions require integration. Thirdly, the approach provides the opportunity to critically express expectations and to think about previous requirements and ways of working (Wright & McCarthy, 2010). In this way, new and unexpected things can come to the surface. Finally, the approach provides a comprehensive solution that is user-

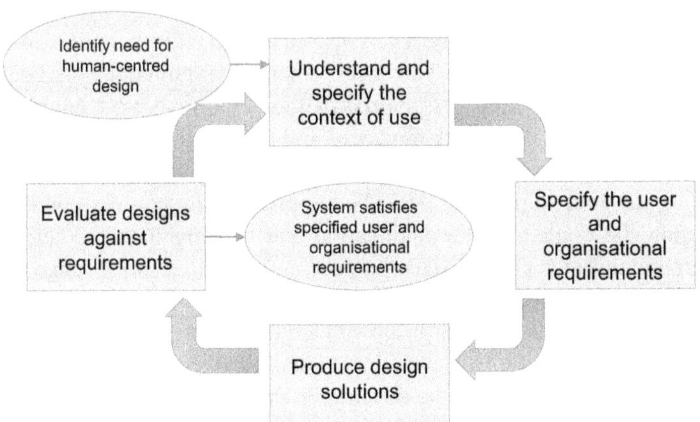


**Fig. 12.2** Human-centred design cycle phases (International Organization for Standardization, 2019)

and action-oriented and spans the design life cycle (Garreta-Domingo et al., 2018). Human-centred design is therefore not intended as an approach that is developed in a design office and then presents a finished product as a solution without having dealt with the reality of users' lives (UNICEF, 2022).

The design approach is based on a specific work process, as already described. This means that user requirements are created based on use cases. Use cases represent, so to speak, the possible scenarios where a particular piece of software can be usefully applied. These cases consist of a description of how users can accomplish certain tasks and how the technical system to be developed behaves in relation to the user's requests. In the context of the technology development of artificial intelligence and especially in relation to the possibilities of collaboration, it is of particular importance to involve employees in the decision-making processes, to develop artificial intelligence and equip them with the appropriate tools and methods, to design work processes with the help of and in interaction with artificial intelligence.

The HCD approach is applied to several pilot demonstrators in the knowlEdge project (Pins, 2021). However, in the following, the process of designing human-artificial intelligence interactions in the context of a company in the food process industry will be described. This use case shows typical supply chain problems that are to be solved with the support of artificial intelligence solutions. Specifically, this case focuses on continuous monitoring and data analysis of food parameters for process quality management and defect detection and prediction for higher production quality and process efficiency.

Identifying problems that ought to be solved is the first step, the trigger for starting the cycle phases in the HCD approach. It shows that there is a need for change and a desire to learn and adapt processes in organisation or production.

The example use case refers to the use of AI software in the context of human-computer interaction. Hence the reference to user interfaces and interaction possibilities. However, since the design resulting from the process has an influence on the organisation of the company, it can of course also refer to other technologies, tools, and methods than software. The basic principles of human-centred design are the same. After all, the guided design process focuses on the work environment of the employees involved, their work processes and tools, which should ideally reflect the needs of the people involved (cf. Pins, 2021).

Following the first step of the described approach, the process continues with understanding the context of the application and relating it to the relevant user groups, relating this step clearly to the purpose and intention in the range of service design building blocks. The partners involved in the project then select several people who correspond to the user group and are therefore among the most suitable ones who know the use case. However, the collaborators should describe the context of use as well as possible so that the developers can later improve this context with their solution. The basis for the context description in this case is the interview.

The interview covers the spectrum of the function, that is, the role description, of the user in his or her working environment, as well as the typical workflow, its prerequisites, and, beyond that, the organisational circumstances. Based on the answers, it is later possible to discover the corresponding possibilities for improving the workflow through technical solutions. To avoid any misunderstanding, open points or ambiguities should be addressed when answering and should also be checked by the respondents. This will lead to the best possible representation of the context of use.

The next step in the HCD approach is the analysis of user needs and organisational requirements, focusing on the actual actors in the service design process. In this step, the results of the surveys of the individual interview participants are divided into individual needs and requirements. Each detailed precisely described context and workflow is joined by corresponding user needs. Here again, it is important that the relevant staff members who were interviewed can identify with the described user needs and thereby approve them. Furthermore, the user needs are subjected to scaling. This means that the needs are further subdivided so that important needs are distinguished from less important needs, and finally separated from needs that are not important at all. In this way, essential elements for the later development of the solutions crystallise.

Of course, this is not only about materials or available infrastructure, but also about the planning of personnel deployment. It is precisely through the comprehensive planning tool that employees can react quickly to changes and initiate internal adjustment processes, at least where there is no automation from the outset. The precautionary options include, for example, corresponding predictive simulations for decision support. These interfaces connect to upstream and downstream areas of a supply chain to make appropriate procurement and sales decisions.

The next step in the elaboration of the HCD process is to structure the user needs in terms of a distinction between main tasks and subtasks. This corresponds to the phase in which a design solution is produced. In this way, the current actual scenarios can be distinguished from the future target scenarios and defined in a later phase. The technologies are naturally developed based on the target scenarios.

The solution found from the analysis of the design process for the use case builds on the functional components of the information technology platform and consists of three main tasks. Following this division into main tasks and subtasks, the process behind each task is graphically represented (Fig. 12.3 for an example process). This representation is based on the user's perspective; the individual steps of the employee are visualised here in a comprehensible way. An important feature is the compilation of the user actions along a timeline, so to speak, or along the work stages. This representation thus characterises an interpretation of the process flow with the respective employees as the main actors. Thus, this represents the process building block in the design process. It also demonstrates the use of materials in the form of an analysis of touchpoints and interfaces between users, the technology, and the work environment.

The individual scenarios of the user contexts are split into different phases. These in turn consist – both from the point of view of the employees and regarding the flow of information – of inputs, actions, and outputs. This analysis is complemented by possible challenges, then the employees involved in the respective phases and possible outcomes or opportunities that result from the work phases. It is precisely

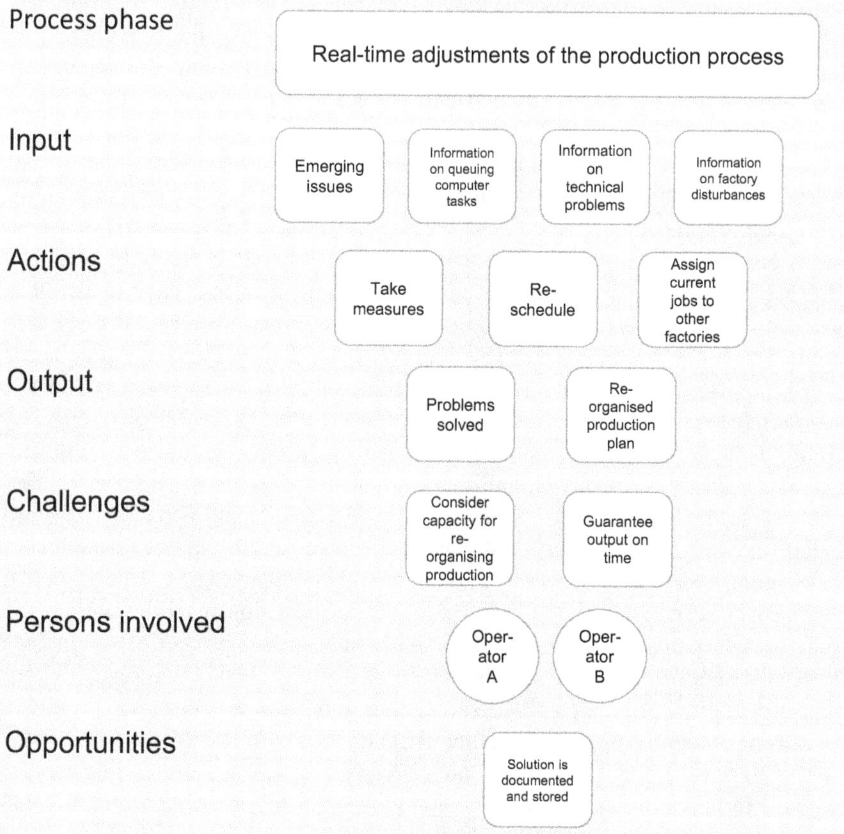


**Fig. 12.3** Visual representation of task specifications. (Adapted from Pins (2021))

this subdivision or focus on inputs, processing, and outputs that makes sense in terms of the development of information technology-based solutions.

Basically, the structured work phases consist of the compiled experiences of the respective employees, as described here based on interviews. However, information provided by the company itself about the company processes can also be considered. These are only intended to supplement the presentation, but can, for example, occur within the framework of the structuring into rough main phases.

Last but not least, in the final step of the HCD approach, evaluating the new solution against the set requirements, individual scenarios are defined in detail in relation to the current situation in the workflow, that is, the actual state, and the future ideal situation, that is, the target state. In other words, the scenarios were taken over from the last step of the graphical representation of the process flows and are now presented here in a narrative manner. Furthermore, the requirements, challenges, and possible solutions are also described in detail here in addition to the future ideal. Specifically, problems and challenges are emphasised in the as-is state,

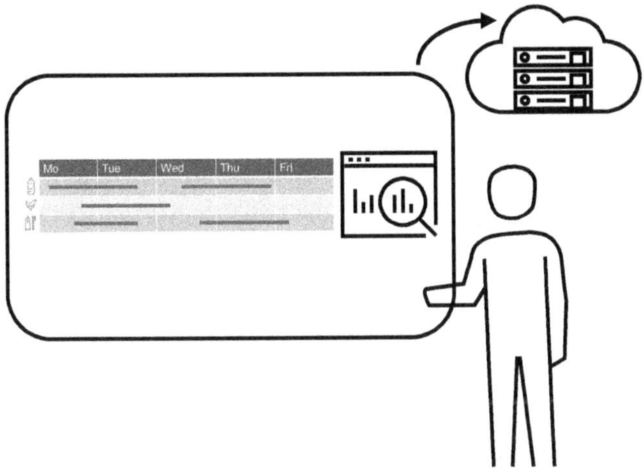


**Fig. 12.4** Graphical and narrative presentation of example scenario. (Adapted from Pins (2021))

whereas in the to-be state, the vision as the result of technology implementation and its requirements as the prerequisite for possible solutions naturally predominate.

In addition to the narrative description, it is possible to use graphical options to depict the respective detailed scenarios. These are once again helpful in promoting the understanding of the scenarios, tasks, and users involved (Fig. 12.4).

The described procedures within the process of HCD result here in the definition of user needs and a corresponding solution. The result, thus, satisfies the specifications of user and organisation. It is possible, and certainly intended within the framework of the design approach, that in one or more further iterations of the HCD, the results will be improved. This demonstrates yet another material block in the design process and shows that the purpose can be defined and refined on various levels, reflecting the learning element, which is so important in the cognitive factory or supply chain.

The main task called "real-time adjustment of the production process", is arguably very interesting from the point of view of the interaction between employees and artificial intelligence. This task consists of continuous notifications of the occurrence of deviations and indications of possible problems in the material flow, production process, available resources, etc., that result from the deviations and that need to be corrected. These, of course, need to be corrected, which requires significant human intervention.

Suggestions for fixing problems come from the AI-driven system, which builds on real-time monitoring data and predictive models of potential emerging problems. Secondly, suggestions come from access to a database of previously documented errors and deviations. The combination of both sources gives an adequate idea of how to proceed.

Furthermore, the staff concerned can intervene in this system based on their own experience, so that the data basis is improved. It is important to know that data collected automatically by the system via sensors, for example, is not always correct or sufficient to support the logic of the artificial intelligence algorithm in such a way that a sensible, satisfactory, or even common-sense result emerges from the system's analysis.

This means that this function, which documents any deviations, errors, or problems that occur in the flow of production and the options for rectifying them, corresponds to a mechanism for capturing human domain knowledge or experience. In this way, the basis of the database, which contains the documented deviation and associated solution paths, is then improved over time, leading to an increase in the quality of the artificial intelligence system that would not have been achieved without human intervention.

It is clear that both directions of information exchange lead to a mutual evolution. Human employees learn from the contextual suggestions provided by artificial intelligence, and the artificial intelligence can learn from human experience. From a management point of view, this illustrates what we mean by a learning organisation and will have a significant impact on a wide range of areas in companies and value networks, ranging from high-quality and zero-defect manufacturing to high resource efficiency, zero waste, and set-up flexibility.

Further iterations of the HCD approach help to straighten out any possible issues, as users involved will undoubtedly realise gaps, inconsistencies, or contradictions and other challenges, such as evolving market needs, that will come to light. It is speculative, though, to which extent such iterations may already represent the transition towards collective design in the sense anticipated by Vink et al. (2021).

## Emergent Properties of the Design Ecosystem

Following the proposition that a service design ecosystem has emergent properties, that is, outcomes, which are the result of co-creation, and that the evolving direction is beyond concrete planning, it becomes interesting to highlight some possible consequences beyond the technical impact. Apart from the technical impacts of a stronger integration of human work and artificial intelligence, such as high-quality manufacturing or flexibility, there are others, of which the focus shall be on the topics of controlling or corporate management and the robustness and continuity of supply chains.

The following impacts are, thus, not intended, but are likely outcomes as part of co-creation activities and emergence of design ecosystems. The characteristics and how they relate to efforts to design interactions between humans and artificial intelligence are described here.

## *Impact on Management*

Management processes in companies should generally be as constant as possible without being too dependent on the respective management. Especially for larger companies, and even more so for much more complex organisations such as supply networks, it would be detrimental if the entire management processes had to be replaced whenever there was a new managing director or board of directors. To ensure continuity, there is controlling. The concept stands for the coordination of planning and controlling activities that steer a company. Planning itself is described by Vahs and Schäfer-Kunz (2015) as a goal-oriented and systematic process that serves a company to describe goals and the ways to reach them and to achieve them in such a way that possible problems can be identified and addressed in advance.

There are several types of planning, such as strategic, tactical, and operational planning (Schmidt & Wilhelm, 2000). However, the coordination of the plans is discussed in more detail here. Regarding the hierarchical coordination of plans, there are several commonly known possibilities. These include top-down planning, bottom-up planning, and circular planning (Kim et al., 2014; Vahs & Schäfer-Kunz, 2015) (Fig. 12.5).

Here it is important to understand that the types presented are certain basic examples in the sense of archetypes. They represent a main pattern of behaviour in the context of management. Of course, the types are only a model, which illustrates an essential characteristic, as it is necessary for a scientific analysis. The model does not show all the details but is used to examine and understand the effects of changes in design.

In terms of coordination levels, coordination often takes place hierarchically. This is the classical coordination in which central and fundamental planning is developed in management. This means that the direction of planning is from top to bottom (top-down). Since the lower levels are usually only insufficiently involved in

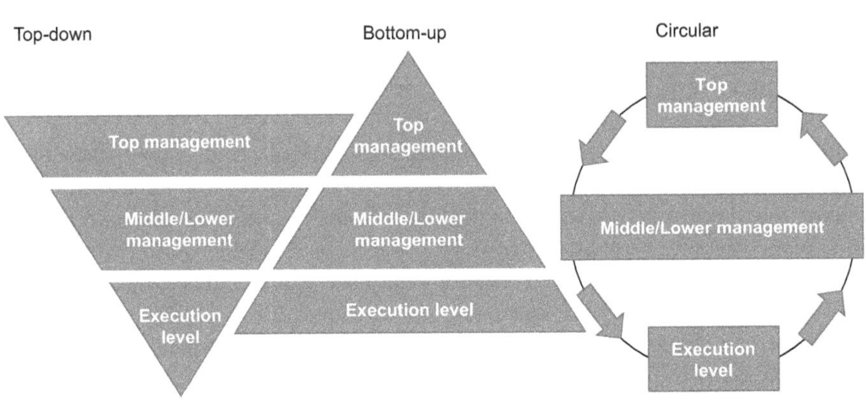


**Fig. 12.5** Planning types. (Adapted from Vahs and Schäfer-Kunz (2015): 343)

planning, this certainly has an impact on motivational performance and often results in less identification with the company and its goals (Vahs & Schäfer-Kunz, 2015).

In bottom-up planning, on the other hand, the process is reversed and planning now takes place from the lower levels of management or the shopfloor to the upper level. Here, top management must coordinate the sub-plans of the lower levels, which entails a high degree of complexity. In addition, it is likely that this coordination will lead to conflicts due to mismatched objectives (Vahs & Schäfer-Kunz, 2015). Therefore, this approach is also more theoretical and conceptual in nature.

The combination of both approaches is reflected in circular planning (Martinich, 2016). Management specifies framework targets, but the planned measures are defined, integrated, and implemented bottom-up. If the result does not meet management's expectations, further runs are carried out. The concept is thus developed as an iterative model. In contrast to the other models of top-down and bottom-up planning, this approach also leads to an uncovering of the interrelationships between individual plan and target ideas of the different levels. This approach usually leads to strong employee participation, acceptance, and motivation.

In the model shown, these remarks do not mean that other communications do not take place in company practice. For example, organisational silos are not an unknown problem in many companies, where employees in one department or work area have no or insufficient contact with other employees in other departments. These silos then lead to a situation where communication and cooperation do not take place to the extent that would be necessary for effective problem-solving. Organisational problems of this kind can clearly pose challenges for management (Lam & Shankararaman, 2007; Fenwick et al., 2009).

Sociologically, goals implemented during planning serve as a way to reduce complexity (Seidl & Mormann, 2015). This makes it possible for organisations to focus on a small number of meaningful areas. At the same time, goal setting enables organisations to manage different and sometimes conflicting expectations.

Simultaneously, wherever knowledge and information gaps exist and need to be managed, there is a need for trust. Trust is the enabling mechanism that increases the number of options available to management or available in the organisation because of its structure. Thus, where trust is present, the options for action increase (Seidl & Mormann, 2015).

In general, it can be argued, the management and planning approaches that require a high degree of complexity and coordination (bottom-up, circular planning) require more knowledge and skills in dealing with the necessities of the workplace. This is increasingly difficult in times of a shortage of skilled workers and the growing complexity of business environment and technological advance. One can equally postulate that the increasing coordination effort, the desire for error avoidance, and adaptive processes will lead to an increasing demand for artificial intelligence for planning and decision support.

Combined with the statement that in a learning organisation employees can learn from artificial intelligence and vice versa artificial intelligence, which is after all also involved in the (automatic) control of sometimes important processes, can learn from human experience; this illustrates a shift from hierarchical to flat structures in which decisions are made in a decentralised manner. It is also important to note here that artificial intelligence can play some role in the education of individual employees. This is because the normative goal of human empowerment also involves giving employees the ability to act independently. In this sense, artificial intelligence is the trigger for participative management.

With regard to organisational problems that are due to the fact that top-down management does not have suitable answers for all challenges in the company, we can therefore state that decentralisation of decision-making capacities has the potential to provide approaches to solutions. As an example, reference is again made to the problem of organisational silos. Employees who are put in a position to make decisions of far-reaching significance on their own can break down such silos by initiating the necessary communication and cooperation on their own initiative. Many problems that can be traced back to a lack of foresight on the part of management can be solved in this way, such as problems in product development caused by traditional, sequential information flows, duplication of tasks, and the feeling of disenfranchisement on the part of employees (Motingoe & Langerman, 2019; Tett, 2015; Werner, 2020).

## *Impact on Resilience*

The concept of resilience refers to the need to make structures more robust against certain disturbances. This applies both to industrial production itself and to critical infrastructures, which undoubtedly include extensive supply chains. These, particularly in the form of supply and value networks, have recently received the most attention in the context of the global COVID-19 experience. Of course, this also includes crises caused by disasters or armed conflicts. In the long run, of course, the resilience of any production necessary for the market is important to ensure prosperity (European Commission et al., 2021).

However, the idea that resilience is significant has existed for much longer than the crises of recent years would suggest. Indeed, the term, which originated in ecological research, has had a long use in climate impact research. Here, it is particularly useful in understanding the interaction between climate change and the ability of systems to maintain structure and function or reorganise to continue to exist in the face of perturbations.

In the context of social systems, such as those represented by businesses and their alliances and networks, according to O'Brien et al. (2012), resilience should be understood as a dynamic process that is directly related to people's ability to know how to cope with change. This includes both current challenges or disruptions and future changes, which are sometimes not known at all or not precisely known, that is, uncertain. These capabilities include the ability to learn and innovate. In addition, we can speak of the ability to develop leadership qualities. This is particularly important because, for example, machines or artificial intelligence systems cannot lead themselves.

And in fact, the concept of resilience goes beyond conventional technical systems that have some degree of shock absorption and that have been developed with the help of engineering (Folke, 2006). Resilience or resilience enhancement is not about simple re-engineering.

From an information technology point of view, it is essential to keep in mind that it is not only an engineering issue. Technology is essential, but technology alone is not enough to make a company or a supply chain resilient. Especially in the context of the use of artificial intelligence, we can assume that a company should have a secure foundation of infrastructure and also a meaningful and effective organisation and planning (Baskerville & Dulipovici, 2006; Janssen et al., 2020; Oracle, 2018). Both, in combination, help to reduce risks that can arise from multiple factors and also cushion unexpected shocks.

The infrastructure should be built on an architecture that is agile, flexible, open, and adaptable and can also be integrated into existing systems. The latter is often a bigger challenge than building a new system from scratch. Legacy technology is therefore often problematic in terms of exploiting new opportunities, innovation, and the application of cognitive, self-learning, or learning-support technologies (Arevian et al., 2018). Nevertheless, building a corresponding infrastructure is very cost-intensive and will therefore also lead to a development that promotes the use of open platforms (Oracle, 2018).

Secondly, a robust foundation in terms of the organisation consists of a suitable corporate culture that is designed in such a way that it promotes new things, that is, is innovation-friendly (Claver et al., 1998; Stacho et al., 2016). This means that the organisation must structurally allow and promote human agency.

The effects of human agency on the continuity of a complex organisation will be briefly illustrated here using the example of a supply chain. In Fig. 12.6, the supply chain is shown in the form of a network. The network consists of nodes, such as certain facilities, warehouses, factories, companies or the like, and arcs, that is, connections between the nodes. These can be roads, for example, but in the model, they rather represent an abstract connection that exists between two companies (Fig. 12.6).

One can now imagine that all possible nodes in the network also have workflows that are characterised by cooperation between humans and artificial intelligence. It has already been explained above that this collaboration in the context of management and planning coordination leads to greater autonomy for employees. This means that the relevant processes are controlled in a decentralised manner. Considering that hierarchical organisation cannot accommodate all contingencies or the unexpected, the decentralisation that results from greater staff autonomy is a prerequisite for rapid adaptation to changing conditions. Artificial intelligence, or rather the interaction between humans and artificial intelligence, thus leads to greater resilience.

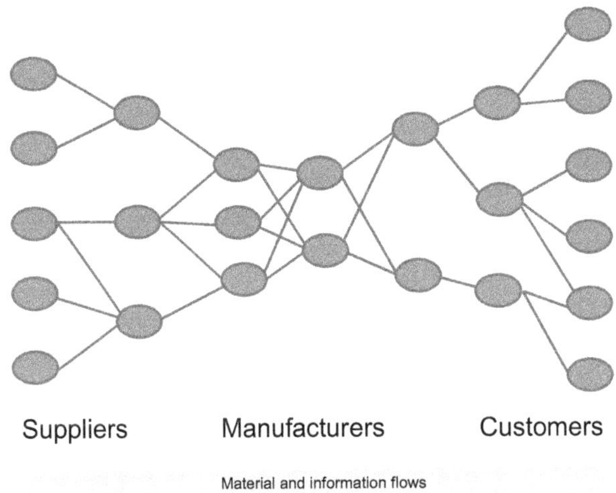


**Fig. 12.6** Nodes and arcs of a generic supply chain. (Adapted from Schulte (2013): 17)

# Conclusion

This research illustrated the design of services in the form of interaction between humans and artificial intelligence, the specifics of its design, and the way it is embedded in companies and supply chains. From a rationale, or purpose, point of view, the chapter also described the obvious benefits of applying or embedding such interaction capabilities. These include enabling industry to move towards adaptive and flexible strategies with the support of integrated supply chains, thereby significantly increasing competitiveness. Secondly, such an approach can address the need for holistic integration of different perspectives, experiences, and backgrounds. The latter already targets the capacity of individual actors to actively shape and design opportunities, leading to the emergence of service design ecosystems.

The multiple possibilities of human engagement in a supply chain dominated by cognitive technologies were discussed here from the perspective of human-centred design and knowledge capture. The work highlights that human involvement leads to impacts that represent shared value creation, but seemingly beyond hierarchical planning efforts. Future research on methodology is likely to address the problem of how to capture expertise and human experience in different contexts. This may also, for example, expand the opportunities for professional design experts to initiate the basic design process.

The designer's original goal may have been to introduce artificial intelligence in a way that users do not perceive it as a black box. This includes that technology in general and artificial intelligence in particular should be used in a way to improve work, which may translate into higher efficiency, speed, or fewer errors. Artificial intelligence systems should nevertheless be explicable and enable workers, that is, also non-experts of information technology, to use appropriate tools in their work processes. However, the underlying logic of designing service ecosystems has led to an outcome that has self-organised and will continue to self-organise in the sense of the function of social systems and co-creation. The author referred to this in the context of circular management and robust supply chains.

The ideal of continuous competence development of employees that goes hand in hand with interaction also shows that the central theme of a cognitive supply chain is learning. Learning at the same time reflects the idea of Industry X as continuous evolution, also in the sense of the Service Design Ecosystem. This forms the important basis for the design capacity of all actors as well as the ability to adapt to changing conditions in a decentralised way. It is in the nature of the Service Design Ecosystem that a changed understanding of management is to be expected, as well as almost inevitably a stronger resilience of companies and supply chains.

**Acknowledgement** The research leading to these results has received funding from Horizon 2020, the European Union's Framework Programme for Research and Innovation (H2020/2014-2020) under grant agreement no 957331 (www.knowlEdge-project.eu).

## References


Adobor, H. (2020). Supply chain resilience: An adaptive cycle approach. *The International Journal of Logistics Management, 31*(3), 443–463. https://doi.org/10.1108/IJLM-

Anderson, J., Viry, P., & Wolff, G. B. (2020). Europe has an artificial-intelligence skills shortage, Bruegel-Blogs, 27 Aug 2020. Gale Academic OneFile, link.gale.com/apps/doc/A634244496/AONE?u=anon~84177555. Accessed 7 Apr 2022.

Arevian, A. C., O'Hora, J., Jones, F., Mango, J., Jones, L., Williams, P. G., Booker-Vaughns, J., Jones, A., Pulido, E., Banner-Jackson, D., & Wells, K. B. (2018). Participatory technology development to enhance community resilience. *Ethnicity & Disease, 28*(Suppl 2), 493–502. https://doi.org/10.18865/ed.28.S2.493

Bak, O. (2021). Understanding the stimuli, scope, and impact of organizational transformation: The context of eBusiness technologies in supply chains. *Strategic Change, 30*, 443–452. https://doi.org/10.1002/jsc.2466

Baskerville, R., & Dulipovici, A. (2006). The theoretical foundations of knowledge management. *Knowledge Management Research & Practice, 4*, 83–105. https://doi.org/10.1057/palgrave.kmrp.8500090

Bowersox, D. J., Closs, D. J., Cooper, M. B., & Bowersox, J. C. (2016). *Supply chain logistics management* (4th ed.). McGraw-Hill Education.

Boy, G. A., & Narkevicius, J. M. (2014). Unifying human centered design and systems engineering for human systems integration. In M. Aiguier, F. Boulanger, D. Krob, & C. Marchal (Eds.), *Complex systems design & management*. Springer. https://doi.org/10.1007/978-3-319-02812-5_12

Claver, E., Llopis, J., Garcia, D., & Molina, H. (1998). Organizational culture for innovation and new technological behavior. *The Journal of High Technology Management Research, 9*(1), 55–68. https://doi.org/10.1016/1047-8310(88)90005-3

Coulton, P., & Lindley, J. G. (2019). More-than human centred design: Considering other things. *The Design Journal, 22*(4), 463–481. https://doi.org/10.1080/14606925.2019.1614320

Del Vecchio, C., & Paschalidis, I. C. (2006). Enforcing service-level constraints in supply chains with assembly operations. *IEEE Transactions on Automatic Control, 51*(12), 2000–2005. https://doi.org/10.1109/TAC.2006.884961

Delpla, V., Kenné, J.-P., & Hof, L. A. (2022). Circular manufacturing 4.0: Towards internet of things embedded closed-loop supply chains. *The International Journal of Advanced Manufacturing Technology, 118*(9–10), 3241–3264.

European Commission. (2019). *The European Green Deal*, https://eur-lex.europa.eu/legal-content/EN/TXT/?uri=COM:2019:640:FIN

European Commission. (2021). *2030 Digital compass: The European way for the digital decade*, https://eur-lex.europa.eu/legal-content/en/TXT/?uri=CELEX:52021DC0118

European Commission, Directorate-General for Research and Innovation, Breque, M., De Nul, L., & Petridis, A. (2021). *Industry 5.0: Towards a sustainable, human-centric and resilient European industry*. Publications Office. https://data.europa.eu/doi/10.2777/308407

Fenwick, T., Seville, E., & Brunsdon, D. (2009). *Reducing the impact of organisational silos on resilience: A report on the impact of silos on resilience and how the impacts might be reduced.* Resilient Organisations Research Report, University of Canterbury, 28 pp.

Folke, C. (2006). Resilience: The emergence of a perspective for social-ecological systems analyses. *Global Environmental Change, 16*(3), 253–267.

Forbrig, P. (2016). Continuous software engineering with special emphasis on continuous business-process modeling and human-centered design. In *Proceedings of the 8th international conference on subject-oriented business process management (S-BPM'16)* (pp. 1–4). Association for Computing Machinery. https://doi.org/10.1145/2882879.2882895. Article 11.

Frederico, G. F. (2021). From supply chain 4.0 to supply chain 5.0: Findings from a systematic literature review and research directions. *Logistics, 5*, 49. https://doi.org/10.3390/logistics5030049

Garreta-Domingo, M., Sloep, P. B., & Hernández-Leo, D. (2018). Human-centred design to empower "teachers as designers". *British Journal of Educational Technology, 49*(6), 1113–1130.

Glock, C. H., Grosse, E. H., Neumann, W. P., & Sgarbossa, F. (2017). Editorial: Human factors in industrial and logistic system design. *Computers & Industrial Engineering, 111*, 463–466. https://doi.org/10.1016/j.cie.2017.08.014

Gorlach, I., & Wessel, O. (2008). Optimal level of automation in the automotive industry. *Engineering Letters, 16*(1). http://www.engineeringletters.com/issues_v16/issue_1/EL_16_1_21.pdf

Gupta, S., Modgil, S., Bhattacharyya, S., & Bose, I. (2022). Artificial intelligence for decision support systems in the field of operations research: Review and future scope of research. *Annals of Operations Research, 308*, 215–274. https://doi.org/10.1007/s10479-020-03856-6

Hashizume, A., & Kurosu, M. (2013). Understanding user experience and artifact development through qualitative investigation: Ethnographic approach for human-centered design. In M. Kurosu (Ed.), *Human-computer interaction. Human-centred design approaches, methods, tools, and environments. HCI 2013* (*Lecture notes in computer science*) (Vol. 8004). Springer. https://doi.org/10.1007/978-3-642-39232-0_8

Hayat, K., JianJun, Z., Ali, S., & Khan, M. A. (2021). Exploring factors of the sustainable supply chain in the post-COVID-19 pandemic: SWARA approach. *Environmental Science and Pollution Research*. https://doi.org/10.1007/s11356-021-16908-6

Heylighen, F. (1989). Self-organization, emergence and the architecture of complexity. In *Proceedings of the 1st European conference on system science* (pp. 23–32). AFCET.

Heylighen, F. (1991). Design of a hypermedia interface translating between associative and formal representations. *International Journal of Man-Machine Studies, 35*(4), 491–515. https://doi.org/10.1016/S0020-7373(05)80089-6

Iarovyi, S., Lastra, J. L. M., Haber, R., & del Toro, R. (2015). From artificial cognitive systems and open architectures to cognitive manufacturing systems. In *2015 IEEE 13th international conference on industrial informatics (INDIN)* (pp. 1225–1232). https://doi.org/10.1109/INDIN.2015.7281910

International Organization for Standardization. (2019). *Ergonomics of human-system interaction – Part 210: Human-centred design for interactive systems*. International Organization for Standardization.

Janssen, M., Brous, P., Estevez, E., Barbosa, L. S., & Janowski, T. (2020). Data governance: Organizing data for trustworthy Artificial Intelligence. *Government Information Quarterly, 37*(3), 101493. https://doi.org/10.1016/j.giq.2020.101493

Ketchen, D. J., Jr., & Hult, G. T. M. (2007). Bridging organization theory and supply chain management: The case of best value supply chains. *Journal of Operations Management, 25*, 573–580.

Khan, I., Lim, H., Jemai, J., & Sarkar, B. (2019). Effect of electrical energy on the manufacturing setup cost reduction, transportation discounts, and process quality improvement in a two-echelon supply chain management under a service-level constraint. *Energies, 12*(19), 3733.

Kim, Y. H., Sting, F. H., & Loch, C. H. (2014). Top-down, bottom-up, or both? Toward an integrative perspective on operations strategy formation. *Journal of Operations Management, 32*(7–8), 462–474. https://doi.org/10.1016/j.jom.2014.09.005

Kurtmollaiev, S., Fjuk, A., Pedersen, P. E., Clatworthy, S., & Kvale, K. (2018). Organizational transformation through service design: The institutional logics perspective. *Journal of Service Research, 21*(1), 59–74.

Kuure, E., & Miettinen, S. (2017). Social design for service. Building a framework for designers working in the development context. *The Design Journal, 20*(sup1), S3464–S3474. https://doi.org/10.1080/14606925.2017.1352850

Lam, W., & Shankararaman, V. (2007). Dissolving organisational and technological silos: An overview of enterprise integration concepts. In W. Lam & V. Shankararaman (Eds.), *Enterprise architecture and integration: Methods, implementation and technologies* (pp. 1–22). IGI Global.

Li, S., Wang, R., Zheng, P., & Wang, L. (2021). Towards proactive human–robot collaboration: A foreseeable cognitive manufacturing paradigm. *Journal of Manufacturing Systems, 60*, 547–552. https://doi.org/10.1016/j.jmsy.2021.07.017

Lii, P., & Kuo, F. I. (2016). Innovation-oriented supply chain integration for combined competitiveness and firm performance. *International Journal of Production Economics, 174*, 142–155.

Luhmann, N. (1984). *Soziale Systeme. Grundriß einer allgemeinen Theorie*. Suhrkamp.

Luhmann, N. (1987). *Rechtssoziologie* (3. Auflage). Westdeutscher Verlag.

Maguire, M. (2001). Methods to support human-centred design. *International Journal of Human-Computer Studies, 55*(4), 587–634.

Maozhu, J., Wang, H., Zhang, Q., & Zeng, Y. (2020). Supply chain optimization based on chain management and mass customization. *Information Systems and e-Business Management, 18*(4), 647–664.

Martelaro, N., & Ju, W. (2018). Cybernetics and the design of the user experience of AI systems. *Interactions, 25*(6), 38–41. https://doi.org/10.1145/3274570

Martinich, L. P. (2016). Excellent execution: Balancing top-down and bottom-up management. *IEEE Engineering Management Review, 44*(2), 20–22. https://doi.org/10.1109/EMR.2016.2568720

Masood, T., Cherifi, C. B., & Moalla, N. (2021). A machine learning approach for performance-oriented decision support in service-oriented architectures. *Journal of Intelligent Informations, 56*, 255–277. https://doi.org/10.1007/s10844-020-00617-6

Mavrikios, D., Papakostas, N., Mourtzis, D., & Chryssolouris, G. (2013). On industrial learning and training for the factories of the future: A conceptual, cognitive and technology framework. *Journal of Intelligent Manufacturing, 24*, 473–485. https://doi.org/10.1007/s10845-011-0590-9

McDougall, N., Wagner, B., & MacBryde, J. (2022). Leveraging competitiveness from sustainable operations: Frameworks to understand the dynamic capabilities needed to realise NRBV supply chain strategies. *Supply Chain Management, 27*(1), 12–29.

Miettinen, S. (2017). *An introduction to industrial service design*. Routledge.

Miettinen, S., & Vuontisjärvi, H.-R. (2016, August). Rethinking the marginal – Service design for development. Design for all – A publication for Design for All Institute of India. *Designing for the Public Good, 11*(8), 82–95. http://www.designforall.in/newsletteraug2016.pdf; Download 14.10.2022.

Min, H. (2019). Artificial intelligence in supply chain management: Theory and applications. *International Journal of Logistics: Research and Applications, 13*(1), 13–39.

Motingoe, M., & Langerman, J. J. (2019). New organisational models that break silos in organisations to enable software delivery flow. In *2019 International conference on system science and engineering (ICSSE)* (pp. 341–348). https://doi.org/10.1109/ICSSE.2019.8823257

Nguyen Ngoc, H., Lasa, G., & Iriarte, I. (2022). Human-centred design in industry 4.0: Case study review and opportunities for future research. *Journal of Intelligent Manufacturing, 33*, 35–76. https://doi.org/10.1007/s10845-021-01796-x

O'Brien, K., Pelling, M., Patwardhan, A., Hallegatte, S., Maskrey, A., Oki, T., Oswald-Spring, U., Wilbanks, T., & Yanda, P. Z. (2012). Toward a sustainable and resilient future. In: Managing the risks of extreme events and disasters to advance climate change adaptation [Field, C.B., V. Barros, T.F. Stocker, D. Qin, D.J. Dokken, K.L. Ebi, M.D. Mastrandrea, K.J. Mach, G.-K. Plattner, S.K. Allen, M. Tignor, and P.M. Midgley (eds.)]. *A Special Report of Working Groups I and II of the Intergovernmental Panel on Climate Change (IPCC)* (pp. 437–486).

Oracle. (2018). *Oracle Government 360 – Autonomie und Resilienz*. ORACLE Deutschland B.V. & Co. KG.

Penin, L. (2018). *An introduction to service design: Designing the invisible*. Bloomsbury Publishing.

Pins, D (Ed.). (2021). D2.1 user need analysis and scenario definition, deliverable of knowlEdge Project Consortium, WP2 System Engineering, Specifications and External Collaboration, www.knowledge-project.eu.

Pyykkö, H., Suoheimo, M., & Walter, S. (2021). Approaching sustainability transition in supply chains as a wicked problem: Systematic literature review in light of the evolved double diamond design process model. *Processes, 9*(12), 2135. https://doi.org/10.3390/pr9122135

Rupp, M., Schneckenburger, M., Merkel, M., Börret, R., & Harrison, D. K. (2021). Industry 4.0: A technological-oriented definition based on bibliometric analysis and literature review. *Journal of Open Innovation: Technology, Market, and Complexity, 7*(1), 1–20. https://doi.org/10.3390/joitmc7010068

Sakuramoto, C., Di Serio, L. C., & de Vicente Bittar, A. (2019). Impact of supply chain on the competitiveness of the automotive industry. *RAUSP Management Journal, 54*(2), 205–225.

Sarkar, M., & Chung, B. D. (2019). Flexible work-in-process production system in supply chain management under quality improvement. *International Journal of Production Research, 58*(13), 3821–3838. https://doi.org/10.1080/00207543.2019.1634851

Schipper, E. L. F., Dubash, N. K., & Mulugetta, Y. (2021). Climate change research and the search for solutions: Rethinking interdisciplinarity. *Climatic Change, 168*, 18. https://doi.org/10.1007/s10584-021-03237-3

Schmidt, M., & Schwegler, R. (2008). A recursive ecological indicator system for the supply chain of a company. *Journal of Cleaner Production, 16*(15), 1658–1664.

Schmidt, G., & Wilhelm, W. E. (2000). Strategic, tactical and operational decisions in multi-national logistics networks: A review and discussion of modelling issues. *International Journal of Production Research, 38*(7), 1501–1523. https://doi.org/10.1080/002075400188690

Schulte, C. (2013). *Logistik: Wege zur Optimierung der Supply Chain* (6. Auflage). Verlag Franz Vahlen.

Seidl, D., & Mormann, H. (2015). Niklas Luhmann as organization theorist. In P. Adler, P. du Gay, G. Morgan, & M. Reed (Eds.), *Oxford handbook of sociology, social theory and organization studies: Contemporary currents* (pp. 125–157). Oxford University Press.

Sgarbossa, F., Grosse, E. H., Neumann, W. P., Battini, D., & Glock, C. H. (2020). Human factors in production and logistics systems of the future. *Annual Reviews in Control, 49*, 295–305. https://doi.org/10.1016/j.arcontrol.2020.04.007

Snowdon, A. W., & Saunders, M. (2021). COVID-19, workforce autonomy and the health supply chain. *Healthcare Quarterly, 24*(2), 15–26.

Stacho, Z., Potkány, M., Stachová, K., & Marcineková, K. (2016). The organizational culture as a support of innovation processes' management: A case study. *International Journal for Quality Research, 10*(4), 769–784.

Stickdorn, M., & Schneider, J. (2011). *This is service design thinking. Basics – Tools – Cases*. BIS Publishers.

Tett, G. (2015). *The silo effect: The peril of expertise and the promise of breaking down barriers*. Simon & Schuster.

UNICEF. (2022). *Human-centred design*, https://www.unicef.org/innovation/hcd, United Nations Children's Fund (UNICEF), Download 04.04.2022.

Vahs, D., & Schäfer-Kunz, J. (2015). *Einführung in die Betriebswirtschaftslehre* (7. Auflage). Schäffer-Poeschel Verlag.

van der Bijl-Brouwer, M., & Dorst, K. (2017). Advancing the strategic impact of human-centred design. *Design Studies, 53*, 1–23. https://doi.org/10.1016/j.destud.2017.06.003

Vink, J., Tronvoll, B., Edvardsson, B., Wetter-Edman, K., & Aguirre, M. (2017). Service ecosystem design: Doing institutional work through design. In E. Gummesson, C. Mele, & F. Polese (Eds.), *Service dominant logic, network and systems theory and service science: Integrating three perspectives for a new service agenda: Proceedings of the 2017 Naples forum on service*. Tricase, 2017 Naples Forum on Service, Naples.

Vink, J., Koskela-Huotari, K., Tronvoll, B., Edvardsson, B., & Wetter-Edman, K. (2021). Service ecosystem design: Propositions, process model, and future research agenda. *Journal of Service Research, 24*(2), 168–186.

Wahlström, M., Walter, S., Salonen, T.-T., Lammi, H., Heikkilä, E., & Helaakoski, H. (2020). *Sustainable Industry X – A cognitive manufacturing vision*. VTT Technical Research Centre of Finland.

Wan, J., Li, X., Dai, H.-N., Kusiak, A., Martínez-García, M., & Li, D. (2021). Artificial-intelligence-driven customized manufacturing factory: Key technologies, applications, and challenges. *Proceedings of the IEEE, 109*(4), 377–398. https://doi.org/10.1109/JPROC.2020.3034808

Wang, Q., Sun, Y., Yuan, X., Cao, D., Zuo, J., & Gao, Z. (2017). Addressing the efficiency of the core ecological industrial chain: A DEA approach. *Journal of Cleaner Production, 156*, 235–243.

Werner, H. (2020). *Supply chain management: Grundlagen, Strategien, Instrumente und Controlling* (7. Auflage). Springer Gabler.

Wesselink, A. (2008). Interdisciplinarity, problem focused research and normativity. In *SRI Papers No. 11*. Sustainability Research Institute, University of Leeds.

Wetter-Edman, K. (2012). Relations and rationales of user's involvement in service design and service management. In S. Miettinen & A. Valtonen (Eds.), *Service design with theory: Discussion on change, value and methods*. Lapland University Press.

Wohlgezogen, F., McCabe, A., Osegowitsch, T., & Mol, J. (2020). The wicked problem of climate change and interdisciplinary research: Tracking management scholarship's contribution. *Journal of Management & Organization, 26*(6), 1048–1072. https://doi.org/10.1017/jmo.2020.14

Wright, P., & McCarthy, J. (2010). *Experience-centered design – Designers, users, and communities in dialogue*. Springer.

Young, R. A. (2012). Refocusing the practice of service design in socially responsible contexts. In S. Miettinen & A. Valtonen (Eds.), *Service design with theory. Discussions on change, value and methods*. Lapland University Press.

Zhang, J., Xu, J., & Liu, Y. (2009). Complex adaptive supply chain network: The state of the art. In *2009 Chinese control and decision conference* (pp. 5643–5647). https://doi.org/10.1109/CCDC.2009.5195204

Zheng, P., Xia, L., Li, C., Li, X., & Liu, B. (2021). Towards Self-X cognitive manufacturing network: An industrial knowledge graph-based multi-agent reinforcement learning approach. *Journal of Manufacturing Systems, 61*, 16–26.

# Part IV
# Case Studies on the Service Design for Emerging Technologies Applications

# Chapter 13
# Service Design Methods for the Design of Smart Surfaces

**Jonna Häkkilä, Markus Löchtefeld, Damien Brun, and Ashley Colley**

## Introduction

Our everyday environments are becoming increasingly embedded with technology. This includes mobile devices, Internet, and social media, which have taken a significant role in our daily routines and interactions, but also includes intelligence embedded in the physical environment around us. Smart environments including sensors, displays, and connected artifacts are an integral part of today's building planning. This includes homes, offices, public spaces, and smart building infrastructures. Public displays are invading both indoor and outdoor urban spaces (Kostakos & Ojala, 2013) and span from the resolution of one pixel to building-sized media walls and high-resolution installations. In the emerging domain of smart surfaces, rather than being add-on elements, displays and user interaction mechanisms are seamlessly integrated into our everyday environments. As well as conventional touch interaction, such smart surfaces can utilize a variety of sensors for interaction, e.g., to react to the proximity or gestures of the user.

In this chapter, we focus on smart surfaces, as interactive surfaces that can include both input and output elements. In particular, we explore electrochromic materials as the enabling technology and discuss its feasibility for end-user applications. Electrochromic materials change their optical properties, such as color, when an electrical voltage is applied. When formulated as an electrochromic ink and applied to a surface, an electrochromic display can be created, which follows the contours

J. Häkkilä (✉) · D. Brun · A. Colley
University of Lapland, Rovaniemi, Finland
e-mail: jonna.hakkila@ulapland.fi; damien.brun@ulapland.fi; ashley.colley@ulapland.fi

M. Löchtefeld
Aalborg University, Aalborg, Denmark
e-mail: markus.lochtefeld@aau.dk

 
U. Z. A. Hamid, M. Suoheimo (eds.), *Service Design for Emerging Technologies Product Development*, Springer Series in Design and Innovation 29,
https://doi.org/10.1007/978-3-031-29306-1_13

of the applied surface. Our research has been conducted in close collaboration with other research disciplines, e.g., chemistry and industry partners, and with end-user applications and product design in mind as the final research goal. The research is part of the DecoChrom project (DecoChrom, 2022), which addresses different aspects of electrochromic display development, including the sustainability of the technology. This chapter focuses on service design and introduces how service design methods have been applied in our research on smart surfaces and discusses different service design methods and their strengths in this context.

During the recent decade, service design has been adopted as a new approach for consumer-centric service product creation (Miettinen, 2017). Building essentially on the tradition of industrial design and user-centered design, it applies methods from design and user research, considering the holistic user experience with the service. Service design also takes into account different stakeholders, and the design process considers the interplay between them, beyond designing for an individual user. This means taking into account the viewpoints of the consumer, service provider, and those interfacing to the consumer in the service interface. For instance, when designing a mobile sport coaching app for young athletes, in addition to the teenage mobile app user, the service design process takes into account stakeholders such as the sports coach, medical support, and the service platform handling the data (Häkkilä et al., 2016). As a consequence, the design process needs to consider a more complex contextual framework of users and a wider set of other stakeholders than traditionally considered in human–computer interaction design cases (Lee et al., 2022).

Fundamentally, interactive systems designed for today's world are integrated services, rather than individual products (Forlizzi, 2018). Thus, it is important to consider service design aspects already in the early phases of the product design process. In the following, we present how service design thinking and methods have been applied in our research on smart surfaces. As design cases, we present the concept of an interactive wallpaper (Brun et al., 2021) and two cases demonstrating the use of interactive surfaces to support the smart social distancing required during the Covid-19 pandemic, i.e., interactive surfaces in a smart building and a smart cafe table (Colley et al., 2021). In each of the three design cases, design probes with different levels of physicality were used (Fig. 13.1).

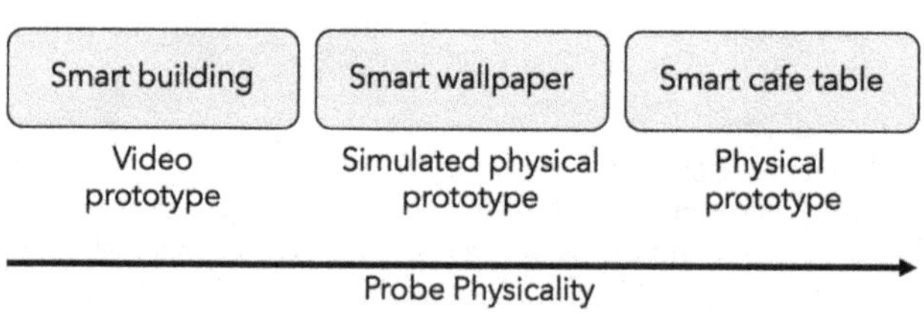


**Fig. 13.1** The 3 presented case studies demonstrate the use of prototypes with different levels of physicality as probes in the service design process

## Electrochromic Smart Surfaces

Mark Weiser presented a vision of twenty-first century computers as computers that will "find their way invisibly into people's lives" (Weiser, 1999). This means that we will no longer carry around devices such as smartphones, but we will be surrounded by an environment containing smart surfaces and objects that we can interact with. However, current technology is not suited for this, and new material developments are needed to realize this vision (Abowd, 2020). In the future, the very surfaces that we use to construct products or buildings will become inherently intelligent. In this chapter, we focus on one specific technology that has the potential to form a building block of smart surfaces—electrochromic materials.

### *Electrochromic Technology*

Electrochromism is the capability of some materials to change their optical properties through chemical oxidation or reduction when an electric current is applied to them (Granqvist, 2015). This means that electrochromic (EC) materials can change their optical absorption characteristics or color when an electrical voltage is applied. This behavior is similar to the more well-known thermochromic materials that change color based on temperature. A variety of different electrochromic materials have been explored that can switch between different color combinations and intensities. One of the most commonly used is PEDOT:PSS (the chemical name poly(3,4-ethylenedioxythiophene) polystyrene sulfonate), which can change its color from nearly transparent to a dark blue in less than two seconds when a small voltage is applied (Jensen et al., 2019).

EC materials have a variety of different properties that are extremely favorable when designing large interactive smart surfaces. First of all, EC displays operate at low voltages (usually between 1 and 3 V) and do not require complex control electronics, such as those needed for E-Ink or OLED displays. Furthermore, these displays are bi-stable, meaning once the change in color has been performed, the display will remain in this state for several hours without the need for additional power. EC materials are not light emitting, making them suitable for environments such as the bedroom, where artificial light may have negative impact on human sleep patterns (Czeisler, 2013). Finally, EC displays are very thin (around 3 mm), flexible, and can be produced in any shape (Jensen et al., 2019), which enables their integration into a variety of different products as user interfaces (Colley et al., 2021) or soft robots (Löchtefeld & Milthers, 2021), and are specifically suited for interactive smart surfaces.

### *Design with Electrochromics*

Electrochromics also have several advantages when it comes to designing and prototyping EC displays. Prototyping can be done using off-the-shelf printing techniques such as screen printing or a slightly modified ink-jet printer (Jensen et al., 2019). The process of assembling the displays is easy and does not require any special requirements such as a clean room, making it relatively easy for designers to test their ideas in practice (Jensen et al., 2022).

Given that the display itself can be transparent, a wide variety of application scenarios are possible. The simplest being is to add an EC display on top of a standard printed product. For example, in Jensen et al. (2020), EC displays are integrated onto the printed graphics of a classic board game to create new interactive experiences. The fact that EC displays are flexible and can be produced in every shape has also been enabled their integration into clothing and wearable devices such as shoes (Jensen et al., 2019), jackets (Genç et al., 2020), and face masks (Genç et al., 2020). These examples demonstrate the potential of the electrochromic technology and the ways in which designers can leverage it.

## Service Design as an Approach

Service design has its roots in design research. Whereas the field of product design has addressed physical artifacts, service design has shifted the focus beyond the traditional design scope. Instead of focusing on tangible design artifacts, service design applies the design methods and design thinking in creating and improving services. It aims to create and improve services around both immaterial and material products (Miettinen, 2012). In the area of human–computer interaction (HCI), service design has commonly been interpreted as designing digital instead of physical products, and with a focus on the consumer journey and system-level design (Yap et al., 2021). Even though digital applications are today commonplace in all sectors of life, service design can also address cases that do not have digital components, and often it concerns hybrid products with both physical and digital elements.

Service design applies methods based on those from the tradition of user-centric design. In essence, a user-centric design approach means applying a design process, which keeps the user in its core focus, and involves target users through different study methods throughout the whole design cycle. From a historical perspective, design principles for creating interactive systems first focused on human factors and ergonomics, and after that, started addressing the wider concept of usability, i.e., the efficiency, effectiveness, and user satisfaction, with the system. It was, however, realized that these aspects are not enough when investigating users' interaction with the system and motivations behind the use. The development in the field thus proceeded to consider factors beyond usability and to design for a holistic user experience that includes both utilitarian and hedonic aspects (Hassenzahl & Tractinsky, 2006). User experience design emphasizes aspects that do not focus only to the instrumental value of the interactive artifact but include aspects such as engagement, emotion, and affect when interacting with the system (Law et al., 2008). Service design considers yet a larger framework, as it broadens the focus from an individual user and takes into account also other stakeholders (Forlizzi, 2018). Lee et al. (2022) have presented four dimensions for service design from interaction level to system level, which are: (1) the customer experience, (2) backstage work, (3) collaborative network, and (4) ecosystem and infrastructure. The viewpoint of the service provider is important to take into account in order to enable creating services, which result in functional and seamless experiences for the end user. Typically, there also exist other stakeholders, who have dependencies with or influence over the service. Also, it is important to remember that services do not operate in isolation but can be seen as part of a larger ecosystem and utilize some infrastructure. It has been pointed out that HCI-oriented research on service design often focuses on the interaction dimension, whereas the system-level dimension includes tools and methods that are more business-oriented (Lee et al., 2022).

Service design differs from conventional interaction design also in the manner in which it considers the holistic consumer journey during which the user interacts with the system. The interaction with the system is not anymore studied as an isolated phenomenon, as is often the case with usability studies and interaction design research, but interaction is considered as part of a longer path, during which the user interacts with the service at several touchpoints. This includes not only the actual core transaction or task completion, but also the pre- and post-service experiences. Methodologically, this means that a service design study must be designed to cover a larger sequence of events, during which the user interacts with the system. For instance, the designers can organize a study that looks at searching and finding the service, taking it into use as a first time user, and completing the actual transaction with the service.

A variety of tools and methods commonly applied in service design have been charted by Alves et al. (2013). In the analysis, gathering 164 methods and tools used in service design, the most common ones included scenarios, storyboarding, prototyping, consumer journeys, shadowing, and service blueprints (Alves & Jardim Nunes, 2013). These reflect the general approach in service design, which emphasizes understanding the users and the use cases, and investigating the service as a holistic experience. Service design is also experimental and agile in its very nature, as reflected through service prototyping and scenario workflows.

To organize studies covering a significant part of the consumer journey or to concept a holistic service, different methods to simulate the service and user interaction with it have become especially useful. Service prototyping is a method to where the service is simulated or set up as a demo in a place or condition resembling the actual use context. This way, the service can be tested and its features iterated before taking it into real use. An example of a service prototyping tool is a service design laboratory space, where different use contexts can be staged with the help of back-projected screens. For example, two screens placed at a 90° angle can form a corner, where different use scenarios can be staged and acted out (Rontti et al., 2012). Projected presentations of service spaces have been found to facilitate collaboration (Colley et al., 2020). With a portable set of equipment, the stage can be set up at different locations, for instance, at a project collaborator's or service provider's office space for a joint service design session (Häkkilä et al., 2015). Wizard-of-Oz techniques can also be used to simulate a functional service (Meurer et al., 2020). In addition, virtual reality (VR) can be utilized to investigate the service context and allow remote access to it, although this may limit the collaborative and participatory nature of the design process, and thus is not optimal in all design phases (Colley et al., 2020).

In the following, we present case studies where we have used a service design approach to investigate the potential for smart surfaces. The examples apply different methods for service prototyping, each presenting different maturity levels and allowing a different amount of interactivity.

## Service Design Cases with Smart Surfaces

### *Case 1: Interactive Wallpaper*

As the first design case adopting service design methods, we present the case of interactive wallpaper. The concept of interactive wallpaper has already been explored with the aims of providing either useful, engaging, or entertaining ambient experiences (Huang & Waldvogel, 2005; Campbell et al., 2014; Hoare et al., 2015). However, despite showing the potential applications of using interactive wallpapers, the previous works were mostly agnostic regarding the technical implementation, and many years or decades later they remain too far-fetched to be implemented. With this background, we addressed the topic of interactive wallpaper based on the technical possibilities of electrochromic inks (Section "Electrochromic Technology").

An exploratory study (Brun et al., 2021) was conducted to collect user perceptions and qualitative feedback while emulating the opportunities (e.g., low consumption with solar panels integrated into the design) and limits (bi-state monochromatic graphical elements) of this technology in five interactive wallpaper concepts, each aiming to address different ideas:

- *Timer*, in our case taking place in the bathroom, with slowly moving graphics to emphasize the required duration for hand washing
- *Notification*, taking place in the office with the visual form of portholes notifying the user of either an upcoming meeting, the microphone status, or as a reminder for physical activity
- *Voting*, or more broadly the idea of communal expressive walls, visually rendering the results of voting at the corners of a room

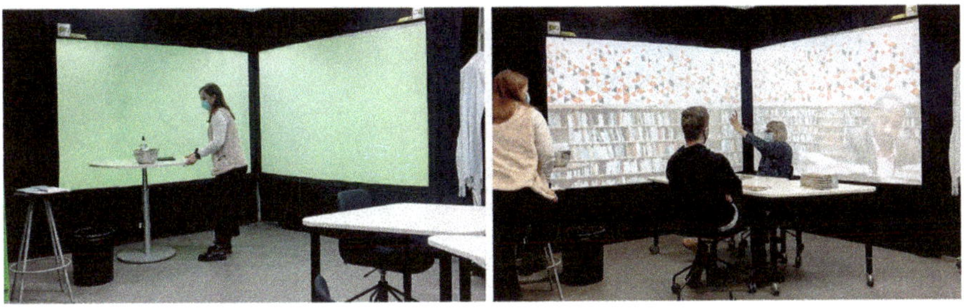

Fig. 13.2 The instructor setting up the additional furniture next to the two back-projected walls (left) and two participants giving feedback to the instructor during the *noise* concept scenario taking place in a simulated library (right)

- *Noise*, with a colored repeated pattern showing the current ambient sound level (in our case situated in a library)
- *Playful*, taking the form of a hide-and-seek game with integrated solar panel elements and a repeated pattern of animals

The concepts were simulated in the service design laboratory, where the wallpaper concepts were visualized on two large screens. Two back-projected walls were used to display the interactive wallpaper concepts, complemented with physical furniture (tables, chairs) and tangible artifacts (keyboards, tablets, books, and a sanitizer dispenser) to improve the realism of the simulation (Fig. 13.2).

In seven sessions, fourteen participants (nine women) were recruited in pairs from the university campus from students and administrative employees. For each session, the participants first gave their informed consent and were then introduced to physical electrochromic display samples. Then, participants were invited to engage with a think-aloud protocol for each simulated concept presented in a set order as part of an imaginary short story. The narrative was such that the participant just arrived at work and had to wash their hands in the bathroom with the *timer* concept. Then, they sat in an office with the *notification* concept and transitioned to a meeting room incorporating the *voting* concept. Finally, they stopped at the library with the *noise* concept (including simulated audio) before going home and engaging with the *playful* concept. One moderator gave explanations of each concept and took notes, while another was responsible for changing the physical structure of the settings according to each text context. The *Wizard-of-Oz* method was used to simulate interactions, such as activity (e.g., washing hands) or touch detection.

Hassenzahl has defined user experience (UX) as "*a momentary, primarily evaluative feeling (good-bad) while interacting with a product or service*" (Hassenzahl, 2008). Thus, after completing the test scenarios, participants completed a questionnaire to assess their feeling toward each wallpaper concept (1–5 scale) and to collect their least and most preferred concept (including explanation of the reasons) as well as demographic data and their additional free form comments. Each session was video-recorded for subsequent analysis and lasted around 45 minutes in total. The methodology allowed us to receive preliminary feedback, both

**Fig. 13.3** Screenshots from the created consumer journey video illustrating a low-fi simulation of a smart door display that checks if a face mask is worn

quantitative and qualitative, and their analysis provided the opportunity to highlight and further develop the two most favored concepts, *Noise (library)* and *Playful (animals)*.

## *Case 2: Smart Building*

As the second case, we present a service design case where the user interacted with smart building infrastructure, supporting precautions for social distancing. Here, the design context was inspired by the Covid-19 pandemic. Digital technologies have been widely harnessed both to fight against Covid-19 in healthcare (Whitelaw et al., 2020) and to support society during periods of lockdown and social distancing (He et al., 2021). In our design case, a smart building was considered as a service, integrating different technologies that enabled the creation of a safe zone, minimizing the infection risk for those moving through the space. With our focus on smart surfaces with integrated display elements, scenarios including safety instructions and indicators were in the main interest.

A consumer journey description was created in the form of a video, which demonstrated different touchpoints with the smart building service. The service concept included display screens that were integrated into the building. An example of one of the concept scenarios is presented in Fig. 13.3, where the user enters the building, and the smart display at the door reminds them to wear a face mask. When the system detects the user is wearing a face mask, the display shows a smiley face icon. In the consumer journey video, the technology is illustrated using a low-fi paper prototype, with overlaid textual explanations. The video collected together several use cases for smart building technology and provided a holistic overview, which could be used to communicate the possibilities of technology to different stakeholders, such as citizens and technology companies, in an easily digestible form.

## *Case 3: Interactive Cafe Table*

The surfaces of tables in cafes and public spaces such as trains and libraries provide a potential mechanism for the spread of the Covid-19 virus. To reduce this risk, the table surface should be cleaned between each user. In the *CleanLeaf* table concept, the table surface is a smart surface that visually indicates its status as either clean or dirty. In this way, arriving users can be directed to clean tables, and staff can directly identify which surfaces require cleaning. Two approaches to integrating indication to the table surface were explored: (a) LEDs under a thin wood veneer and (b) electrochromic displays integrated to the surface. With both approaches, the design target was that the display should form an integral part of the surface design, rather than being simply an added indicator.

To evaluate the concept, functional physical prototypes of the *CleanLeaf* table were developed (Fig. 13.4). Passive infrared (PIR) sensors integrated at the top of the table leg were used to identify if a user was static at the table for an amount of time indicative of actual usage. When usage was detected, the surface display of the tables changed to indicate the table was dirty (red illumination or filled electrochromic leaf motifs). After cleaning the table surface, staff could reset the surface indication by placing an NFC tag in the center of the table, where an NFC reader was located under the table surface. A clean table was indicated by green illumination or empty electrochromic leaf motifs. A delay of five seconds was provided such that the cleaner could exit from the range of the PIR sensors after resetting the table's state.

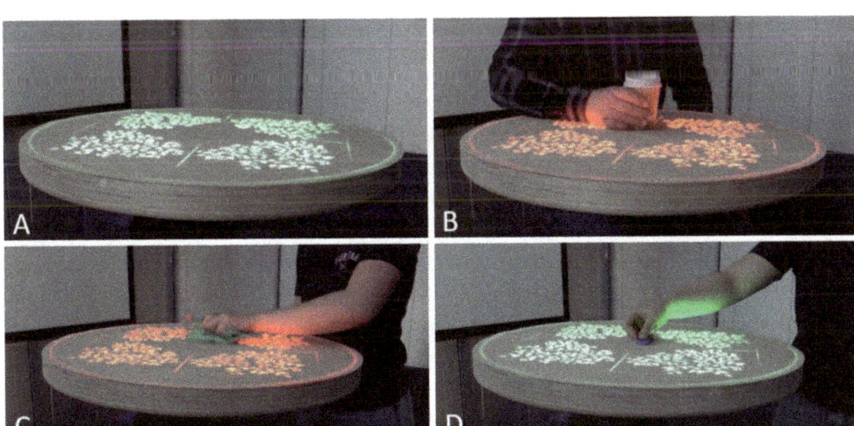


**Fig. 13.4** CleanLeaf table, demonstrating different patterns in a simulated version

## Discussion

The design cases described in this chapter provide insights into how different service design methods have been used in developing service concepts for smart surfaces with integrated displays. These design cases have emphasized service prototyping and showcase different approaches for it. In the consumer-journey-based case, the technology is illustrated using very simple low-fi prototypes, but the video method presents them in an authentic use context. The smart wallpaper concepts were visualized in a simulated context, which allowed a relatively immersive and easily configurable setup in which to organize a user study. The interactive cafe table concept includes a functional, interactive prototype. Compared to two other design cases, the cafe table is at a high maturity level. It offers a plausible demonstration of the concept and provides an almost end-product-like user experience, but on the other hand it demanded significantly more effort to design and implement. This makes the prototype more costly and harder to modify when design iterations are needed.

Both the interactive cafe table prototype and the smart building consumer journey video have been used for communication purposes when discussing the concepts with companies producing related products and services. Prior research has pointed out that service designers use visualizations to articulate insights, communicate insights, and maintain empathy, and they tend to use more polished visualization when communicating externally compared to internal communications (Segelström, 2012). This is also valid in our work.

Forlizzi has called for a more service-oriented approach to replace the conventional way of thinking in human–computer interaction research (HCI), which has emphasized a single user and taken solely the user's perspective, rather than also considering other stakeholders (Forlizzi, 2018). Reflecting against the framework presented by Lee et al. (2022), the service design cases presented here are rather interaction-design-oriented and consider less the system-level design aspects. However, they already take a wider perspective to the use of smart surfaces, including several actors in the use scenarios, such as waiter and cafe customer, or following a sequence of user actions during a consumer journey. Studying the interactive wallpaper concepts on a simulated stage, where a bodystorming technique (Oulasvirta et al., 2003) was applied, emphasizes the participatory design aspects that are essential in service design. In bodystorming, participants engage to the scenarios with their whole body presence and can act them out. In its overall approach, our research follows a similar path to that identified by Yap et al., who chart the interface between HCI and service design in general (Yap et al., 2021). System-level thinking becomes more important when developing the concepts targeting the use of particular technologies, here, electrochromic displays. However, differing from most HCI research utilizing service design (Yap et al., 2021), we have focused not on digital services but on digital–physical systems. In our experience, integrating this approach has functioned well in the design process.

## Conclusion

In this chapter, we have presented three example cases of how service design methods have been applied to study and communicate concepts exploring smart surfaces. The service design cases demonstrate three different fidelity levels of prototypes, from paper prototyping presented on a video, to a simulated service scenario, and finally a functional prototype. From an industry perspective, these methods provide tools to explore and evaluate applications of technology from the user viewpoint at an early development phase. The service design approach provides a broader picture of the user experience than a single user study and helps in orientating toward system-level thinking.

**Acknowledgments** This project has received funding from the European Union's Horizon 2020 research and innovation programme under Grant Agreement No. 760973 DecoChrom.

## References

Abowd, G. D. (2020). The Internet of materials: A vision for computational materials. *IEEE Pervasive Computing, 19*(2), 56–62.
Alves, R., & Jardim Nunes, N. (2013). Towards a taxonomy of service design methods and tools. In *International Conference on Exploring Services Science* (pp. 215–229).
Brun, D., Lehtinen, S., & Häkkilä, J. (2021). Discreet interactive wallpaper concepts study. In *IFIP Conference on Human-Computer Interaction* (pp. 390–394).
Campbell, R., Felton, R., & Hoarse, C. (2014). Smart wallpaper. In *Proceedings of the 2014 ACM International Conference on Interactive Experiences for TV and Online Video, Industry Track-TVX* (Vol. 14). Citeseer.
Colley, A., Genç, Ç., Löchtefeld, M., Mueller, H., Jensen, W., & Häkkilä, J. (2021). Exploring button design for low contrast user interfaces. In C. Ardito et al. (Eds.), *Human-Computer Interaction – Interact 2021* (pp. 411–415). Cham: Springer International Publishing.
Colley, A., Gröhn, W., & Häkkilä, J. (2021). CleanLeaf Table: Preventing the spread of covid-19 through smart surfaces. In *20th International Conference on Mobile and Ubiquitous Multimedia (MUM 2021), December 5–8, 2021, Leuven, Belgium. ACM.*
Colley, A., Suoheimo, M., & Häkkilä, J. (2020). Exploring VR and AR tools for service design. In *19th International Conference on Mobile and Ubiquitous Multimedia* (pp. 309–311).
Czeisler, C. A. (2013). Perspective: Casting light on sleep deficiency. *Nature, 497* (7450), S13–S13.
DecoChrom. (2022). *DecoChrom project*. https://decochrom.com/. Accessed 10 October 2022.
Forlizzi, J. (2018). Moving beyond user-centered design. *Interactions, 25*(5), 22–23.
Genç, C., Colley, A., Löchtefeld, M., & Häkkilä, J. (2020). Face mask design to mitigate facial expression occlusion. In *Proceedings of the 2020 International Symposium on Wearable Computers* (p. 40–44). New York, NY, USA: Association for Computing Machinery. Retrieved from https://doi.org/10.1145/3410531.3414303
Genç, C., Kantola, V., & Häkkilä, J. (2020). DecoLive Jacket with battery-free dynamic graphics. In *19th International Conference on Mobile and Ubiquitous Multimedia* (p. 338–340). New York, NY, USA: Association for Computing Machinery. Retrieved from https://doi.org/10.1145/3428361.3431192
Granqvist, C.-G. (2015). Electrochromic metal oxides: an introduction to materials and devices. In *Electrochromic materials and devices* (pp. 3–40). Germany: Wiley-VCH Weinheim.

Häkkilä, J., Alhonsuo, M., Virtanen, L., Rantakari, J., Colley, A., & Koivumäki, T. (2016). MyData approach for personal health-A service design case for young athletes. In *2016 49th Hawaii International Conference on System Sciences (HICSS)* (pp. 3493–3502).

Häkkilä, J., Colley, A., Inget, V., Alhonsuo, M., & Rantakari, J. (2015). Exploring digital service concepts for healthy lifestyles. In *Design, User Experience, and Usability: Design Discourse* (pp. 470–480). Springer.

Hassenzahl, M. (2008). User experience (UX) towards an experiential perspective on product quality. In *Proceedings of the 20th Conference on l'interaction Homme-Machine* (pp. 11–15).

Hassenzahl, M. & Tractinsky, N. (2006). User experience-A research agenda. *Behaviour & Information Technology, 25*(2), 91–97.

He, W., Zhang, Z. J., & Li, W. (2021). Information technology solutions, challenges, and suggestions for tackling the covid-19 pandemic. *International Journal of Information Management, 57*, 102287.

Hoare, C., Campbell, R., Felton, R., & Betsworth, L. (2015). Hide and seek: Exploring interaction with smart wallpaper. In *Proceedings of the 2015 Annual Symposium on Computer-Human Interaction in Play* (pp. 129–133).

Huang, J., & Waldvogel, M. (2005). Interactive wallpaper. In *ACM SIGGRAPH 2005 Electronic Art and Animation Catalog* (pp. 172–176).

Jensen, W., Colley, A., Häkkilä, J., Pinheiro, C., & Löchtefeld, M. (2019). TransPrint: A method for fabricating flexible transparent free-form displays. *Advances in Human-Computer Interaction, 2019*, 14, Article ID: 1340182. https://doi.org/10.1155/2019/1340182.

Jensen, W., Colley, A., & Löchtefeld, M. (2019). VitaBoot: Footwear with dynamic graphical patterning. In *Proceedings of the 23rd International Symposium on Wearable Computers* (pp. 279–283). New York, NY, USA: ACM. Retrieved from http://doi.acm.org/10.1145/3341163.3346937

Jensen, W., Craft, B., Löchtefeld, M., & Bjørn, P. (2022). Learning through interactive artifacts: Personal fabrication using electrochromic displays to remember Atari women programmers. *Entertainment Computing, 40*, 100464.

Jensen, W., Streubel Kristensen, T., Sand Kirk, C., Hameed, H. A., Bergmann Villadsen, D., & Löchtefeld, M. (2020). Hybrid settlers - integrating dynamic tiles into a physical board game using electrochromic displays. In *Extended Abstracts of the 2020 CHI Conference on Human Factors in Computing Systems* (p. 1–7). New York, NY, USA: Association for Computing Machinery. Retrieved from https://doi.org/10.1145/3334480.3382857

Kostakos, V., & Ojala, T. (2013). Public displays invade urban spaces. *IEEE Pervasive Computing, 12*(1), 8–13.

Law, E., Roto, V., Vermeeren, A. P., Kort, J., & Hassenzahl, M. (2008). Towards a shared definition of user experience. In *CHI'08 Extended Abstracts on Human Factors in Computing Systems* (pp. 2395–2398).

Lee, J.-J., Yap, C. E. L., & Roto, V. (2022). How HCI adopts service design: Unpacking current perceptions and scopes of service design in HCI and identifying future opportunities. In *CHI Conference on Human Factors in Computing Systems* (pp. 1–14).

Löchtefeld, M., & Milthers, D. (2021). ChromaBot - prototyping soft robotic actuators with integrated electrochromic displays. In *Proceedings of the 20th International Conference on Mobile and Ubiquitous Multimedia*.

Meurer, J., Pakusch, C., Stevens, G., Randall, D., & Wulf, V. (2020). A wizard of Oz study on passengers' experiences of a robo-taxi service in real-life settings. In *Proceedings of the 2020 ACM Designing Interactive Systems Conference* (pp. 1365–1377).

Miettinen, S. (2012). Discussions on change, value and methods. In *Service design with theory: Discussions on change, value and methods* (pp. 6–10). Lapland University Press.

Miettinen, S. (2017). *An introduction to industrial service design*. Routledge.

Oulasvirta, A., Kurvinen, E., & Kankainen, T. (2003). Understanding contexts by being there: Case studies in bodystorming. *Personal and Ubiquitous Computing, 7*(2), 125–134.

Rontti, S., Miettinen, S., Kuure, E., & Lindström, A. (2012). A laboratory concept for service prototyping-service innovation corner (SINCO). In *Proceedings of ServDes*.

Segelström, F. (2012). Communicating through visualizations: Service designers on visualizing user research. In *Conference Proceedings ServDes. 2009; DeThinking Service; Rethinking Design; Oslo Norway 24–26 November 2009* (pp. 175–185).
Weiser, M. (1999). The computer for the 21st century. *ACM SIGMOBILE Mobile Computing and Communications Review, 3*(3), 3–11.
Whitelaw, S., Mamas, M. A., Topol, E., & Van Spall, H. G. (2020). Applications of digital technology in covid-19 pandemic planning and response. *The Lancet Digital Health, 2*(8), e435–e440.
Yap, C. E. L., Lee, J.-J., & Roto, V. (2021). How HCI interprets service design: A systematic literature review. In *IFIP Conference on Human-Computer Interaction* (pp. 259–280).

# Chapter 14
# A Product-Service System Design Approach for the Frame Innovation of Civil Airliners Catering

**Paola Maria Trapani, Ke Ma, and Mo Jiao**

## Introduction

The research project is a collaboration between Boeing and the College of Design and Innovation at Tongji University in Shanghai. Specifically, it involved Boeing's two groups of cabin technology experts located in Beijing and Seattle, respectively. Tongji contributed to a multidisciplinary group of faculty and students in product-service systems design, industrial design, open design, business, artificial intelligence and algorithms, and data visualization.

The multidisciplinary nature of the research team allowed the research topic to be approached from diverse perspectives but simultaneously created difficulties in finding common ground for dialogue given the disciplinary differences. The section below Frames and Metaphors describes how the joint search for metaphorical imagery for developing creative concepts provided a powerful dialogic device for accommodating diversity within a framework of coherence.

Boeing approached the Tongji research group with the idea of investigating the future of the dining experience on-board airliners. The investigation started from the observation of intense turbulence in the market, the push of new lifestyles, the

P. M. Trapani (✉)
AHO Oslo School of Architecture and Design, Oslo, Norway
e-mail: paola.trapani@aho.no

K. Ma
School of Artificial Intelligence and Automation, Huazhong University of Science and Technology, Shanghai, China
e-mail: suta@tongji.edu.cn

M. Jiao
College of Design and Innovation, Tongji University, Shanghai, China
e-mail: jmo@tongji.edu.cn



U. Z. A. Hamid, M. Suoheimo (eds.), *Service Design for Emerging Technologies Product Development*, Springer Series in Design and Innovation 29,
https://doi.org/10.1007/978-3-031-29306-1_14

increasing food deliveries at the airport, the gradual spread of digital payments, the connection between consumers and the sharing of real-time experiences, the consciousness of healthy nutritional needs for well-being, the pressing competition from fast trains, the requirement to adopt new practices such as, for example, the classification of cabin waste, and vending machines installed at the airport. The starting point was not a precise brief but an overview of a sociotechnical system evolving rapidly under the effect of very different forces.

Indeed, Boeing was looking for innovation, but it was not clear at first what kind. Focusing on incremental innovation of the on-board galley could have been the goal of the research. In this case, the toolkit of Human-Centered Design could have been used, applying several cycles of observation, concept development, prototyping, and testing to obtain improved versions of the product concerning desired performance aspects. However, this approach would soon lead to a plateau such that even applying new cycles would not result in appreciable improvements. In fact, according to Norman and Verganti (2014), the iterative process of HCD is highly effective and productive if one tries to do better at what one already does yet remains within the same paradigm or regime. Indeed, the limitations of available technologies and knowledge ultimately place an insurmountable limit on the degree of incremental innovation possible.

Boeing's aspirations went beyond that: as a multinational company that builds aircraft for business customers, that is, airlines, it does not operate directly in the food and beverage market. Therefore, Boeing intended to foreshadow possible evolutionary trajectories in the industry that could help airlines create value for their end customers, that is, passengers, in the medium and long term. This awareness justified the adoption of product-service system design methodologies and tools capable of offering a holistic approach to creating scenarios that would tell new meanings of the on-board dining experience (Verganti, 2009). The project was framed as a radical business-to-business innovation case for the F&B product-service system (Verganti, 2017).

The research produced four distinct scenarios, each populated with a bundle of products and services that answer the three aforementioned questions (see Abstract). The following paragraphs will explain the process that led to identifying these three frames to radically innovate the on-board dining experience and its meaning.

The designated scenarios make the future tangible but do not aspire to predict what will happen 10 or 15 years from now. They provide the opportunity for discussion for Boeing's management in search of future strategies and opportunities while keeping an eye on risk avoidance in the medium and long term.

## Theoretical Background

Past research has shown that the dining experience is a strategic lever for an airline's customer retention (Han et al., 2020). Then, one might wonder why people eat so poorly on airplanes, especially in economy class. The paradox might appear to be a simple one, but it is not. The factors at play that determine the palatability of the experience are the most diverse: the type of menu, flavor, texture, personal dietary regimen, physical condition, level of satiety, portion size, and even the pleasantness of the cabin environment influence perceptions and experiences, at mealtime (You et al., 2020). On the other hand, the F&B service is subject to a strict protocol that controls its entire life cycle. As a result, it is almost impossible to escape standardization: one can choose from a pool of menus the one closest to personal taste. Figuring out how to personalize the experience without incurring unintended side effects of redesign, such as increasing cabin waste or crew workload, is a big conundrum. Immediately, we are faced with a problem with four characteristics (Dorst, 2015) discussed as being open, complex, dynamic, and networked.

## *Open*

First, it is not easy to know where to establish the system's limits to be investigated. Such limits do not exist and thus must be artificially established by the research team as a necessity burdened with unexpected consequences in the later stages of the research. From the initial interviews with the actors in the system (Fig. 14.1), it quickly became apparent that it would be challenging to point the finger, for example, at the catering company. They closely cooperate with the airline, which imposes precise budget limits. Both, in turn, must observe strict policies and rules designed not so much to ensure the pleasantness of the experience but to protect public health and the safety of food and beverages throughout the production and distribution chain. The aspiration to improve the on-board dining experience does not seem to be a real issue among industry insiders. Indeed, it is a legitimate aspiration but something that can be safely placed on the backburner. There are more concrete issues to be addressed in the business' everyday chores.

Moreover, the system of actors operates within an enlarged sociotechnical complex that, although external to the field in the strict sense, influence it at the level of trends and dynamic forces. Therefore, it is unclear where to draw that circle necessary to understand what to include and exclude from the field of attention.

## *Complex*

A product as apparently simple as an economy-class meal results from the collaborative work of many actors in different industries. In daily practice, they exchange continuous flows of information, materials, and money that can be represented in the system map (Fig. 14.2), that is, a model that is still reductive of reality but "good enough" to get an idea of the system functioning.

Looking at the intricate mesh of arrows, it is cumbersome to understand which relationships to touch to improve the quality of the dining experience. The poor

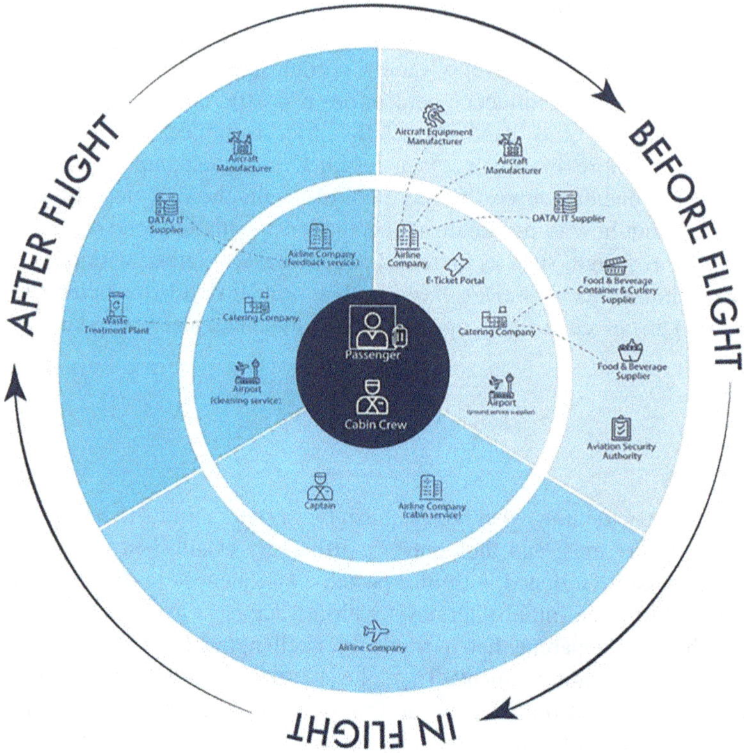


**Fig. 14.1** Stakeholders map. The actors involved in the production and distribution of on-board F&Bs are linked by interdependent relationships. They are arranged in concentric rings according to the distance from the main actors placed in the center. (Image by HUANG Jiawen)

result could be caused by many different factors: the quality of ingredients, the production process, distribution, packaging, the time between preparation and consumption, or microwave heating. Changing one area of the system, such as interrupting some flows, or creating new ones where none exist, could have unpredictable effects in areas of the system even far away.

It might be tempting to deal with such complexity by reducing it into limited and more manageable packages, for example, by relying on the interaction between pairs of actors who are strategic in the system, so-called *service encounters* (Manzini, 2015): passenger-cabin staff; airline manager-catering manager; catering distributor-cabin staff; and so on. However, it is recommended to keep the system's wholeness and alternate zooming in/zooming out by jumping levels from the entire to the detail according to the research stage. At one point in the investigation, and as we shall see, we focused on the service encounter between passengers and cabin staff.

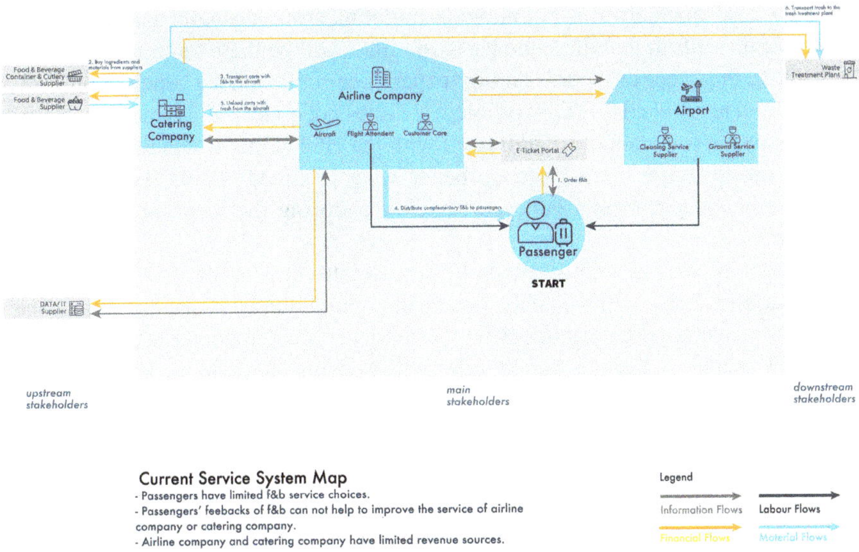


**Fig. 14.2** System map. The product-service system comprises elements and connections of a heterogeneous nature: products, services, networks of actors, and digital and physical infrastructures. These parts exchange information, materials, money, and labor. (Image by HUANG Jiawen)

## *Dynamic*

The system is constantly changing under the impetus of dynamic forces. Some changes are slow, for example, passengers' eating habits and lifestyles; others are sudden, such as the appearance of a new policy or competitor in the market. Changes alter priorities and information, monetary, and material flows between actors. New actors appear on the scene; others disappear. Continuous transformation means that any attempt to capture the system's configuration results in an ambition destined for repeated frustration.

## *Networked*

The production and distribution of meals involve synchronizing the daily activity of thousands of people who do not belong to the same organization. Very often, even within the same company, the food department has no clear idea of what the commercial one is doing. This reflection is enough to marvel that a result is eventually produced, even if of questionable quality. As Dorst writes, interlacing business relationships interweaves opportunities but also problems. A financial problem in one company can reverberate ripple through all the others. A failure to communicate can have disastrous effects, and so can an error in delivering materials.

No matter how much effort is put into consolidating system ganglia and joints, it is impossible to confirm that the same parts and links will be there tomorrow.

If innovating the on-board dining experience falls among open, complex, dynamic, and networked (OCDN) problems, it does not qualify as a typical problem-solving activity, so it cannot be addressed with the HCD toolkit. The research team is not trying to figure out how to do better what is already being done, such as building a better-performing galley or quickly choosing the menu at the time of ticket purchase.

The scope of the research is broader: it investigates how to improve the airplane meal experience holistically, namely, improving other boundary aspects, such as the ability to move around the cabin during the service, reduce food waste and garbage while still enjoying a more varied offering. Arriving at defining these aspects required participatory negotiation involving the principal and its stakeholders. Meanings are always collective constructs and, as such, are not given aprioristically, nor can they be changed unilaterally by the project team. The experience of eating on an airplane goes far beyond choosing the menu online or from a brochure. Therefore, imagination, invention, and dialogue among the parties are necessary to get to the point where the perspective is to be broadened.

Among the many valid methodologies for dealing with the radical innovation of meaning, the research team adopted (Dorst, 2015) *Frame Innovation*, which must be considered a reference for the entire chapter.

## Research Stages and Methods

### *Archaeology*

The Frame Innovation methodology begins with a field research phase to discover the problem's origins. Suppose, for an initial simplification, that the problem is figuring out why passengers eat poorly on airplanes. If the lack of satisfaction is not easy to remedy, they might as well get used to it. Boeing does not operate directly in the in-flight F&B business; therefore, it is not directly identifiable as the problem owner.

The first phase of field research, which Dorst calls *archaeology*, involved the F&B officer of a Chinese airline and a catering company at Pudong Airport. Moreover, the Asia-Pacific region manager of a European airline, the cooperation director of an online travel portal, and two engineers specializing in commercial robotics also got involved.

The data, at this stage, are derived from interviews and a field visit to the production plant, which ended with the tasting of a business-class menu.

All the information gathered is analyzed and leads to the production of the system map, the service blueprint, and several customer journey maps (Fig. 14.3) to reconstruct and clarify the current process.

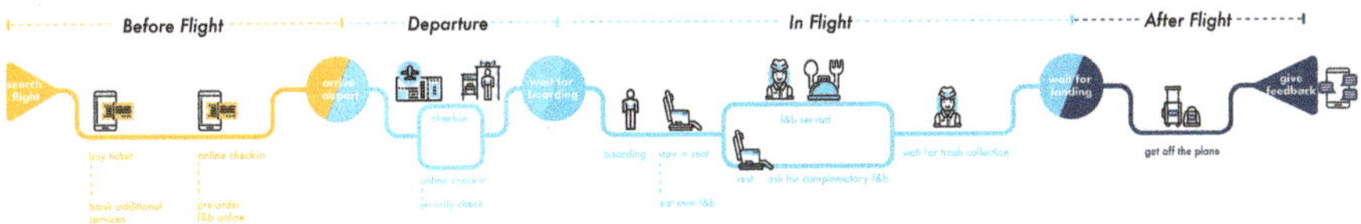


**Fig. 14.3** Customer journey map. (Image by HUANG Jiawen)

## *Paradoxes*

Next, the research team meets to reconstruct, based on the data collected and personal observations as a passenger, the current functioning of the system. The goal is to understand the reasons that may underlie an unsatisfactory dining experience and why it is difficult to solve it. The game at this stage is to formulate a series of paradoxes in the form of pairs of contradictory cause-effect statements. An impasse comes from the contradiction that does not allow movement in any direction. Eleven paradoxes have been formulated; one is given here as an example:

*Statement A: Because of the safety regulation from aviation organizations* (e.g., *FAA, CAAC, ICAO) and the airlines' service protocol, the meal must be served at a specific period of the flight to all passengers.*

*Statement B: Because passengers can have different physical and emotional states, especially during long-haul flights, a rigid food and beverage service often makes passengers dissatisfied.*

The two statements result in a stalemate that can be turned into a research question, namely, a design challenge to be addressed: *How can we guarantee the compliance to policies while increasing F&B variety and personalization?*

## *Context and Field*

Let us set aside the paradoxes and take up the data collected in the field during the archaeology phase. We try to find out why the catering company adopts certain practices, what factors determine the practices, and under what typical perspectives the problem of daily meal preparation is framed. We repeat this exercise for all the stakeholders in the first inner circle and try to discover mutual influences. At this stage, the movement starts by identifying established behavioral patterns in business practices and then discovering significant mutual influences, which can reinforce each other in a loop of resistance to change. We seek to include an investigation of Boeing's role, if not in generating, at least in perpetuating the problem of the limited on-board dining experience.

**Table 14.1** Rich picture table

| Stakeholder | Value/interest | Power | Practice | Frame |
|---|---|---|---|---|
| Passenger | Affordability<br>Good value for the money<br>Taste, experience<br>Lifestyle, diet<br>Safety | Buying or not | Buying ticket<br>Repeating the purchase<br>Leaving feedback<br>Sharing content on SN | Cost<br>Quality/cost ratio<br>Experience |
| Airline (cabin crew) | Cost control<br>Customer care<br>Safety<br>Efficiency and simplicity | Service timing and modality | Receiving meals from catering<br>Preparing and serving meals<br>Collecting garbage and used trays<br>Storing used props away<br>Handing over used trolleys to the catering<br>Selling meals upgrades | Customer experience<br>Safety<br>Efficiency and simplicity |
| Airline (F & B dept.) | Cost control<br>Customer experience<br>Safety | Decision about catering company<br>Menu setup<br>Retaining passengers' data<br>Daily checks on catering | Negotiating menu (including special menu) with catering company<br>Deciding tray setup<br>Daily checks on catering company<br>Menu for sale<br>Reviewing passengers' feedbacks | Cost control<br>Customer experience<br>Safety |
| Catering company | Cost control<br>Passengers' experience<br>Safety<br>Efficiency<br>Relationships with airline | Negotiation with airline about menu<br>Little or no competitors<br>Monopoly to provide F&B on airplanes | Negotiate menu with airline<br>Prepare/deliver/collect meal-related props<br>Buying low value products from third parties<br>Store meal-related props for airline<br>Sort garbage and pay for waste disposal<br>Wash tools<br>Check on hygiene | Cost control<br>Customer experience<br>Safety<br>Efficiency<br>Relationships with airline |

Stakeholders' perspectives on F&B

Later, the group takes a step back to broaden the perspective of secondary stakeholders: for example, it visits a packaging, container, and cutlery supplier and conducts online research on packaging recycling and disposal facilities.

The analysis is later consolidated into a rich context table (Table 14.1) highlighting the currency of exchange among the various stakeholders, for example, the power, interests, values, practices, and frames they bear in the problem arena. The goal is to sort everything and everyone into taxonomic boxes, except immediately afterward, to break the rigidity of the embeddings by identifying cross-cutting themes that ring common values and purposes across stakeholders, which can promisingly direct the development of future creative concepts.

## *Themes*

Since, in reality, no one operates in vitro but is part of a force field that simultaneously "is shaped by" and "sculpts" the practices of others, it is necessary to break the schematic constraint in which the work of individual stakeholders lives until now. The research team extrapolates specific themes that chain motifs, values, practices, and behaviors of different stakeholders into streams of meaning that constitute the common denominator underlying individual storylines. A theme is a red thread that weaves into the same fabric actors, seemingly driven by different agendas. Themes lie on a more abstract level than the immediately experiential and generalize the problem situation into a broader human dimension. Condensing needs on the one hand and constraints and limitations on the other is a process of invention and insights below the surface of appearance, a sort of enlightened exploration that opens to the creative dimension.

For example, the issue of *safety, hygiene, and public health* unites the interest and practice of airlines, catering companies, and passengers, but also, at a higher level, policymakers. The theme of *aspiration to reduce carbon footprint* unites, for different reasons, the airline – which recently in China must pay for waste incineration – a particular portion of the clients and policymakers.

One issue that affects cabin staff and passengers is the *rationalization of service and space* available during service. We have all experienced the discomfort of eating on small trays, constrained by the armrests and seatback in front. The inability to move along the aisle, for example, to reach the toilet during service, is another classic example of a constricting and frustrating situation.

## *Frames and Metaphors*

Although themes help reduce the field's complexity while not permanently losing the richness of the information gained in the field, they remain a yarn too tangled for our cognitive system and its natural limitations. A further round of non-lossy complexity reduction is necessary. We must understand how to segment and organize information on layers that can be turned on or off as appropriate. Frames have been the right tool for the purpose. A frame is a window through which one looks at the problem from a given point of view. Since it represents a particular perspective on the landscape, one must never forget that it represents a reduction of it for cognitive purposes, so it is a device for simplifying reality. However, it works very well if one must reduce the entanglement by setting aside and retaining parts and relations if they are not relevant to the problem under analysis.

The research process moves into the creative phase, where the first concepts are to be created and developed. As mentioned at the beginning, the heterogeneity of the research group could have functioned as a double-edged sword. To avoid getting lost in too abstract layers of reasoning, it was necessary to give the frames concreteness and tangibility to function as an inventive tool in the hands of a multidisciplinary team. Representing the frame through a commonly found metaphor avoids the danger of misunderstandings due to different experiences and specific disciplinary backgrounds failing to converse. A metaphor is based on the similarity between a source term and an end term. The evocative and communicative power of metaphor is the more effective the further apart the two terms are in reality ("Metaphor", 2021).

The first frame (Fig. 14.4) weaves together the themes of cost control, food health and safety, service efficiency, simplification, and sustainable lifestyles. It has been called *garbage and packaging reduction.* Looking at the innovation problem of the F&B experience under this frame means looking at the problem as if it were a problem of reducing food waste and garbage production. Everything else, it can safely be set aside for the time being. If one looks at the problem from this point of view, then consistently, everything must be done to reduce the carbon footprint of the cabin, simplify service, lighten spending, and gain health. Previous studies (Sweet et al., n.d.) show that each passenger on a long-haul flight produces an average of 0.52–1.81 kg of waste, including food waste, depending on the flight class. The use of disposable containers and cutlery should, in theory, lighten the workload of

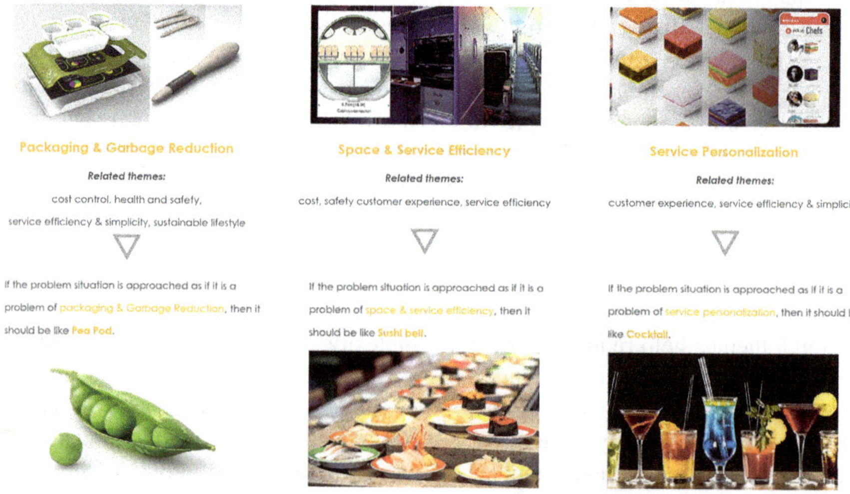


**Fig. 14.4** Frames (above) and metaphors (below). Frames have the function of looking at the problem from a specific point of view, reducing complexity. Their representation through a metaphorical term provides concreteness and facilitates discussion among researchers with different backgrounds. (Image by ZHANG Yiqin)

in-flight staff. However, it shifts the problem to other operators, such as catering company staff or waste disposal workers.

What metaphor could be used if one wanted to find solutions to the problem that could reduce food waste and trash? The project group found a good metaphor for containers in the *peapod*. If the containers were like a pea pod, less trash would be produced. Let us see why this metaphor works well:

- It is organic and protects the pea inside.
- It grows and degrades.
- Its material is homogeneous.
- It exactly fits the content and is modular.
- It goes from the plant to the table.
- It is traceable from origin to destination.
- Its bivalved shape can be a saucer and a lid.

The second frame (Fig. 14.4) weaves the themes of cost control, health and safety, service efficiency, passenger experience, and simplicity. The frame is named *space and service efficiency*. Looking at mealtime from this perspective, one might ask how to optimize the use of limited space. For example, how to optimize the table and tray or allow passengers to pass by the side of the trolley without getting in the way of the crew's service. To concretize this framework, the metaphor of the *sushi belt* was negotiated for the following reasons:

- It is stretched.
- It moves automatically along a track at a steady speed.
- It constitutes a food showcase.
- It caters to an interactive service.
- The price is color-coded.
- It is constantly resupplied.

The last framework (Fig. 14.4) that intertwines the themes of service simplicity with the passenger experience is called *service personalization*. In this perspective, the project team's efforts aspire to personalize menus based on individual preferences analyzed from previous purchases. The attitude of not finishing the meal can have many reasons, including the presence of disliked ingredients or even allergenic. Serving meals and refreshments that match individuals' preferences could certainly be a way to reduce waste. The metaphor that represents this frame is the *cocktail* because of the following:

- It offers endless combinations of multiple base liquor and ingredients.
- It has a strong identity and regional diversity.
- It is convivial and spectacular.
- The preparation requires a simple mixer.

At this point, the process has entered the more concrete phase of developing creative concepts. The frames and the metaphorical images associated with them functioned as a dialogue device that allowed anyone to reason around nebulous conceptual cores and dance around them for as long as necessary without losing the path and going astray. The image functioned as a powerful magnet that allowed each researcher to explore its associative power without getting lost in collateral dead ends. Everyone was able, despite their diversity, to play with the metaphorical image by producing different creative associations, which nevertheless remained within the established framework of coherence.

The metaphor inspired the creation of many touchpoints, even though different individuals in the design team developed them. The metaphorical device inspired the different mindsets precisely because it was found during a co-creation session and was not imposed unilaterally by someone in the group.

## *Futures*

At this point, the research process again presents a discontinuity: metaphorical imagery and experimentation with creative associations are left in the corner. With a quick zoom-out, the focus returns to the big picture. Thinking-forward exercises are held to build future scenarios in which different creative ideas live on.

Without considering the suggestions indicated by the frames and metaphors, the team returned to the research question, rephrasing it more precisely in the following way: *how could the on-board dining experience be improved without adding to the workload for the cabin crew and reducing the carbon footprint?*

Focusing on the in-flight dining experience means focusing on passenger and cabin staff encounters that, at present, are highly codified in a choreography common to any airline. However, is this the only possible way to describe the interactive pattern between the two actors? Two distinct driving forces, varying from a minimum to a maximum, were found to explore possible alternative scenarios: the level of passenger involvement and the collaboration between passengers and cabin staff. Crossing the two axes results in a two-by-two matrix that gives rise to four distinct scenarios (Fig. 14.5):

*DIY* Passengers are involved to the fullest in the production of the service, even to the point of doing it themselves without the cooperation of the flight attendants. For example, the passenger may go to the vending machine when allowed to take a few steps in the cabin.

*Being served* is the usual scenario in which the passenger is seated and served by the cabin crew. His involvement is minimal, limited to exchanging a few words and receiving or delivering the tray. The burden of service is almost totally on the shoulders of the staff.

*Co-managed* The passenger cooperates with the stewardess in that they can actively order snacks and drinks from a screen but are not directly involved in their preparation.

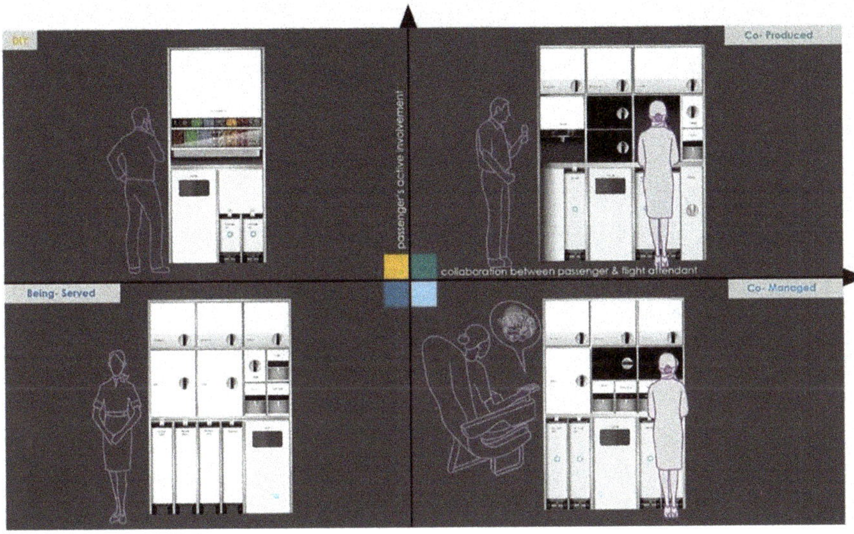


**Fig. 14.5** Scenarios. The four scenarios are generated by crossing the two axes representing the critical variables of the service encounter between passengers and cabin crew. (Image by ZHANG Yiqin)

*Co-produced* Passenger involvement and cooperation with cabin crew are at their maximum. In this scenario, which is reserved only for the few passengers in the topflight classes, they can independently go to the galley and prepare a cocktail or a snack, with or without the assistance of a steward.

## Result

Having discussed how the team constructed the four scenarios, this section addresses how frames and metaphors returned to the scene to populate each quadrant with distinctive products and services (Fig. 14.6).

Packaging and garbage reduction, space and service efficiency, and service personalization are organized from the out-ringer to the core. Each quarter-circle takes on the characteristics of the scenario in which it is located, reflecting the specific kind of passenger's involvement and collaboration with the cabin crew.

To illustrate the user scenarios with a vivid picture, a suggested cabin layout (for a Boeing 787 Dreamliner) is prepared. The co-production galleys serve the first and business class at the front of the aircraft. The low ratio of passengers per flight attendant can assure an interactive, quality service. The fittings for the co-managed service are in the front and middle of the cabin and provide complementary and paid F&B. The DIY equipment is at the back of the aircraft since it serves low-cost

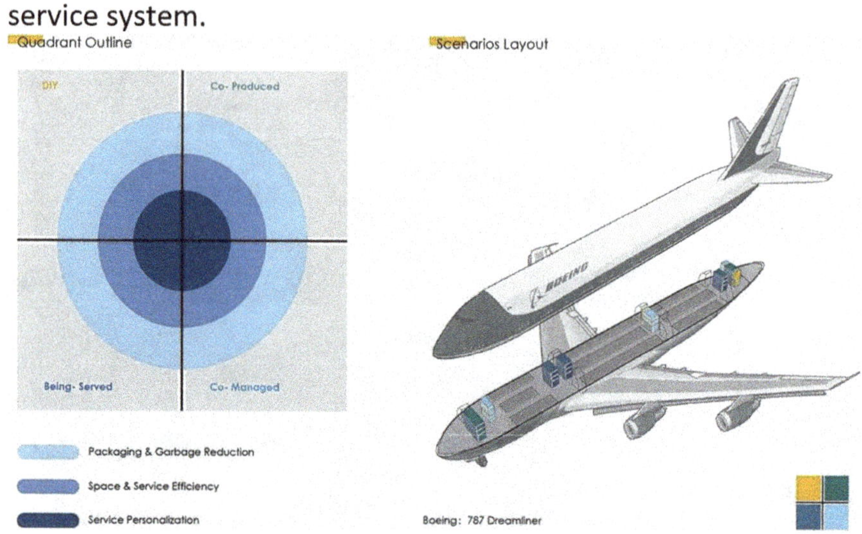


**Fig. 14.6** Scenarios X Frames. (Image by ZHANG Yiqin)

meals and beverages for high-frequency purchases. The being-served galley stays in the usual middle cabin section to bridge the current and the new service system.

*DIY Scenario* The installed vending machines have an intuitive graphic interface. The dispenser refills recyclable containers delivered by the robotic cart. For those who do not want to get up, it is possible to order snacks from the seat via the armrest screen. A robotic cart will distribute them to seated passengers and collect the used containers (Figs. 14.7 and 14.8).

The vending machine is the fundamental concept of the DIY scenario. The possibility of food tracking displayed on the machine interface guarantees freshness and quality and pushes the promotion of food and beverage. A redeemable purchase system encourages the container's reuse.

However, the robotic cart that can distribute food and beverage to sitting passengers and collect the used packaging and containers requires a high investment in the Radio-Frequency IDentification (RFID) and automotive technology. Passengers can place orders on their InFlight Entertainment (IFE) screen without standing and moving.

From the tangible touchpoints to the invisible backstage services, this service scenario is populated with reusable packaging, automatic cart, and on-board vending machines.

*Being served scenario* In this quadrant, the innovation is primarily about the product touchpoints. For instance, the reduced width of the distribution cart allows sideways passage (Figs. 14.9 and 14.10). The foldable tray and the modular system of reusable cups rationalize the limited space and reduce waste production.

**Fig. 14.7** DIY: high passenger's involvement × low collaboration. (Image by ZHANG Yiqin)

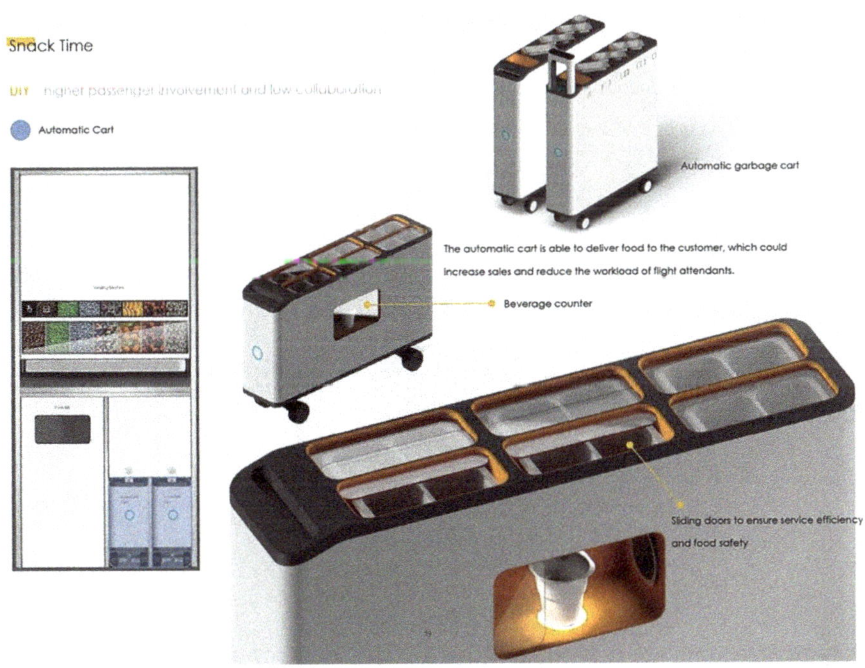


**Fig. 14.8** DIY: the vending machine and the robotic carts. (Image by ZHANG Yiqin)

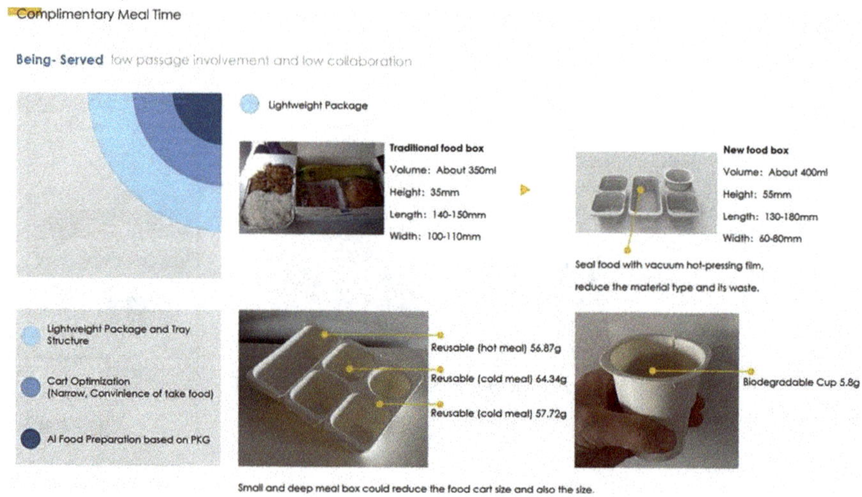


**Fig. 14.9** Being served: low passenger's involvement × low collaboration. (Image by ZHANG Yiqin)

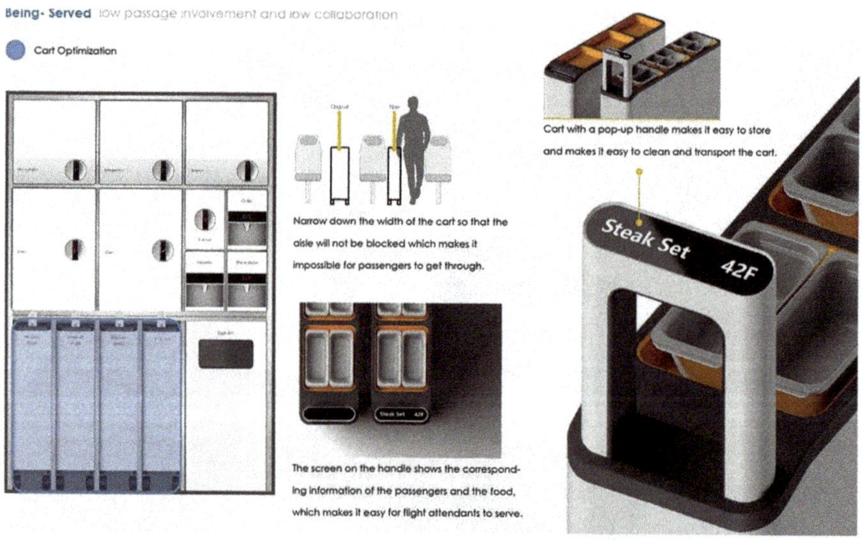


**Fig. 14.10** Being served: galley and narrow robotic cart. (Image by ZHANG Yiqin)

Regarding the service, it is possible to offer a personalized menu based on a predictive AI algorithm besides a meal-swapping opportunity if unwanted meals are still wrapped.

Being served, among all scenarios, is the one that involves just incremental innovation. A careful study led to optimizing single food portion weight versus

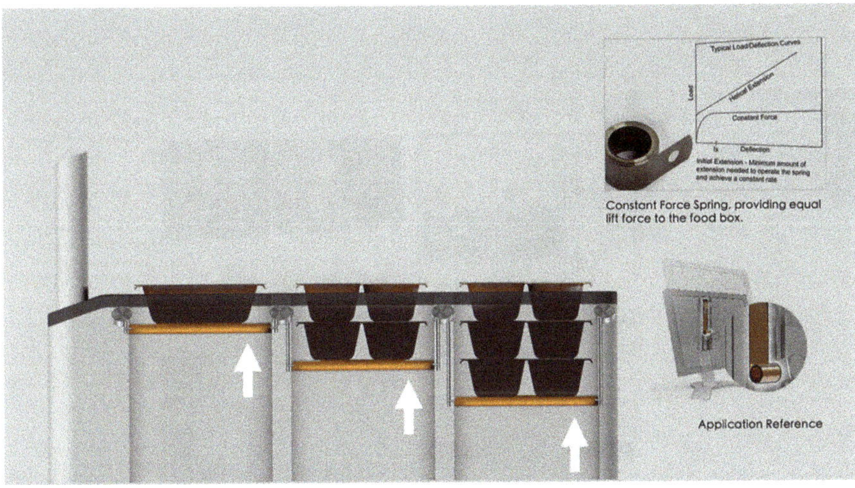


**Fig. 14.11** Automatic lifting springs. (Image by ZHANG Yiqin)

container size. Observing the proportion between an average human body and the alley's width, a new cart 30% narrower than the regular one is proposed to allow easy passage during the service.

The new cart design fits the peapod food and beverage packaging, while a foldable tray installed in the armrest can host the standardized package. Thanks to constant-force springs (Fig. 14.11) present in the cart mechanism, the flight attendant does not need to bend to extract the tray from the cart as in the current system, which increases the cart's user-friendliness.

From the tangible touchpoints to the invisible backstage services, this service scenario is populated with a lightweight package and tray structure, optimized cart, and AI food preparation based on PKG.

*Co-managed scenario* In this scenario, passengers can order food, drinks, and snacks on the spot. They can also decide the type of cooking; in fact, the smart galley is equipped with an air frying and vacuum machine. The food, though, is prepared expressly by the flight attendant.

This scenario is the most challenging in terms of service intensity.

Based on the in-flight stock, the entertainment system presents an "a la carte" menu. Modern cooking appliances make express food preparation possible: dry-frier, sous-vide machines, and traditional ovens with temperature and time controller displays on the handle (Figs. 14.12 and 14.13).

The optimized ergonomic galley design ensures that operation causes less fatigue to the flight attendant.

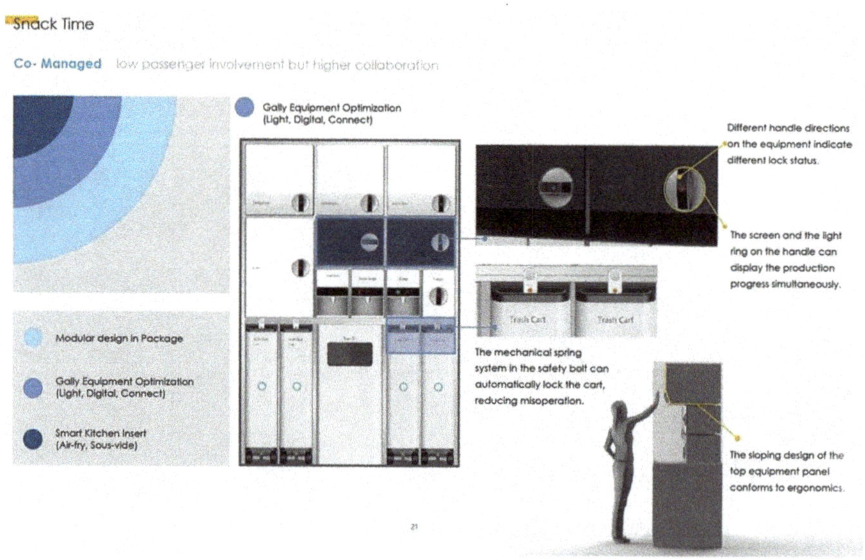


**Fig. 14.12** Co-managed: low passenger's involvement × high collaboration. (Image by ZHANG Yiqin)

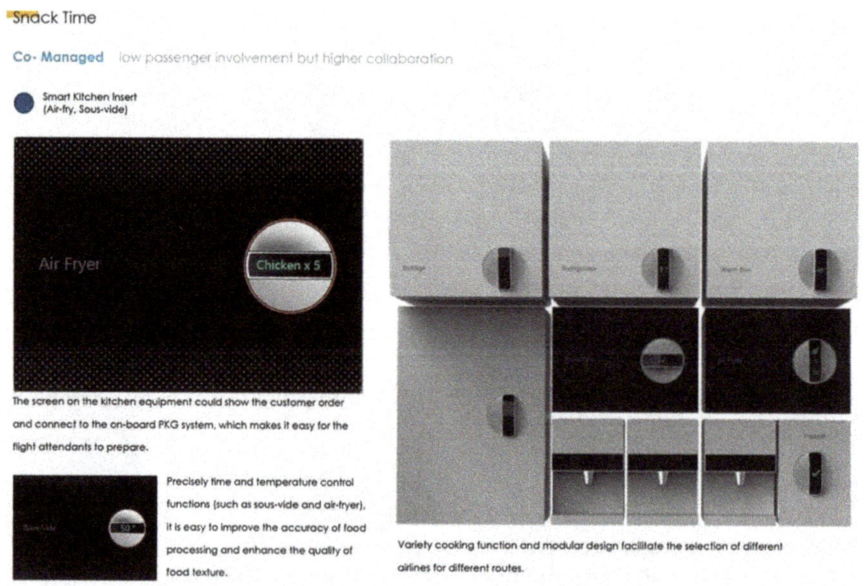


**Fig. 14.13** Co-managed: smart kitchen appliances. (Image by ZHANG Yiqin)

In this service scenario, the design concept from the tangible touchpoint to backstage services is the packaging's modular design, the smart kitchen inserts, and the galley equipment optimization (Fig. 14.14).

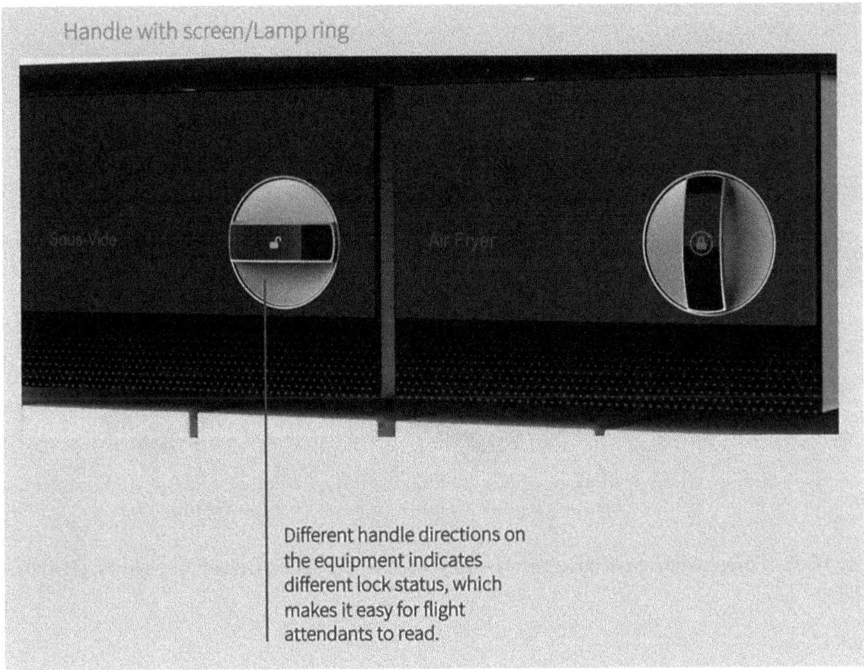


**Fig. 14.14** Co-managed: handle design detail. (Image by ZHANG Yiqin)

*Co-produced scenario* In this scenario, passenger involvement is at a maximum. The passenger can prepare snacks and cocktails with the optional assistance of the stewardess. This galley is provided in the top classes and is free to access.

The service encounter can be more interactive, thanks to the generous space allowance. The design concept of this service scenario focuses on providing ample temperate controlled storage cabinets, ranging from cooler to warmer cells.

Although the "a la carte menu" based on PKG and on-board food stock should accurately predict the passengers' preferences, an AI-food swapping algorithm is developed to meet at best individual needs.

The RFID technology robotic cart collects wasted packaging.

In this scenario, the design concepts from the tangible touchpoint to backstage services are the recycling package cart with side instruction, counter and storage with temperature control, AI food preparation, and swap system (Figs. 14.15 and 14.16).

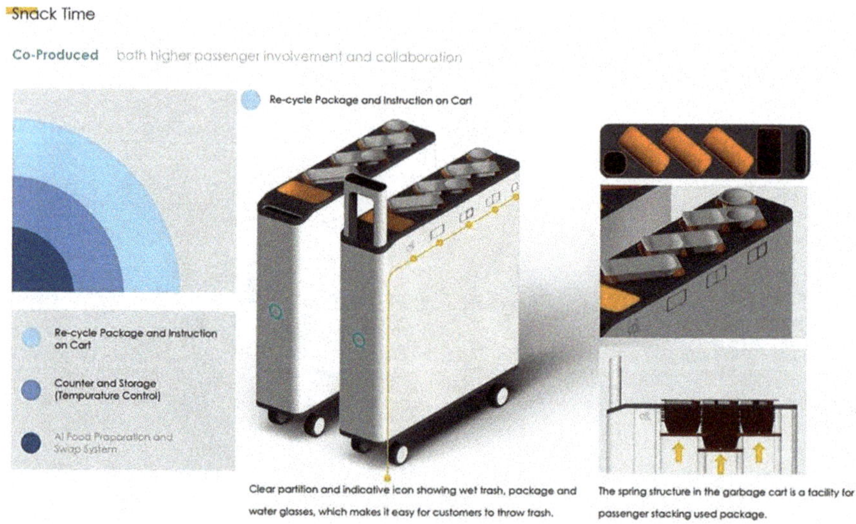


**Fig. 14.15** Co-produced: high passenger's involvement × high collaboration. (Image by ZHANG Yiqin)

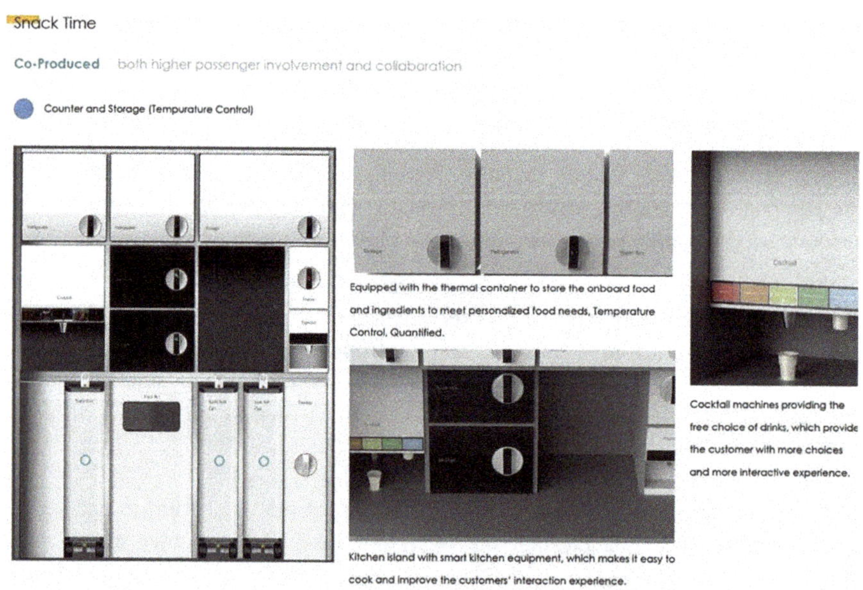


**Fig. 14.16** Co-produced: the cocktail galley for the highest class. (Image by ZHANG Yiqin)

## Prototyping and Evaluation

Changing the meaning of an experience is a very ambitious goal. Creating a new in-flight culinary experience combining taste and sustainability involved expanding the initial intention to redesign the galley.

The reasoning required a lengthy and complex design process instead of a simple one-off choice between a few components from a catalog. There was neither a dominant nor right design for guaranteeing an in-flight dining experience, so imagination was needed to share a definition. The design situation required *learning devices* such as prototypes and simulation techniques. Designing the social interactions between passengers and cabin crew within the confined space of the cabin was part of the design process. The research team needed to make the solution tangible, share the view, judge the answer, and decide which way to go.

A prototype is a product-service concept demonstration instrumental in collecting feedback on the user experience. With the joint effort of the design team, engineers, and manufacturer, a typical galley for each of the four service scenarios is built, equipped with functional touchpoints. Some modifications and design iterations occur during the prototyping process. The final prototypes are installed on the exhibition floor for visitors and peers to review.

A final evaluation workshop included all the initially interviewed participants. Boeing representatives participated online, while one author was absent due to pandemic restrictions. The event was held on the college's campus in Shanghai and lasted three and a half hours. The schedule started with a tour through a 1:1 scale galley prototype (Fig. 14.17), including demonstrations of the robotic cart, the foldable tray, the reusable food package, and the interactive prototype of the AI-aided user interface.

After that, the activity continued with the presentation of the product-service system articulated in the four scenarios, each containing three metaphor-inspired

**Fig. 14.17** Galleys' prototypes, scale 1:1. (Image by ZHANG Yiqin)

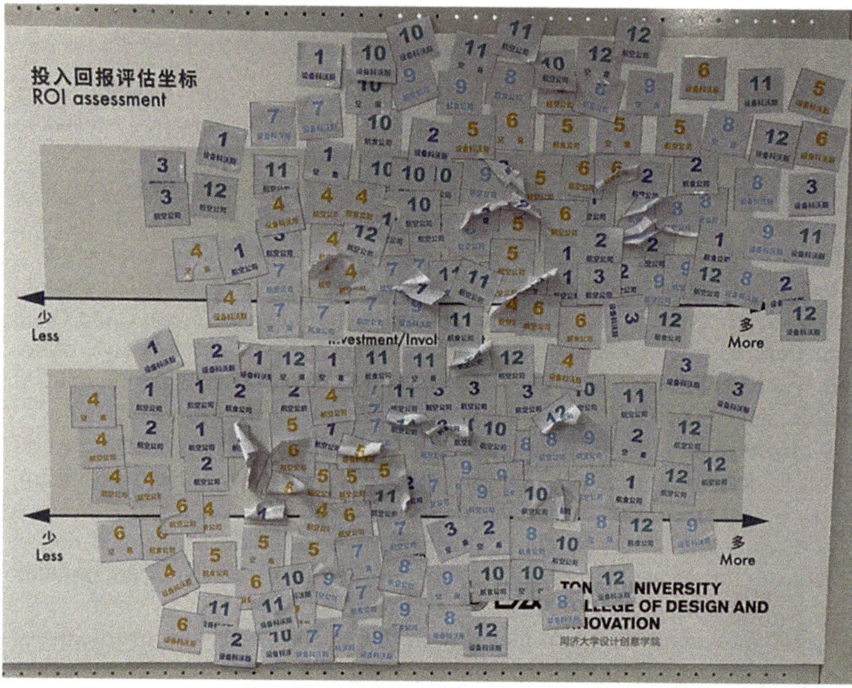


**Fig. 14.18** Common exercise to evaluate the return on investment of each of the 12 design ideas. (Image by ZHANG Yiqin)

solutions. A brief Q&A session cleared up any doubts, followed by a plenary activity that assessed the return on investment for each of the 12 design ideas (Fig. 14.18).

On the wall was a poster with two axes, "investment" (above) and "return" (below), which vary between "low" (left) and "high" (right). Each received a set of 12 stickers corresponding to the ideas. As seen from the final image, the stickers were distributed along the investment axis toward the right, reflecting the propensity to judge the ideas as expensive to implement. The result on the return axis was more evenly distributed. The exercise confirmed industrial participants' tendency to understand "return" based on market revenues rather than on a broader meaning that the future vision might suggest.

Finally, each participant received a table for the evaluation of the 12 ideas according to the criteria tabulated in Table 14.2.

The evaluation was individual, and the results were analyzed and shared soon after the collection. Designs 3, 7, 8, and 9 did not present problems. All other ideas raised concerns about safety, cost of maintenance, or doubts about return on investment. The most criticized was the robotic cart: the reduced width could compromise its stability during turbulence. Robotization was considered a pale substitute for the personal encounter between the passenger and cabin crew.

**Table 14.2** Evaluation sheet

Name, Company

| | | Safety | | | Operation | | | | Custom Loyalty | | Sustainability | | |
|---|---|---|---|---|---|---|---|---|---|---|---|---|---|
| Score | 0=Pass, 1=Poor, 3=Average, 5=Good | Equipment | Food | Personnel | Return of Investment | Efficiency | Maintenance | Quality | Options | Sales Potential | Energy | Trash | Feedback Collection |
| Score Weight | 1=Less Important, 3=Average, 5=Important | | | | | | | | | | | | |
| General Concept | | | | | | | | | | | | | |
| Pea Pod - Package System | | | | | | | | | | | | | |
| Sushi Belt - Automatic Cart, Smart Kitchen Inserts<br>Cocktail - AI Food and Beverage Preparation System | | | | | | | | | | | | | |
| Service Scenarios | Related Designs | | | | | | | | | | | | |
| Being Served | 1. Lightweight Lunch Box and Tray Structure | | | | | | | | | | | | |
| | 2. Cart Optimization (Width, Meal Pick Up) | | | | | | | | | | | | |
| | 3. Intelligent food preparation (PKG based) and reservation system | | | | | | | | | | | | |
| DIY | 4. Reusable package | | | | | | | | | | | | |
| | 5. Automatic Cart | | | | | | | | | | | | |
| | 6. On-board Vending Machine | | | | | | | | | | | | |
| Co-Manage | 7. Modular box design | | | | | | | | | | | | |
| | 8. Smart Kitchen insert (Air fry, Sous vide) | | | | | | | | | | | | |
| | 9. Galley Equipment Optimization | | | | | | | | | | | | |
| Co-Produce | 10. Meal box packaging recycling and instruction on the Cart | | | | | | | | | | | | |
| | 11. Counter and Storage (Temperature Control) | | | | | | | | | | | | |
| | 12. Intelligent recommendation catering system (PKG based) and Swap System | | | | | | | | | | | | |

## Technology

Technology responds to the development of the product-service system. Specifically, applied technology adapted practical methods for incremental design while emerging technology motivated intelligent inventions for radical innovation. The team implemented service and product touchpoints aligned with the four scenarios.

### *Applied Technologies*

Applied technologies include the following.

*IoT sensors and the network* Sensors that capture changing quantities are embedded into the product structure. For instance, the tray's frame integrates a sensitive weight sensor to measure the volume of leftovers in the food box, contributing to analyzing individual passenger food waste. Additionally, the armrest implants an RFID digital tag inside to pass the passenger's identity and order to the delivery cart.

*Tangible user interfaces (TUIs)* Presented product touchpoints, such as the vending machine, require information exchange with passengers for transparent service realization. The team developed specific TUIs for all touchpoints that involve user interactions. TUIs generally incorporate multitouch surfaces, multimodal input (speech, gestural, tactile), and multimedia output (sound, light, motion). Therefore, the implementation of TUIs advances both user experience and service efficiency. Herein we list out TUI-aided touchpoints regarding four scenarios.

- Scenario DIY: vending machine, robotic cart, and tray
- Scenario being served: galley and trolley
- Scenario co-managed: cooking machines in the galley
- Scenario co-produced: cocktail galley

*Robotic Simultaneous Localization and Mapping (SLAM)* Robotic carts present technical challenges. Their movement in the cabin aisle demands robust mechanical control to prevent accidents. Localization and navigation to a precise seat are even more complex tasks. However, the SLAM algorithm (Khairuddin et al., 2015) can overcome the difficulties by providing a semantic map of the aircraft cabin for the robot operating system (ROS) to navigate to each appointed passenger's seat. For instance, the robotic task of garbage recycling in the co-managed scenario requests the trash cart to move afront a passenger who has finished the served F&B. With the utilization of SLAM, the cart can spontaneously localize the current position and set off to destination coordinates enabled by the laser radar; still, the initial construction of the map requires human guidance.

## Emerging Technologies

Other emerging technologies include the following.

*Personalized knowledge graph (PKG)* Service personalization, as the third frame, proposed the requirement of customizing the served F&B. The team sought a comprehensive data model to organize and understand passengers' preferences around F&B. The so-called Knowledge Graph (KG) is a novel technology that employs graph structure to manage knowledge-based relationships between entities (Chen, 2020). Google initiated the Google KG for the optimization of search results. Well-known KG applications, such as Wikipedia, provide public knowledge services obtaining millions of data entities. However, PKG (Passenger Knowledge Graph) has revealed a superior knowledge representation of individual users rather than worldwide things (Safavi et al., n.d.). Accordingly, the team considered PKG an emerging practice to uplift radical innovation for managing domain-specific knowledge of passengers' F&B preferences. The PKG undergoes gradual and adaptive generation by collecting and clustering relational metadata on three key entities: passenger, F&B preference, and flight.

The relationships between the three entities answer three questions:

1. What F&B choices does the passenger prefer?
2. Which flight does the passenger expect to take?
3. What types of F&Bs will the flight offer?

Each entity connects a set of nodes and underlying secondary nodes, reserving, and updating the value or property of each node metadata. The PKG leveraged crowdsourcing technology to passively search for metadata from multiple sources of

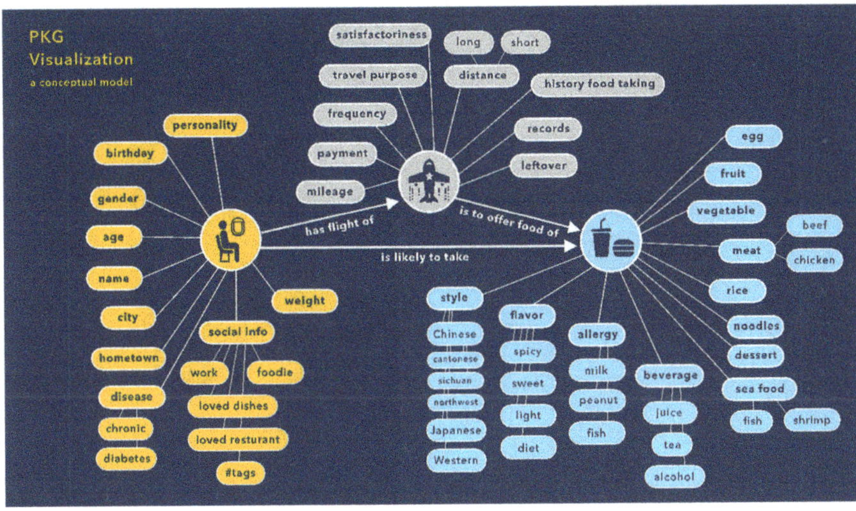


**Fig. 14.19** A conceptual data model of the PKG generation. (Image by Ke Ma)

social networks, airline platforms, and open knowledge graph databases following the privacy-preserving policy (Belguith et al., n.d.). Besides, built-in sensors (biosensors, RFID, leftover weight sensors) and TUIs performed guided learning of the required metadata from the passenger and flight attendants during the flight. Moreover, PKG collaboratively acquired the metadata from passengers by pushing questionnaires, games, and confirmation UI kits from the airline company's official customer service website or APP. The PKG runs like a neural network capable of forgetting false memories but refreshing with novel inputs. The generation of the PKG is an iteration of the entire customer journey and the subsequent coming journeys.

To better understand the structure of PKG regarding an individual passenger, the team visualized the PKG from a conceptual data model (Fig. 14.19) to a graphic representation. The circular network presents the structure, sources of data collection, and metadata of each node (Fig. 14.20). The outside circle presents nodes of three key entities, using color to encode the entity and bandwidth to encode the node's weight. Inside the band, texts, icons, and marks characterized the property of each node. The inside circle shows the composition of sources of data collection and the relationship referring to connected nodes. By switching to a specific customer journey, the visualization network shows different time modalities of the PKG.

The catering company decides the F&B offerings based on a group of passengers. Therefore, the team further developed the PKG matrix to synthesize the PKGs of selected passengers taking the same flight. PKG matrix can analyze PKG units and transform individual knowledge into clustering knowledge of F&B preferences. The circular annulus diagram visualizes the distribution feature of crucial subjects (Fig. 14.21). A slice of annulus represents a featured entity, and its inside ladder

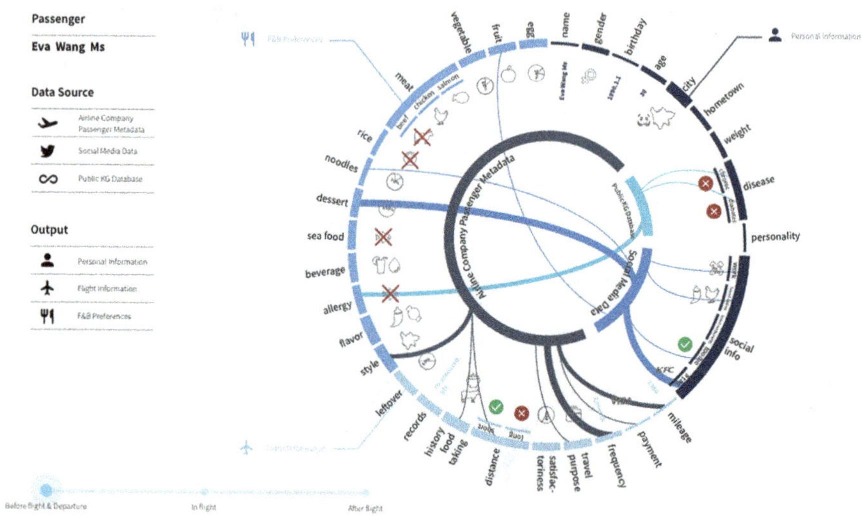


Fig. 14.20 Visualization of the PKG of one passenger. (Image by Ke Ma)

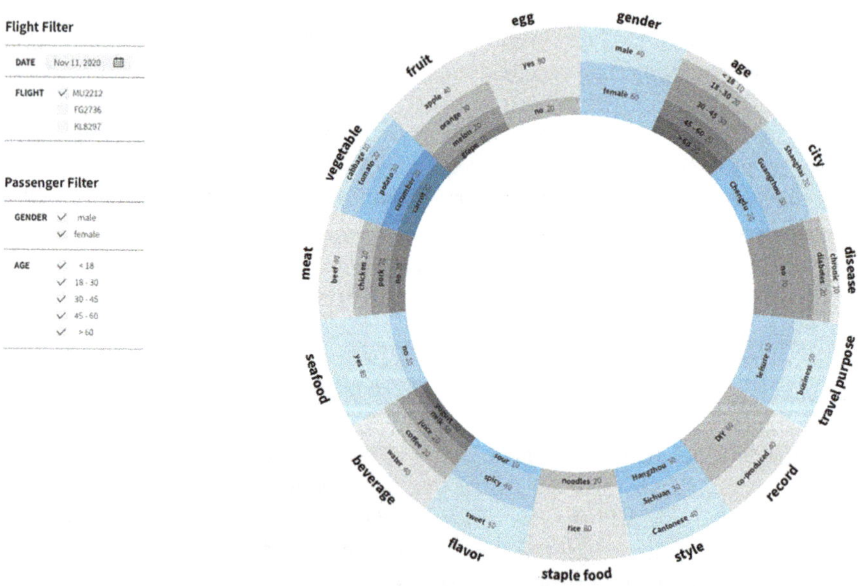


Fig. 14.21 Visualization of the PKG Matrix. (Image by Ke Ma)

reveals the quantitative ratio of different counterparts. The broader bars suggest the distributions preferred by the target group. PKG matrix allows the interactive and exploratory analysis by modifying the conditions, including passenger group, flight date, and flight number.

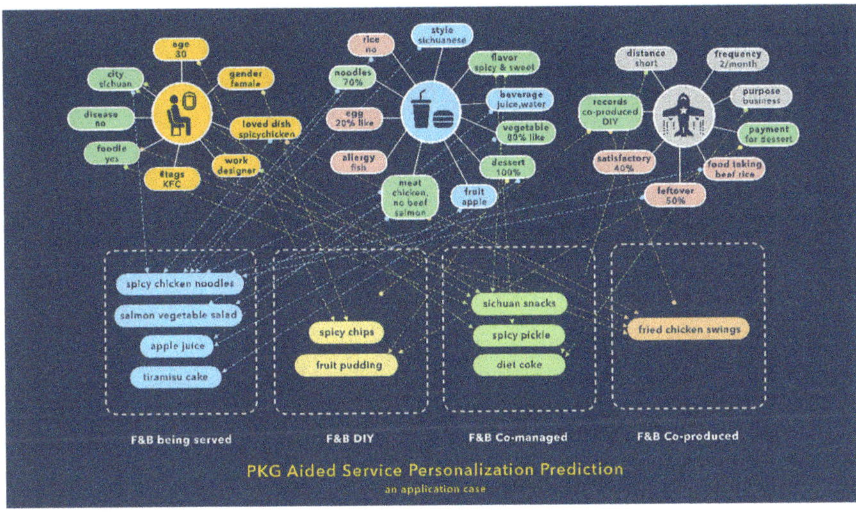


**Fig. 14.22** The conceptual model of PKG-aided service personalization prediction. (Image by Ke Ma)

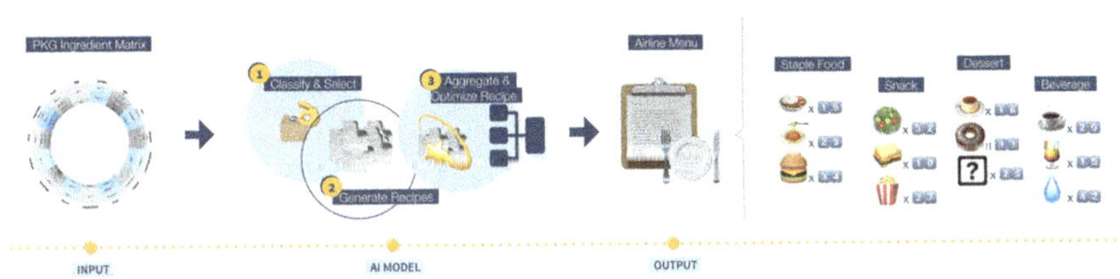


**Fig. 14.23** The workflow of the predictive AI model. (Image by Ke Ma)

*Predictive AI model* Customizing personalized meals demands ingredients and preferred combinations' prediction. The team structured the AI model using a Deep Neural Network (DNN) (Shrestha & Mahmood, 2019) to assist in design decision-making. Predictive computational inferences are obtained by extracting learning from the given input data (Fig. 14.22). DNN accepts the PKG matrix as input vectors; then, it predicts the airline menu with the maximum probability by selecting F&B ingredients and generating recipes. The result of airline menu includes a set of staple food, snack, dessert, and beverage for each passenger. The Cargo company receives the F&B offering suggestions for further decision-making and trade-off before the flight. Figure 14.23 illustrates the technical workflow of the AI prediction model.

*Analytical simulation model* The AI model can make mistakes due to incorrect basis or changing priorities. The team developed the analytical simulation model for service recovery in case of failure. The model can suggest the most reasonable meal swapping between passengers or from the cabin inventory. The simulation model

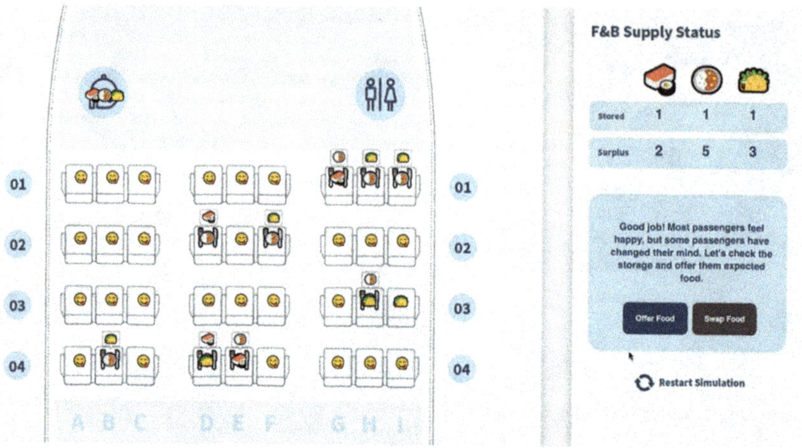


**Fig. 14.24** The user interface of the "food swap" simulation model. (Image by Ke Ma)

runs on the web or mobile app installed on the galley. Figure 14.24 shows the user interface. Readers can manipulate the functional prototyping of the user interface via the link (https://editor.p5js.org/hexinzhang/full/iipHs-bPm) by actions of "offer food" and then "swap food."

## Discussion and Conclusion

Table 14.3 summarizes the new product-service system benefits for the dining experience on-board Boeing airliners.

Ensuring a passenger-centered dining experience is difficult because of differences in nationality, age, lifestyles, purpose of travel, physical and psychological conditions, etc. On the other hand, the menu is designed months in advance by the airline and catering under stringent policies and regulations for safety. The service proceeds according to a formalized script to ensure that the front and back end are synchronized for quality and safety purposes. This standardized operating mode prevents the provision of personalized service and creates dissatisfaction that leads to food and cabin waste.

In our innovative PSS, the use of AI to manage the ingredients' inventory and always recipes' variations, assisted by the autonomous operation of robotic carts for the delivery, could cater to a service granularity pushed to the limit of individual customization.

If in the early days the application of AI was limited to product manufacturing and distribution to cut costs, now applications have expanded to the upstream design stage (Verganti et al., 2020). For example, menu design no longer requires a labor-intensive process involving many people from different organizations. It does not

**Table 14.3** Comparison of traditional and innovative F&B PSS

| Essential difference between the F&B system and our upgraded PSS | |
|---|---|
| Traditional | Innovative PSS |
| Product-service: served meal<br>Function: feed the passenger | Solution: nourish and hydrate according to one's needs and tastes + cutting waste + optimizing staff's workload and cabin space |
| The passenger buys a ticket and decides before the journey on a meal option from among those in the class of ticket | The passenger buys a ticket and can decide before the journey, at the airport, or during the flight on a meal option that is not necessarily conditional on the class of ticket |
| The passenger has a limited choice of menu. It may happen that the preferred meal is sold out | The menu is customized by analyzing previous purchases |
| The meal is prepared at least 24 hours before and reheated in the galley's microwaves. | The galley is equipped with various smart inserts including sous-vide cooker and air fryer. The meal can also be prepared in-flight, in higher ticket's classes |
| The passenger waits for the time of service at the seat. The meal is served by the flight attendants according to the booking, sometimes made long before. | Depending on the class of travel, besides being served, the passenger can order meals, drinks, and snacks at the airport (open catering platform), from the seat screen (co-management), from the vending machines (DYI), prepare snacks and drinks with the help of the cabin crew in the high class galley (co-produced). Food can be distributed at customized times thanks to robotic trolleys without increasing the staff's workload |
| The passenger receives a tray with food packed in aluminum foil and disposable tableware. Average cabin waste amounts to 1.46 kg per passenger, on long-haul flights. The airline must pay for the disposal | Passengers are rewarded for the tableware's reuse, which is collected by robotic carts to ensure proper recycling. Food and drinks can be also supplied from refillable dispensers, further reducing packaging<br>The airline retains the ownership of trays and tableware. The catering company is responsible for its storage and cleaning<br>All tableware is made of renewable or recycled materials. Tableware and tray are modular and adaptable, and the design makes the dining more stable |
| If the passenger does not consume the meal, it is thrown away even though it is sealed | The passenger who wants a different meal can put the still sealed, unwanted one on the virtual swap shelf, reducing the chance of food waste |

need to occur months in advance and remain "frozen" in time. The PKG constantly learns and refines details about individual passenger preferences and can make real-time suggestions, reducing human guessing. As a result, the necessary inventory of perishables is also reduced, decreasing economic risks and waste.

One might wonder what role remains for human creativity. We agree with Verganti et al. (2020): sense-making will remain the prerogative of humans. The fundamental decision to offer new meanings and value of an entirely new offering will not be outsourced to a machine any time soon. Human creativity will also write the machine learning loops that refine the infinite variations of the offering to approximate them more closely to personal needs, moving away from tedious standardization.

**Acknowledgments** We thank the colleagues Costantinos Terzidis, Hyejin Lee, Jianjun Zhang, and Zhenyuan Liu. We express gratitude for the work of the students Yiqin Zhang and Jiawen Huang, who also prepared the iconographic material for this chapter, Sean Charz, Zhengyue Chu, Yuyuan Jin, and Jiaqing Zhang. Cumulative thanks are due to the students of Industrial design (2018), PSSD (2019), and Open Design (2019) of Tongji University, College of Design and Innovation.

**Conflict of Interest Declaration** The author(s) declare that there are no conflicts of interest.

**Funding** This work was supported by Boeing (Research grant n. 2019-GT-328).

## References

Belguith, S., Kaaniche, N., Hammoudeh, M., & Dargahi, T. (n.d.). *PROUD: Verifiable privacy-preserving outsourced attribute based SignCryption supporting access policy update for cloud assisted IoT applications*, p. 24.

Chen, X. (2020). A review: Knowledge reasoning over knowledge graph. *Expert Systems with Applications, 21*, 112948.

Dorst, K. (2015). *Frame innovation: Create new thinking by design*. MIT Press.

Han, H., Lee, K.-S., Chua, B.-L., & Lee, S. (2020). Contribution of airline F&B to passenger loyalty enhancement in the full-service airline industry. *Journal of Travel & Tourism Marketing, 37*(3), 380–395. https://doi.org/10.1080/10548408.2020.1757563

Khairuddin, A. R., Talib, M. S., & Haron, H. (2015). Review on simultaneous localization and mapping (SLAM). In *2015 IEEE international conference on control system, computing and engineering (ICCSCE)* (pp. 85–90). https://doi.org/10.1109/ICCSCE.2015.7482163

Manzini, E. (2015). *Design, when everybody designs: An introduction to design for social innovation*. MIT Press.

Metaphor. (2021). In *Wikipedia*. https://en.wikipedia.org/w/index.php?title=Metaphor&oldid=1056376836

Norman, D., & Verganti, R. (2014). Incremental and radical innovation: Design research vs. technology and meaning change. *Design Issues, 30*, 78–96. https://doi.org/10.1162/DESI_a_00250

Safavi, T., Mottin, D., Belth, C., Muller, E., Faber, L., & Koutra, D. (n.d.). *Personalized knowledge graph summarization: From the cloud to your pocket*, p. 10.

Shrestha, A., & Mahmood, A. (2019). Review of deep learning algorithms and architectures. *IEEE Access, 7*, 53040–53065. https://doi.org/10.1109/ACCESS.2019.2912200

Sweet, N., Morris, E., Roberts, M., & Patterson, K. (n.d.). *IATA cabin waste handbook* (p. 101). WRAP. https://www.iata.org/contentassets/821b593dd8cd4f4aa33b63ab9e35368b/iata-cabin-waste-handbook%2D%2D-final-resized.pdf

Verganti, R. (2009). *Design-driven innovation: Changing the rules of competition by radically innovating what things mean*. Harvard Business Press.

Verganti, R. (2017). *Overcrowded: Designing meaningful products in a world awash with ideas*. MIT Press.

Verganti, R., Vendraminelli, L., & Iansiti, M. (2020). Innovation and design in the age of artificial intelligence. *Journal of Product Innovation Management, 37*(3), 212–227. https://doi.org/10.1111/jpim.12523

You, F., Bhamra, T., & Lilley, D. (2020). Why is airline food always dreadful? Analysis of factors influencing passengers' food wasting behaviour. *Sustainability, 12*(20), 8571. https://doi.org/10.3390/su12208571

# Chapter 15
# Toward Adaptive Homes Through Transdisciplinary Co-design: Case SmartLab

**Harri Hahkala, Toini Harra, and Leila Lintula**

## Background

### *Digitalization and Aging as a Complex Challenge*

Finland is the fastest aging society in Europe. It is also one of the leaders in implementing digital technologies in social and health care services. Older adults associate digitalization with both advantages and drawbacks. Successful adoption of digital technology facilitates everyday activities, whereas inability to utilize technologies results in feelings of alienation and being out-of-touch. The digital divide has been found to occur both between generations and between different socioeconomic groups of older adults (Pirhonen et al., 2020).

According to a digital survey commissioned by the Ministry of Finance, 76% of Finns have good or excellent digital skills. For example, in 2019, about 80% of Finns aged 16–89 used the Internet several times a day (Finnish Government, 2020). This information provides a good starting point for fully exploiting the digital potential of housing.

Digitalization and aging of population are examples of complex future-oriented challenges that concern concrete life. In this case, there are different solution options for each challenge, and each option has several different consequences. The consequences and effects of digital solutions therefore depend on who uses them, how they are used, and in which environments. Therefore, it is necessary to

H. Hahkala (✉) · T. Harra
Metropolia University of Applied Sciences, Helsinki, Finland
e-mail: harri.hahkala@metropolia.fi

L. Lintula
Emerita of Metropolia University of Applied Sciences, Espoo, Finland



U. Z. A. Hamid, M. Suoheimo (eds.), *Service Design for Emerging Technologies Product Development*, Springer Series in Design and Innovation 29,
https://doi.org/10.1007/978-3-031-29306-1_15

combine reality-based and imagination-based discussions and to make joint plans between different stakeholders to identify not only the issues related to use but also the possible consequences and effects.

Various solutions have been sought to safeguard well-being, health, and meaningful life of the aging population, with a wide range of consequences. For example, Finland aims to be recognized as a front-runner that develops and introduces new solutions enabled by digitalization and technological advances across administrative and sectoral boundaries (Finnish Government, 2019). The purpose of universities is to bridge, research, and develop such transdisciplinary research and innovation activities. In addition, regional programs and higher education strategies emphasize network-like collaboration for innovative solutions. National and international innovation collaboration between companies can increase companies' skills and enhance their competitiveness in the market. The importance of cross-border, transdisciplinary collaboration has thus been recognized, but it is not yet universally known what it requires in practice.

## *Individuality of Daily Life*

The importance of home and the living environment is changing with aging and changes in life conditions. Home is more than just walls, electricity, water, and heat. Security, well-being, digitalization, and ecology, among others, have become key factors in housing. The requirements for housing are increasingly influenced by individual's personal needs and daily activities, as well as their need for services. The needs of each resident are individual, which poses challenges for construction. For some, lifestyle and the coziness of the home, unobstructed movement, and the size of the rooms are important, while for others, the opportunity to integrate technology into living, ecological and sustainable construction, and health are important.

Individual housing solutions can improve the flexibility of home and adapt it to meet the changing needs of the user. The flexibility of home conversion is strongly linked to sustainable and responsible construction. It considers the ecological, social, and economic factors of both construction and building, without forgetting solutions related to safety, health, comfort, and versatility of home, from both a user and environmental perspective. The factors described above are strongly linked to construction life cycle thinking, where the environmental and cost impacts of a building are considered not only over the construction period but also over its entire life cycle.

Individual housing solutions support the needs of users and, at the same time, make meaningful life a reality. A meaningful life includes making every day experiences meaningful. The key factors are other people as well as health and situations where you can realize yourself the way you want. Achieving one's own values such as autonomy, community, the common good, and one's own goals is considered important.

Housing solutions are therefore central to a resident's well-being. When considering solutions, attention must be paid to the factors that enable the realization of things that are relevant to the resident. Among other things, safe and unhindered movement at home and outside the home, independence in daily activities, social contacts, and services available at home maintain well-being and functional capacity. In addition to these, various digital solutions related to building technology have become important factors in increasing well-being. They can be used, for example, to regulate the temperature and air conditioning of the resident's apartment to match their daily rhythm. This requires individual information about the resident's daily life so that the equipment can be adapted to the user's needs.

Digitalization and new technology can make the daily lives of people easier. However, digitalization involves threats that are primarily related to the loss of control of one's own life. Thus, the advantages and disadvantages of digitalization always go hand in hand, either encouraging or discouraging commitment to their use. When developing technological and digital solutions, care must be taken to ensure that people have an equal opportunity to benefit from their use, as well as to prevent alienation from their use (Pirhonen et al., 2020).

## *SmartLab as a Smart Adaptive Home*

The novelty of the Myllypuro campus of the Metropolia University of Applied Sciences is the SmartLab (Picture 15.1), a genuine apartment building designed for the development and testing of housing technology. The idea for the SmartLab came from partners Skanska and ABB, and it was built as a student work by the vocational school of Helsinki, Stadin AO.

SmartLab is a platform for learning and a simulation environment. As a research, development, and innovation environment, it can provide research, service design, and testing services. For example, we have utilized it as a platform for projects, such as HIPPA-Remote project (2021–2023). The objective is to help the housing production industry with the near future solutions research and testing. Skanska and ABB want to explore different ways to build a building automation system and the possibilities of connecting different third-party (SMEs) systems, sensors, and consumer products, such as refrigerators, robotic vacuum cleaners, or speakers (Picture 15.2). In a smart home, lights and heating adapt in a user-friendly, cost-effective, and resource-efficient way to the pace of the resident's life. Home will then learn to react to a variety of situations such as home appliance accidentally left on after going to bed, and the next day's hot weather. Smart and proactive home reduces energy consumption, costs, and everyday adjustments and allows the resident to live in one's own home through changing life situations. Metropolia students can familiarize themselves with the building technology and building automation technology of the near future, including the devices, and learn how they operate.

**Picture 15.1** Testing, development, and experimental environment of SmartLab (ABB)

**Picture 15.2** Smart adaptive home (ABB)

The focus of Metropolia's research activities is usability, adaptability, and predictability, that is, how different user groups are able to utilize new technology and what they want to use. In addition, the compatibility of devices is examined to ensure that they operate as part of a building automation (including Internet of Things, IoT) ecosystem and that no separate and overlapping systems exist.

Homes are digitizing at a rapid pace and different systems need to play seamlessly together. We obtain plenty of information about the different parameters of the system, but also about the user's operation. This generates data which machine learning and algorithms can utilize to optimize the operation of devices in advance. As for usability, the question is to what extent people are willing to accept anticipation. We are also particularly interested in what kind of technology people want and accept in general. The most important target is that digital solutions make everyday life easier and that they learn from their users just as much as the resident wants.

A common and substantial ecosystem opens up opportunities for more agile and co-creative development for all its users (Rakennusteollisuus, 2022). As an ecosystem, the SmartLab allows users from different target groups to perform user interface testing. It offers many opportunities for companies in the building services industry to bring their own products for testing and use the expertise of students and staff to help with testing. In addition, the SmartLab offers a unique testbed for the surrounding society with its challenges, to test ecological and social solutions which must now be developed so that the carbon-neutral goals shared by cities, municipalities, and companies can be achieved in the coming years (Kaihovaara et al., 2016).

## *Transdisciplinary Co-design*

In modern product development, more and more attention is paid to the user perspective. Future-oriented product development and research invites users to participate in its processes, as a result of which numerous new collaboration processes, approaches, models, and perspectives have already emerged. An example of this is the variety of concepts used for cooperation, such as co-creation, co-planning, co-production, and co-management. Hakkarainen et al. (2021) propose, based on their co-concept analysis, an integrative understanding of the concepts to facilitate collaborative modes and to enable the transformative aims of research processes.

Co-design is usually seen in front-end, before co-production and co-dissemination. It is a jointly conducted knowledge production process by academic and non-academic partners. In general, it refers to developing a research project and defining collectively interesting research questions, research agendas, and implementation plans. It refers to engagement at the beginning of a participatory project but may also extend beyond the initial phase (Hakkarainen et al., 2021; Moser, 2016).

Service design is a holistic approach that takes into account the multiple points of contact between the user and the product and/or the environment over time. It is suitable for the design of public services, where users interact with a variety of stakeholders, physical and short-term products, and environments over time to achieve their goals. A key part of service design is the direct involvement of users in the process (Parker et al., 2013).

From the perspective of service planning, therefore, all stages of a person's life, from birth to the end of life, are encountered in housing planning. As the design of good and sustainable housing meets not only the needs, wishes, and preferences of the inhabitants but also architecture, construction technology, digitalization, and the promotion of well-being, it is clear that planning is best carried out as a joint design involving all parties. In addition, one should understand how the individual needs of people living in different homes vary, not only from person to person, but also over time at different stages of life.

We describe the first phase of service design as a transdisciplinary co-design aiming to better understand the challenges and different needs of the users. According to Stock and Burton (2011), transdisciplinarity is the highest form of an integrated project, involving not only multiple disciplines, but also multiple non-academic participants in a manner that combines interdisciplinarity with participatory approaches. It maintains a clear emphasis on developing a holistic approach to problem-solving, involving stakeholders and scientists in a joint project. In particular, our interest in transdisciplinarity is based on the fact that, in addition to solving problems alone, it helps to synthesize new sets of data that can be used to better understand and address more complex system problems.

In our previous project, Hippa – Wellbeing and Better Service Housing through Digitalization (2018–2021), we conducted transdisciplinary co-design processes ($n = 25$) with entrepreneurs, testing services, and user trials ($n = 24$) for enterprises and testing in authentic service housing environments ($n = 20$). These actions reinforced the perception that companies do not have enough information about the operating environments or culture of service housing, nor the needs and preferences of users. These experiments showed that the challenges of smart service housing and living at home are complex problems that require a transdisciplinary approach combining social and health care sectors, information and communication technologies, housing and construction, and industry. Co-design requires that older residents and their relatives, care and rehabilitation staff, and other actors, such as cities, companies, and university R&D actors, are involved.

Hippa projects, through co-design of digital technology, aim to better understand the wishes, needs, and preferences of an aging person so that they can live easily and feel at home in their own home. In transdisciplinary co-design, construction and technology companies, design offices, university actors, and current and future residents bring out their own perspectives so that they can be considered in the design. The goal is to find a wide range of technical and digital solutions that promote sustainable and ecological living, health, and well-being. Co-design can be used to identify and create the best solutions together.

Despite its strength, there have been challenges in transdisciplinary collaboration that must be taken into account if it is to succeed. According to Pohl (2005), university researchers collaborate within their own discipline rather than outside it, and cooperation in a problem-based research environment is also usually a form of division of labor rather than real co-working. Therefore, successful transdisciplinary co-design requires careful preparation. Co-design begins with clarifying the focus of the activity and assessing the need for co-design. The next step is to identify the key stakeholders whose expertise is needed in designing user-friendly housing solutions. The key factors in defining stakeholders are the information needed to design housing solutions, and to establish to whom and what solutions are planned. Participants are expected to have new perspectives as well as new thinking, skills, and courage to even break traditional notions. At the same time, it is needed to value others' viewpoints and to be open to new insights. Therefore, it is necessary to agree on common working methods and methods for continuous evaluation to ensure that the desired result is achieved. Co-design requires a commitment to long-term work to achieve impressive results.

The purpose of co-design is to help the industry to produce multi-perspective and user-friendly innovations that are easier to commercialize and to take to proper use and, therefore, also easier to market to the customer. Statistics Finland defines technological innovation as

> A new or improved product or process whose technological characteristics are significantly different from before. Implemented technological product innovations are new products (product innovations) or processes in application (process innovations) that have been brought to market. The product or process is considered to be an innovation if it achieves specified advantages for the enterprise concerned; these need not be new from the point of view of other companies or the market. (Statistics Finland, 2022)

The starting point for innovation is multi-perspective, in which the know-how capital of partners is utilized and expanded and made available for all in a new way. Transdisciplinary collaboration provides multifaceted information to understand digitalization, housing needs, and everyday life, and to support the product and service development of companies. Multidimensionality might produce something unpredictable that is important in improving the usability of products and services.

## *Lessons Learned from Our Previous Projects*

Within Hippa and HIPPA-Remote projects, we have conducted several trials of transdisciplinary co-design processes including ICT, real estate, and building construction and well-being expertise of Metropolia UAS, as well as company representatives and building and service users.

In our previous Hippa project, co-design was an important step in transdisciplinary collaboration between higher education, industry, aging people, service housing staff, and NGOs. In co-design, a multifaceted group of participants innovated together, for example, new housing solutions and opportunities to use technology. The companies received concrete feedback from end users to support the development work. At the same time, users' and the other participants' understanding of the usability of products and services for different user groups expanded. The general and ultimate goal has been to make products and services more responsive to user needs.

Successful co-development requires good preparation. The TUTTUnet website (https://www.tuttunet.fi/en/front-page.html) supports the product and service development of companies by including a wide range of materials from the co-design, co-implementation, and co-reporting forms created in the first Hippa project. The openly accessible website helps product development designers, facilitators, participants, and others interested in the topic. It provides tools, canvases, contract, and information templates, and provides a wealth of information and insights in the form of process descriptions, articles, podcasts, and videos.

Co-design should be planned as a collaboration between the company and the facilitators, as this will best integrate product development issues from a company perspective and identify who should be involved. The co-design plan shall be drawn up on a form developed for that purpose, which may also describe the purpose of the event, the time and place of implementation, the persons to be invited, the procedure, the facilitators, and the necessary methods and tools.

Many other factors should also be taken into consideration in the implementation of co-design. They are not directly related to the issue to be developed, but they can strengthen the positive experience of the participants and their long-term commitment to the joint planning process (Harra & Lintula, 2020):

- Creating an inspiring atmosphere
- Increasing opportunities for people to participate
- Strengthen the experience of participation
- Balancing the interaction of quiet and loud participants
- Encourage all participants to bring up their expertise
- Sharing participants' experiences, knowledge, and skills for common use
- Intensify equal work and decision-making toward the goal

Facilitators from universities recruit participants and send them an invitation and a privacy statement. Our facilitation procedure includes discussion and co-creation of at least the following steps (Harra & Lintula, 2020) (Fig. 15.1).

The company's task is to present the product and the goal of the development work. The facilitators ensure that co-design proceeds according to plan and that everyone has the opportunity to participate in the work. After the implementation, facilitators fill in the reporting form, in which they collect the results of co-design and the next steps for further development. Further development may focus on obtaining additional information, further product development and testing, designing a user experiment, or strengthening commercialization and marketing.

## Research Question and Method

Based on our experience of co-creation and its results and impacts, we suggest more wider use of user-involved co-design in the field of real estate and building construction. However, wider consideration of the user perspective in both design

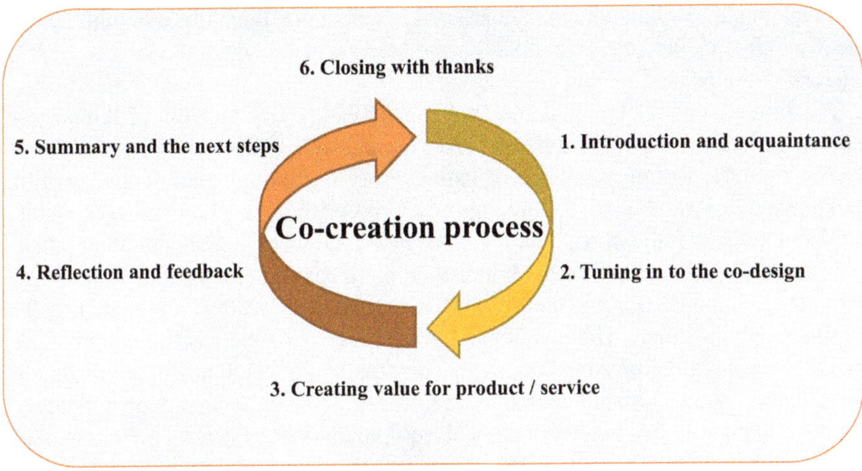


**Fig. 15.1** Co-creation process in HIPPA-Remote project

and building maintenance is still relatively small and operations have still been modeled a little.

The purpose of this chapter is to answer the question:

- How can the user perspective be strengthened in intelligent/smart new and renovation construction?

This is a case study that helps to understand more broadly the importance of a user perspective and multidisciplinary co-development in intelligent/smart new and renovation construction. A case study is well suited to situations where researchers want to re-conceive and better understand practical challenges (Salminen et al., 2006). In this case, it is a kind of "collective case study" that can also be called participatory research partnership in which the participants in the event produce and analyze the co-produced material and also draw conclusions from it together (Sipari et al., 2022).

The case study always examines the phenomenon in its real-life context and is more deeply interested in both using a variety of evidence and different data collection methods (Salminen et al., 2006).

## Case: Co-design for the SmartLab

The context of the case study was received from the HIPPA-Remote project, in which one aim is to co-create a storybank based on end users' everyday life and housing stories. The co-creation event was organized in the SmartLab, and was planned together with the developing experts from Skanska, ABB, Metropolia, and actors of HIPPA-Remote project. The project actors also co-created a facilitation plan that established that one of the Masters' students of Rehabilitation will act as one of the two facilitators. Participants were personally invited to the session by the network of the project.

The first part of background material for the case study consists of three real-life experience stories of older people and one real-time data on the day of a single woman over the age of 80, collected using sensor technology. The second part of background materials consists of the introductions of the SmartLab and visit into it.

The activity of the SmartLab was defined on the co-design planning form as follows: the SmartLab smart home is a platform for testing and developing technologies, interfaces, and practices that can be used to develop user-friendly solutions for housing. The solutions are intended to determine challenges of different stages of the housing life cycle. The broader goal is to utilize the results in the design of next-generation systems and housing, and to scale solutions in housing design in industry. So far, the technical equipment and systems environment of housing are separate from each other and communication between them is very limited. The SmartLab project enables the integration of devices and systems from different actors into one single ecosystem (Hahkala & Tähkävuori, 2022).

Together the four groups developed four imaginary timelines of one person's daily life. The timelines were presented and analyzed together as a first phase analysis. Finally, there was a reflective discussion on the implementation and significance of schedules, as well as on ideas for future actions and feedback on joint planning. The materials were further analyzed by researchers through content analysis, which resulted in the identification of key topics for discussion and their significance for further development (Table 15.1).

The SmartLab's two-hour co-design was held at Myllypuro Campus of Metropolia University of Applied Sciences in early March 2022. The session was attended by 14 people. They represented social and health care housing services, HIPPA-Remote project, Masters' students of Metropolia University of Applied Sciences, companies developing sensor technology, and the SmartLab research staff (Skanska, ABB, and Metropolia). Elderly people and care staff had been invited to the event from the Senior Center Myllypuro in Helsinki, but unfortunately, they had to cancel

**Table 15.1** Implementation of research in co-design context in the SmartLab

| Research implementation | |
|---|---|
| Research question | *How can the user perspective be strengthened in intelligent/smart new and renovation construction?* |
| Method | *Collective case study* |
| Introduction to the subject | *Presentations about the SmartLab*<br>*Visit to the SmartLab* |
| Data collection | *Imaginary timelines by four groups*<br>*Sharing timelines through "exhibition walk"* |
| Data analysis | *Participatory and reflective discussion*<br>*Qualitative content analysis by researchers* |
| Results of the research | *Identification of the key topics for further development* |

at the last minute due to necessary duties of the nursing staff. Participation in co-design was voluntary and the participants' personal data were not disclosed. By their signatures, the participants confirmed their participation in co-design on a voluntary basis, allowing the use of their data in the development work and photography at the session. Participants were not required to keep secret the ecosystem being developed at the SmartLab. The implementation of the co-design, led by facilitators, proceeded according to the next six phases.

**Phase 1. Introduction and Acquaintance**
The host of the session welcomed the participants and told them about the purpose of co-design. The aim was to create an image of an elderly resident's activities during the day to support the development of the SmartLab's automation technology. Participants were then asked to introduce themselves by recounting one thing they did at home at about the same time each day.

**Phase 2. Tuning into Co-design**
ABB, Skanska, and Metropolia presented their reasons, preferences, and benefits for which the SmartLab was built. These introductions serve as input for joint development. After that, the participants had a possibility to visit and feel the SmartLab smart home also inside. Already that round uncovered new ideas about the smart home that arose from the participants.

The main topic of discussion was how digitalization makes it possible to control and anticipate the activities of a resident. For example, by regulating lighting, it is possible to focus the interest and support the control of activities of a person with memory problems to maintain activity. With the help of voice control, it is possible to make it easier for a person with a physical disability to turn the lights on and off, to lower and raise the blinds, and to open and close the television. Also switching electrical devices to standby mode increases the safety of the resident.

**Phase 3. Creating Added Value for Product/Service**
Participants were then divided into four groups. The groups were given the task of discussing what things/activities an elderly person's day can consist of. The real senior stories were shared to groups to stimulate and enrich the discussion. Each group recorded the activities of the imaginary person on post-it notes and placed them on the timeline of the day (Picture 15.3). One group worked based on the activity data of a real elderly person. The other three groups filled the daily timeline mainly based on the activities of a fictive elderly resident.

The Exhibition walk was used as a method to share the results produced by the groups (Picture 15.4). It allowed the whole group to get acquainted with the course of all four days of the elderly. Each group presented their output, followed by a multidisciplinary discussion of the activities that had emerged and their implications for the life and well-being of the elderly (Picture 15.5). The issues raised in the discussion were recorded.

At the end of the Exhibition walk, the participants collected and shared what kind of ideas they had during the task and the results produced by each group.

In the discussion, participants highlighted the following issues:

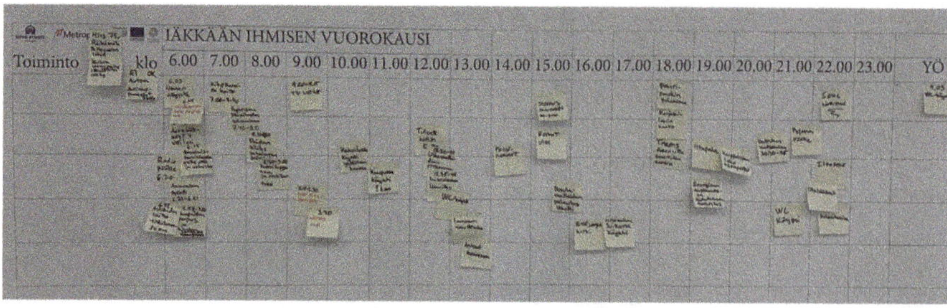


**Picture 15.3** The timeline of an elderly person's day (Harri Hahkala)

**Picture 15.4** The Exhibition walk was used as a nice way to take a peek at the differences of the days of elderly persons (Harri Hahkala)

- *Groups that built a timeline based on an imaginary client story reflected more on their own thoughts and wishes about the day of the elderly* than those who used a genuine client story.
- The lives of the elderly are very different. So, *it is difficult to define a standard day of the elderly.* Older people also appreciate a variety of things. Someone enjoys being alone doing nothing, another lives an active and social life with others, and for the third relatives are important. For this reason, *it will be very important to obtain additional information from the elderly themselves.*
- *Each of us should mentally anticipate our own aging and prepare for it in different ways.*
- *The rhythm of life reflects the state of well-being and changes in it indicate disorder.* The causes should be clarified and fixed as soon as possible.
- New construction should take better account of the life cycle of people and the changes in activity that take place during it. *Housing should be adaptive.* It *should adapt to the different needs of users, but also inspire residents to take care of things that maintain and improve well-being, such as mobility.*

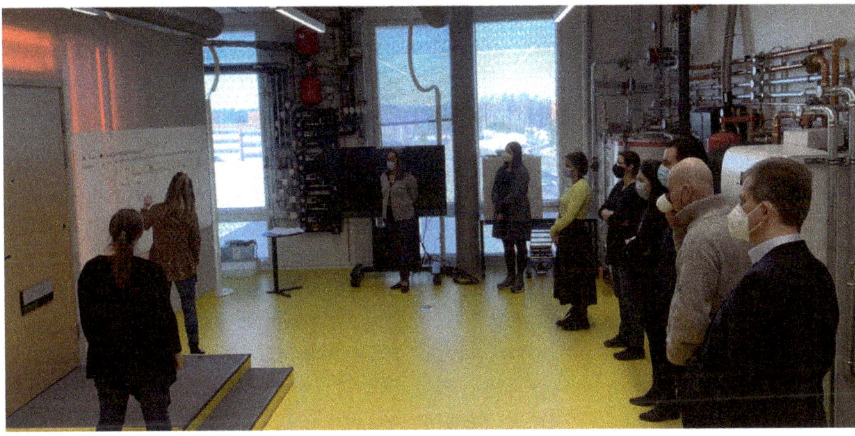

**Picture 15.5** Transdisciplinary discussion (Harri Hahkala)

- *With the help of automation technology,* for example, by changing and focusing light, *it is possible for the resident to continue doing things and maintain their own rhythm.*
- *Services should also be proactive rather than reactive.*
- Different housing-related data are already collected and combined in many ways and from different sources. However, *the biggest challenge is in utilizing that information.*

**Phase 4. Reflection and Feedback**

In the last phase, participants gave feedback about the factors of success, interests, and issues they would like to discuss more. The participants' responses provided valuable information for the implementation of interdisciplinary collaborative development and what we need to consider when we want to strengthen the user perspective in smart/smart new and renovation construction.

Participants' feedback emphasized the importance of elderly people's participation in development work. The reason for this is that what works for one is not always suitable for another. Therefore, it is important to find the best possible solutions in terms of suitability, significance, and impact of digital solutions as a factor that makes an individual's daily life easier, but at the same time, maintains activity and develops functional capacity.

Participants' knowledge and understanding of the possibilities of a smart home to support the daily routines of elderly people increased as a result of the discussions and a visit to the SmartLab smart home. The detailed results of the co-design feedback are in Appendix 1.

**Phases 5–6. Summary, the Next Steps, and Thanks**

At the end of the session, the SmartLab researchers discussed how they will utilize the information generated in co-design. The results promote the development of smart home functionality for different inhabitant groups, for example, the elderly, the disabled, singles, students, families, or others. How the smart home should support and activate their daily life to ensure longer, better, and stimulated life span in their own home?

More knowledge and information about the activities of daily life and living of the elderly is needed. Collaboration with HIPPA-Remote project and the elderly continues. The participation of elderly people in the development work and getting more information from them became the key follow-up measure in co-design. The HIPPA-Remote project will build a digital search service, HIPPA-SEARCH, related to the daily lives of the elderly. It is based on interviews with people over the age of 65. The purpose of the search service is to increase awareness and understanding of the lives, needs, and wishes of elderly people, for example, in building construction. In addition, it is possible to co-design with elderly people as described above, if necessary. These alternative ways of generating knowledge and using the results of this co-design session will be discussed with the SmartLab staff.

The HIPPA-Remote project has good connections with companies that provide housing services and care for the elderly, as well as with NGOs that promote the affairs of the elderly. It is inevitable that for ecosystem building, more companies with different solutions are needed. HIPPA-Remote constantly invites Finnish companies to look for new opportunities to further develop their digital solutions in transdisciplinary collaboration and connect suitable companies with the SmartLab. At the end of the event, the hosts thanked the participants for their active involvement in co-design.

## Discussion

The purpose of this chapter is to answer the question: How can the user perspective be strengthened in intelligent/smart new and renovation construction?

With the increase of freedom of choice and the number of wealthy consumers living in their own homes, the share of private services and digital solutions in housing can be expected to increase. For example, new digital solutions are needed to make everyday life easier, to promote exercise and meaningful activities, to support interaction and participation, and to increase security.

In the future, cooperation and data generated by home-installed sensors, resident biodata, and IoT and service robotics can also be crucial for ecologically sustainable construction. Easy-to-use or unobtrusive and accessible technology can be produced in a resource-wise and scalable way, provided that user orientation is taken into account in product development at an early stage in the development work.

Digital solutions can be used to create meaningful and safe spaces for residents to operate freely, move, and act. Security can be improved, and foresight information can be increased from a variety of data sources (environment, structures, actors) using digital data (open data, big data, sensor data) that is collected and analyzed. However, it requires coordination of information from different sources, which is still a big challenge at the moment.

The development of smart service housing and living at home requires transdisciplinary data collection. The integration of intelligent solutions and digital platforms for the utilization of large data sets in different fields, IoT, and location services are of interest to larger cities in provinces. Related smart solutions can, for example, save energy and promote the vibrancy, safety, and well-being of residents. Especially, in large-scale service and work environments, the use of data sets is of interest, as it allows different user groups to be considered in choices, encounters, and real-time interactions. Obtaining comprehensive data also guarantees actions based on timely and real information. Products and services that cross industry boundaries can be combined to each other on information platforms. However, this is still completely new in service housing and supported home living sector in Finland.

We usually talk about elderly as one large group, but this is not the reality. Instead of one large group, there are at least three groups that differ in their ability to function. These three groups are as follows: third phase, preparative phase, and fourth phase. According to Pirhonen et al. (2020), when the elderly retires in the West, they are still in good condition. Their financial situation is good, and their free time allows them to do things that matter to them. The preparative phase is a kind of transitional phase in which the elderly person begins to consider how long they are able to live in their current home. In the fourth phase, the elderly become dependent on services and assistance of others for a living.

In the third phase, it is important to design digital solutions that nurture and strengthen well being and health promotion so that the next phases would be delayed. In the preparative phase, it is important to have digital ecosystems at home. There is a clear need for active transdisciplinary collaboration with several companies looking for versatile ecosystems for both adaptive living and enabling independent living at home. In the fourth age, digitalization enables the world to enter the elderly's home when their own resources and ability to move around are reduced. It would be important to develop digital solutions so that aging people could live a safe and meaningful life with other people for the rest of their lives.

It is therefore time to break the development bubble and actively involve aging people in co-design. In order to be able to prepare for aging in different ways, information on user-friendly solutions must be disseminated to a wide target group. Likewise, proactivity should be emphasized in the design of services for the elderly people, rather than current reactivity. In addition, the "automation dilemma" must be taken into account in the development of automation technology. While automation can be used to activate, guide, control, and maintain an elderly person's daily rhythm, automatic reminders can work in the opposite direction, causing frustration and irritation, for example.

Co-development has brought for industry a wide range of competitive advantages, which strengthen the importance of service design in development work. Co-design, testing, and user trials have helped companies develop their products to better meet the needs of the elderly. The user experience is improved as the industry better understands customer needs, preferences, and the environments in which products and services are used. By participating in co-design session, companies also gained new knowledge about the regulation of social and health care in the EU. The co-design sessions also increased the awareness of participants of the digital solutions available and their usability. In addition, transdisciplinary co-development strengthened the compatibility of different companies and their ability to collaborate in ecosystems; companies gained marketing references through user trials; and small businesses discovered new partnership possibilities with other companies and the residential industry. All these activities require financial and human resources, and a desire to make changes to the solutions and to the operational culture and mode of the industry.

## Conclusion

The co-design session was carried out in collaboration with the SmartLab and HIPPA-Remote project. It proved to be an important phase of co-development. Its benefits are reflected in the speeches and comments of those involved in co-design as increased knowledge and understanding of the potential smart technology and digitalization. During co-design, participants were encouraged to look at future housing with curious, open, and critical eyes. With the help of questions, thoughts, and ideas from the participants, the SmartLab is able to develop the technology of a smart home in a direction that guides, activates, and anticipates the activities of a resident. The importance and significance of interdisciplinarity is reflected in the results of co-design and expansion of collaboration.

Co-design showed that older people need to be bravely involved in the development work. This is important because older people do and appreciate very different things in their lives. Second, in the collection and planning of the use of digital information, more attention should be paid to the identification of changes in the daily rhythm of the elderly. They indicate a change in functional capacity. Therefore, collecting information about the apartment or the activity of the resident alone is not enough. The information must also be used in proactive planning and organization of the services needed by the resident. Third, the focus of development work should be more strongly on supporting the resident's activities. Examples of this are lighting control, voice control, or large-scale interactive displays that can be used to guide and support resident activity. With such solutions, older people are helped to maintain their own daily rhythm and ability to function and to live longer in their own homes.

Companies in the construction industry are increasingly interested in the lives and wishes of different residents. Interdisciplinary cooperation creates added value. Through interdisciplinary cooperation, the construction industry obtains extensive information on the factors that affect the well-being of different residents. User data helps companies build long-lasting homes that meet the needs of different residents and are economically and ecologically sustainable. Digital solutions that anticipate residents' actions are also gradually renewing the design and production of welfare services.

Collaboration with the SmartLab and HIPPA-Remote project continues. Based on this SmartLab case, we can state that the SmartLab smart home provides a good development environment and co-design platform, where it is possible to utilize transdisciplinary expertise and coordinate building services systems, to study the importance of digital information from different devices and its benefits and usability as a component of well-being in new and renovation construction. Co-design, as it was co-created and conceptualized in the previous Hippa project, is an effective, participatory, and transdisciplinary method for co-development of adaptive homes in the residential industry. The SmartLab smart home is also suitable as an environment where future potential residents can test the functionality of the developed solutions. The benefits to residents increase as knowledge and information increase. "Homes of the future will learn from their inhabitants and the home will beat at the pace of the resident," as Hahkala and Tähkävuori (2022) aptly state.

**Acknowledgement** We would like to acknowledge and give thanks to HIPPA-Remote, HYTKE, FREE projects and SmartLab R&D&I platform which made this article possible. We also want to give our warmest thanks to Leila Lintula who even from her retirement gave us her knowledge and guidance to this article work.

# Appendix

## *Appendix 1: Feedback of Co-design Implementation*

Feedback of co-design

| Feedback questions | Answers |
|---|---|
| I think we succeeded today... | We noticed that the daily profiles of elderly people are different.<br>We recognized that we live in a bubble and we need elderly people to participate in the discussion.<br>We realized that automation involves dilemmas such as the balance of reminders and services, activity and passivity, or pleasant and irritating. |
| I thought it was interesting... | We got to see the SmartLab's smart home.<br>We got to know how to maintain and support everyday routines.<br>We got more information about the importance of small reforms.<br>We heard about the importance and necessity of physiological factors.<br>We discussed how to avoid the frustration of automatic reminders for things that are not important to themselves. |
| I would have liked to think more... | How automation technology could be used to activate a resident?<br>How could automation be developed in the direction that it would help to develop itself as well?<br>Where can you find services that promote self-development?<br>It is important to develop together with the elderly!<br>It is good to continue co-design on how to make the current home data and user data better to serve the future? |

## References

Finnish Government. (2019). *Inclusive and Competent Finland – a socially, economically and ecologically sustainable society*. Programme of Prime Minister Sanna Marin's Government 10 December 2019. Finnish Government 2019: 33. Helsinki. Available on: https://julkaisut.valtioneuvosto.fi/handle/10024/161935

Finnish Government. (2020). *Digitaitokartoitus – Digitaalinen kysely. Digitaitokartoituksen sähköisesti kerätyn aineiston yhteenveto*. Ministry of Finance & Digital and Population Information Agency. 30.8.2020. Accessible on Finnish: https://valtioneuvosto.fi/-/10623/suomalaisten-digitaidot-ovat-suurimmaksi-osaksi-hyvalla-tasolla-digitaitokartoitus-nosti-esiin-myos-huolenaiheita

Hahkala, H., & Tähkävuori, A.-S. (2022). *SmartLab – Älykotien kehitysalusta*. Accessible in Finnish: https://www.metropolia.fi/fi/tutkimus-kehitys-ja-innovaatiot/yhteistyoalustat/smart-lab

Hakkarainen, V., Mäkinen-Rostedt, K., Horcea-Milcu, A., D'Amato, A., Jämsä, J., & Soini, K. (2021). Transdisciplinary research in natural resources management: Towards an integrative and transformative use of co-concepts. *Sustainable Development, 2021*, 309–325. https://doi.org/10.1002/sd.2276. https://onlinelibrary.wiley.com/doi/10.1002/sd.2276

Harra, T., & Lintula, L. (2020). *Developing digital products for service housing through digitalization*. Accessible on: https://www.tuttunet.fi/en/information-and-insights/developing-digital-products-for-service-housing-through-digitalisation.html

HIPPA-Remote. Remote services for product developers to promote the housing of the elderly. (2021–2023). *Project web page accessible in Finnish*: https://www.metropolia.fi/fi/hippa-remote

Hippa-Wellbeing and Better Service Housing through Digitalization. (2018–2021). *Webpage of the project*: https://hippa.metropolia.fi/en/

Kaihovaara, A., Härmälä, V., & Salminen, V. (2016). *Mitä innovaatioekosysteemit ovat ja miten niitä voi kehittää? Valtioneuvoston kanslia*. http://tietokayttoon.fi/documents/1927382/2116852/Mit%C3%A4+innovaatioekosysteemit+ovat+ja+miten+niit%C3%A4+voi+kehitt%C3%A4%C3%A4/feecb2aa-d56e-441d-aa2e-15f5bd18d59b?version=1.0

Moser, S. C. (2016). Can science on transformation transform science? Lessons from co-design. *Current Opinion in Environmental Sustainability, 20*, 106–115. https://doi.org/10.1016/j.cosust.2016.10.007

Parker, C. J., May, A., Mitchell, V., & Burrows, A. (2013). Capturing volunteered information for inclusive service design: Potential benefits and challenges. *The Design Journal, 16*(2), 197–218. https://doi.org/10.2752/175630613X13584367984947. To link to this article: https://doi.org/10.2752/175630613X13584367984947

Pirhonen, J., Lolich, L., Tuominen, K., Jolanki, O., & Timonen, V. (2020). "These devices have not been made for older people's needs" – Older adults' perceptions of digital technologies in Finland and Ireland. *Technology in Society, 2020*, 62. ISSN 0160-791X. https://doi.org/10.1016/j.techsoc.2020.101287

Pohl, C. (2005). Transdisciplinary collaboration in environmental research. *Futures, 37*(10), 1159–1178. ISSN 0016-3287. https://doi.org/10.1016/j.futures.2005.02.009

Rakennusteollisuus. (2022). *Kestävä rakentaminen on vastuullista rakentamista*. Visited webpage on 30th of May 2022. https://www.rakennusteollisuus.fi/Tietoa-alasta/Ilmasto-ymparisto-ja-energia/Kestava-rakentaminen/

Salminen, A.-L., Harra, T., & Lautamo, T. (2006). Conducting case study research in occupational therapy. *Australian Occupational Therapy Journal, 2006*(53), 3–8. https://doi.org/10.1111/j.1440-1630.2006.00540.x. https://onlinelibrary.wiley.com/doi/abs/10.1111/j.1440-1630.2006.00540.x

Sipari, S., Vänskä, N., Lehtonen, K., Helenius, S., Väisänen, S., & Harra, T. (2022). *Participatory research Partnership in Rehabilitation* (Oiva Series 55). https://www.theseus.fi/bitstream/handle/10024/755011/2022%20Oiva%2055%20Participatory%20Research%20Partnership.pdf?sequence=2&isAllowed=y

Statistics Finland. (2022). https://www.stat.fi/meta/kas/tekn_innovaatio_en.html

Stock, P., & Burton, R. J. F. (2011). Defining terms for integrated (multi-inter-trans-disciplinary). *Sustainability, 3*(8), 1090–1113. https://doi.org/10.3390/su3081090

# Chapter 16
# Co-designing Person-Centered eHealth Information Services: The Case of Maternal Health Care in Kenya

**Danny R. Nyatuka and Retha de la Harpe**

## Terms and Key Concepts

| | |
|---|---|
| **Person-Centered Care** | An approach to care which consciously adopts the perspectives of individuals, carers, families, and communities as participants in, and beneficiaries of, trusted health systems that are organized around the comprehensive needs of people rather than individual diseases, and respects social preferences. It requires that people have the education and support they need to make decisions and participate in their own care, and is organized around the health needs and expectations of people rather than diseases. |
| **Service Design Research** | A creative, human-centered, and iterative approach to service innovation. It is an integrative research approach which is context specific, holistic, and multidisciplinary toward innovating and/or improving current services so as to make them more useful, usable, and desirable for consumers, and more efficient and effective for organizations. |
| **Underserved Context** | Communities and/or those groups of people who live primarily in informal settlements, that is, within inner city (slums) and/or rural areas, and who are of lower socio-economic status (SES), hence population groups who frequently experience difficulties due to resource constraints that include lack of, or inadequate, health care coverage, particularly in developing countries. |

D. R. Nyatuka
Strathmore University, Nairobi, Kenya
e-mail: dnyatuka@strathmore.edu

R. de la Harpe (✉)
Cape Peninsula University of Technology, Cape Town, South Africa
e-mail: DeLaHarpeR@cput.ac.za



U. Z. A. Hamid, M. Suoheimo (eds.), *Service Design for Emerging Technologies Product Development*, Springer Series in Design and Innovation 29,
https://doi.org/10.1007/978-3-031-29306-1_16

## Introduction

Health is regarded as one of the fundamental requirements of human capital, and therefore, a critical ingredient for socio-economic development (Bloom et al., 2004; Radojicic et al., 2020). Accordingly, a healthier workforce is one that is energetic, both physically and mentally sound, and thus more productive. More so, people earn better wages as they are less likely to be absent from work for illness-related reasons. This, therefore, justifies the overall objective of Sustainable Development Goal 3 (SDG 3), that is, to "ensure healthy lives and promote well-being for people at all ages" (Nabyonga-Orem, 2017:2). On a global scale, the World Health Organization (WHO) advocates for integrated people-centered health care service model(s), which should be implemented according to the local context to establish holistic trusted health systems organized around the comprehensive needs of the people as well as respect for social preferences (World Health Organization, 2016). It is against this background SGD 9c initially emphasized the need to increase access to information and communication technologies (ICTs) to provide universal and affordable access to the Internet in underserved contexts such as sub-Sahara Africa by 2020 (Clark & Wu, 2016), where the majority of the people are devoid of basic health care hence inequitable distribution of health care (Kaur, 2012).

Particularly, there exists inequalities in the quality of mother and child health (MCH) which needs to be improved. For instance, the complexity of barriers to accessing maternal care services of formal providers indicates that the current narrow-focused interventions are unlikely to improve the situation (Ochieng & Odhiambo, 2019). Existing health information systems (HIS) are deficient in terms of meeting individuals' real needs, and hence they cannot deliver expected results holistically (Boore, 2018; Korpela et al., 2013; Nabyonga-Orem, 2017; Ziraba et al., 2009). Furthermore, the voices of stakeholders, particularly the patients and their families, including those of women have not been heard regarding their health care choices and preferences (Townsend et al., 2013:752). This situation may be attributed to the complex nature of health care ecosystem characterized by overlapping strategic issues and interests among different stakeholders (Manya et al., 2012, 2016), and thus hindering efforts toward the growth and maturity of eHealth in most developing countries. This prompts the need to adopt integrated large-scale ICT innovations to address such challenges, which are prevalent in underserved environments, for example, unequal geographical access to health care, fragmented health systems, inadequate health personnel, and weak supply chain systems (Patrício et al., 2020; Thobias & Kiwanuka, 2018; Lade et al., 2014).

The problem this study seeks to address is that current eHealth interventions have a deficiency to deliver meaningful health and wellbeing services in underserved contexts as they lack holistic contextualization (Boore, 2018; Korpela et al., 2013). The study proposes adoption of principles of service design (SD) Strategy which is both human-centered and context specific to address the research problem, with a focus in establishing maternal health information needs in Kenya, and then design an innovative eHealth service from a person-centered perspective.

## Health Care Context, Opportunities, Challenges, and Trends

Health care is defined as "the maintenance or improvement of health via the diagnosis, treatment, and prevention of disease, illness, injury, and other physical and mental impairments in human beings" (Saravanan et al., 2016:9). It refers to the various service(s) offered by different categories of health practitioners, including physicians, dentists, midwives, nurses, and pharmacists among others, cutting across different levels of health care, that is, primary, secondary, and tertiary care (Saravanan et al., 2016). The ultimate goal of modern health care is to have a continuum of care for the patient(s), integrated on all aforementioned levels. It is against this background that WHO has recently introduced the *people-centered health care policy*, which emphasizes that individuals, families, and communities be treated as active participants in their health care (World Health Organization, 2015, 2016). Specifically, the global policy's key objectives are geared toward achieving the following: (i) to engage and empower people and communities through active participation; (ii) to re-orient and re-design care models; (iii) to coordinate health care services; and (iv) to create a supportive health care environment. The overarching aim is to enhance access to ensure a holistic patient experience through the medical care and treatment process (Grisot & Lindroth, 2019).

This prompts the need to develop globalized non-medical therapeutic strategies and interventions including the use of ICTs to promote health and disease prevention (World Health Organization, 2015, 2016). In pursuit of this agenda, governments across the globe have embarked on efforts to incorporate people-centered health care model(s) into their national health strategies, Kenya included (Ministry of Health, 2016).

Unfortunately, despite the effort, the implementation of the policy is far from the reality in many nations of the world, Kenya included (Malmberg et al., 2019; Ogden et al., 2017). Basically, the nature of relationship(s) between the three thematic categories is that the challenges experienced in health care do influence emerging trends, whereas opportunities offered by design provide solutions to these challenges (Tsekleves & Cooper, 2017).

Literature reveals that design has the potential to contribute immensely to future health care. Consequently, noticeable is that the key agenda going forward and requiring immediate attention is that of preventative health care (Tsekleves & Cooper, 2017). However, design in health care is not without challenges, which could be either more prevalent, recurrent, or demanding (Groeneveld et al., 2018). According to the authors, the key challenges may be categorized into three, namely, practice, managerial, and generic challenges. Specifically, challenges occurring under practice include the processes of conducting fieldwork, involvement of end users, and handling of sensitive situations. Challenges under the managerial cluster, on the other hand, include management of relations, building understanding among project stakeholders, and communicating the value of design to health care. The last cluster, generic challenges, includes time, budget limitations, and building rapport (Groeneveld et al., 2018). Noticeable is that health care is moving from a disease-based approach toward preventive care models, which seek to support and empower patients in various ways. As a result, exciting opportunities have proven to emerge from design such as service prototyping, co-design, evidence-based design, digital design, and holistic design, which contribute a positive experience toward population health and wellbeing among stakeholders, and especially to care recipients and providers (Groeneveld et al., 2018).

## The Nature of SDR and Its Potential for Service Innovation

In the last decade, the art of design has shifted from designing stand-alone products toward building holistic service systems that involve products, interactions, experiences, and services in different areas (Pamedytyte & Akoglu, 2019). Design for services is, therefore, viewed as activities involving the conception, planning, and building of an interactive system(s) or service architecture, with the primary goal to provide resources to support interactions between the users and providers (De Mello Freire & Damazio, 2018). The service concept, therefore, plays a key role in SDR and development, not only as a core component of the design process but also as a means to "concretize" the nature of the service (Goldstein et al., 2002). According to the authors, the service concept does not only define the *how* and the *what* of SDR, but it also ensures integration between the how and the what of SDR while mediating between the needs of both the customer and those of an organization's strategic intent.

It was decided that a design approach is appropriate, and although the proposed design solution would include technology components, these would form part of the proposed service. Service design formed the basis of the design process, but being part of a research study, a service design research (SDR) strategy was selected. In SDR, the focus is on developing theories, using and adapting service design methods and models, and developing new conceptual frameworks for the service based on users' experience for enabling new forms of value co-creation with stakeholder input (Joly et al., 2019; Pekkala & Ylirisku, 2017). SDR is an evolving multidisciplinary field that draws from both design and service research (Teixeira et al., 2019). It is a human-centered, collaborative holistic approach that focuses on real-life problems (Joly et al., 2019; Pekkala & Ylirisku, 2017). When technology components are also included, SDR also draws from human-computer interaction design and information systems research (Teixeira et al., 2019).

In this study, therefore, the research problem is investigated through the lens of the SDR approach as the chosen research strategy to allow for the complexities in health care and the intangible nature of a service. As a demonstration case, an information service in maternal health care in the context of Kenya allowed for the investigation to consider context and situational aspects. The results are not intended to be generalized. In the context of SDR, *service* refers to a series of interactions between providers and users of service, and hence there is participatory design (Thobias & Kiwanuka, 2018; Salgado et al., 2017). Services are intangible and dependent on each individual who use them, and therefore, SDR approach attempts to understand the characteristics of the desired service by engaging the designer(s) and stakeholders as well as the service users during the design and process. This helps to facilitate the understanding of end users' needs and their experiences with a particular service using various design methods, tools, and techniques such as the co-design method, and hence there is active participation (Huang et al., 2013; Hurley et al., 2018; Moritz, 2005).

SDR, therefore, has the capacity to generate new ideas through an iterative process while the designer explores users' experience toward creating new innovative services (solutions), which are then translated into prototypes, tested and implemented as a new service (Trischler et al., 2018). In SDR, knowledge is gained from the designer's intellectual reflection on the design activities and the phenomena representing the service situation being investigated, similar to research through design (Sun, 2020). Generally, much effort is needed to address challenges associated with the design and maintenance of public services and operations which may involve a complex network of actors – hence interdependent – who co-create their value by facilitating many-to-many interactions among themselves (Thorbjørnsen et al., 2009).

In the context of this empirical study, the focus was *maternal health and wellbeing information needs*, whereby the principles of design are applied to investigate and determine the nature of maternal health information needs of stakeholders in an underserved context such as Kenya and then create a new technology-enabled information service to meet those needs. Therefore, the study adopted SDR being a multidisciplinary, human-centered, collaborative, and iterative approach that promotes service innovation (Blomkvist et al., 2010; Ostrom et al., 2015; Sangiorgi et al., 2017), and thus a catalyst for organizational transformation (Kurtmollaiev et al., 2018).

## The Role of SDR in Creating eHealth Innovations

eHealth refers to "the cost-effective and secure use of ICTs in support of health-related operations including service delivery, surveillance, literature, education and literacy, knowledge and research" (Alamri, 2012:448). This is achieved through clinical data storage and exchange of medical information by means of interpersonal communication, computer-based systems support, patient-provider communication, health education/literacy, health service, community health, and telemedicine (Boore, 2018). In the context of this study, therefore, eHealth is regarded as the application of a wide variety of medical information and applications toward promoting health care delivery as well as the distribution and sharing of health information. On a global scale, implementation of eHealth (ICT innovations) by government(s) is one of the strategies to enhance ongoing global health care reforms (Boore, 2018) toward improving population health outcomes. Adoption of eHealth innovations is regarded as a pointer toward economic development particularly in

Kenya, whereby the government has put several initiatives in place to build capacity for eHealth infrastructure, in conjunction with other health care stakeholders as part of ongoing health care reforms (Boore, 2018). Among such initiatives include adoption of the National Health Policy, the District Health Information System 2 (DHIS 2) as the National Health Information System (NHIS), the National eHealth Strategy, and the National eHealth Policy among other strategies (Ministry of Health, 2016; Karuri et al., 2014). Proper utilization of ICT can, therefore, greatly empower individuals, organizations, and communities to take control over their own health and wellbeing (Miah, 2017).

However, this can only be achieved through stakeholder engagement and participation in their care decisions, freedom for individuals to access their health data/information, and also utilizing patient-reported outcomes or feedback to guarantee person-centered care (Girgis et al., 2017). Unfortunately, the process of implementing eHealth innovations is not without challenges. Among the common factors which hinder realization of the set objectives for eHealth in Kenya include social, economic, technical, and political issues (Ministry of Health, 2016). Industry standards, policies, legislation, infrastructure, and ICT capacity building are among other factors which curtail the growth and maturity of eHealth (Boore, 2018; Korpela et al., 2013). As a result, eHealth still continues to remain at infancy stage (Ministry of Health, 2016). Being a catalyst for service innovation and organizational transformation, SDR strategy can be leveraged to co-create eHealth innovations to overcome the aforementioned challenges.

To leverage the stakeholders' potential for value co-creation, networked services such as a national EHR service should be designed in such a way to support interdependent interactions rather than individually (Edvardsson et al., 2014), whereby SDR approach can indeed play a pivotal role. It is on that basis research recommends the involvement of patients through SDR to initiate cultural change (Malmberg et al., 2019) since SDR promotes active participation in the design process, and therefore, it has the capacity to induce change in people's health behavior (Wetter-Edman et al., 2018). This is because SDR approach can engage both the patient(s) and their family perspective(s) in the SD and development process whereby people's voices are heard (Malmberg et al., 2019). Therefore, to demonstrate the value of design in a health care environment, a human-centered design process with a bottom-up approach such as SDR approach could help design researchers to challenge the hierarchical situations to create innovative eHealth solutions, and particularly for patients in vulnerable situations (Pamedytyte & Akoglu, 2019). This study is, therefore, aimed at combining opportunities presented by both SDR and ICT to complement efforts to provide person-centered health and wellbeing health care service innovations (Malmberg et al., 2019). SDR is mainly context focused, holistic, participatory, and multidisciplinary in innovating current services toward a desired situation, that is, to make them more meaningful in the context of use (Nyatuka & de la Harpe, 2021). It engages and probes the relevant stakeholders to gain deeper insight from an individual's lived experience to co-create services, hence enabling participatory health care design (Tindall et al., 2021). Also, the methodology employed involves key phases, that is, field and design research phases, respectively. Similarly, the design and implementation process of the proposed solution involves stages, which include the service, information, and user layers, respectively, hence a blueprint for eHealth innovation process. This work, therefore, defines the service concept and describes how it can be applied to enhance technology-enabled health care service design process.

## Research Methods

The purpose of this study was to establish maternal health information needs and then design technology-enabled integrated person-centered health and wellbeing information services to support delivery of maternal health care in an underserved context. The study adopted both interpretive and pragmatic (co-design oriented SDR) approaches, and hence this section provides a detailed discussion of the research design. This includes research strategies, target population and sampling, data collection methods and tools, and data analysis techniques employed.

### *Research Design*

Since the study used thematic analysis (TA) method, inductive coding approach was adopted, whereby raw textual data was read and interpreted to develop concepts and themes (Elliott, 2018).

### *Initial Field Research Phase*

At this stage, the study, the concept(s) of stakeholder mapping (Johnson et al., 2008), and principles of interpretive field research in information systems (IS) (Klein & Myers, 1999) were applied. Potential stakeholders of maternal health were identified to facilitate contextual inquiry/analysis of the research context whereby individuals, groups, organizations, and community contexts were explored and described. In this study, principles of interpretive field research in IS were applied to facilitate constant movement of understanding from the whole to the part and back to the whole, toward the harmony of all details with the whole as the criterion of correct understanding (Klein & Myers, 1999). Various qualitative methods were used to gather primary data (Amoako & Rivett, 2015) and among key aspects assessed were: stakeholders' information needs, technical and economic feasibility capabilities around proposed solution(s), and the potential effect of the project on the beneficiaries. Potential stakeholders were identified to be engaged as participants.

**Design Research Phase**

This phase involved the application of SDR strategy to explore the health context and maternity journey to create the desired situation (Salgado et al., 2017; Moritz, 2005). This involved identifying and understanding various touch points throughout pregnancy period and childbirth as well as women's experiences, viewpoints, behaviors, and preferences (Salgado et al., 2017) related to information services and needs using co-design-oriented SDR approach. Service prototyping was used whereby various design methods, tools, and techniques were used to design the desired service (Salgado et al., 2017; Moritz, 2005). The aim was to assist the researcher to discover and define concepts relating to design characteristics and components of the proposed cloud-based PHR service for maternal health care within the research context.

## *Target Study Population and Sampling*

The target population for the study was drawn from potential stakeholders of maternal health care in public health facilities located within informal settlements of Nairobi City County (NCC), Kenya. Participants were selected from across six public community health facilities located in the slum areas where majority of the population live. The study specifically targeted the public health sector where majority of the urban-poor population receive health care services since the government pays part of the cost to offer health care services at subsidized rate, unlike in the private sector where only the privileged can afford. The selected facilities include Mathare North H/C, Embakasi Health Centre, Eastleigh H/C, Kibera South H/C, Mutuini Sub-County Hospital, and Pumwani Maternity Hospital, from which the representative sample was obtained. The sampling procedure was conducted in two levels: (i) firstly, identifying potential stakeholders of maternal health care in public health sector; and (ii) secondly, identifying the relevant stakeholders who could be selected to participate in the research by virtue of their role in maternal health care. Snowball sampling technique was used to select potential participants through the help of Facility-In-Charge respondents. The same procedure was repeated with emerging stakeholder groups until a sufficient sample was generated. The relevant stakeholders were identified and their roles, relationships, influence, and interactions in relation to the delivery of maternal health care services in existence. For purposes of this study, personal judgment was also applied to select individuals whose competence level was deemed sufficient, with an awareness of operations in their working environment, and who were willing to share information freely (Kumar, 2011; Guest et al., 2014). A variety of criteria were applied, that is, age, level of education, profession/career, job position/rank, and corresponding experience (Marshall, 1996), which were considered as important variables. Saturation point was achieved with 35 out of 47 participants. The key stakeholder groups (respondents) identified were: pregnant women (mothers), community health workers (CHWs), health practitioners (nurse/midwives, clinical officers, and doctors), health facility administrators, and ICT experts.

Specifically, pregnant women participants helped the researcher understand the information needs of mothers in maternal health care while the health practitioners helped with understanding of information needs of the caregivers. The facility administrators assisted the researcher to understand policy-related information needs, whereas the CHWs helped with understanding of the dynamics of community health services and information needs. ICT experts helped to understand design characteristics and components for technology-enabled PHR service in an underserved context.

## *Data Collection Methods, Tools, and Sources*

In this study, the researcher collected a variety of data, including qualitative data, design data, and secondary data, using a variety of methods as shown in Fig. 16.1 developed by the researcher.

Principles of field research in IS (Klein & Myers, 1999) were applied to evaluate and guide the data gathering process, which involved abstraction and generalization (generating initial data codes, categories and data themes); contextualization (adoption of SDR strategy to facilitate contextual inquiry); interaction (engaging the

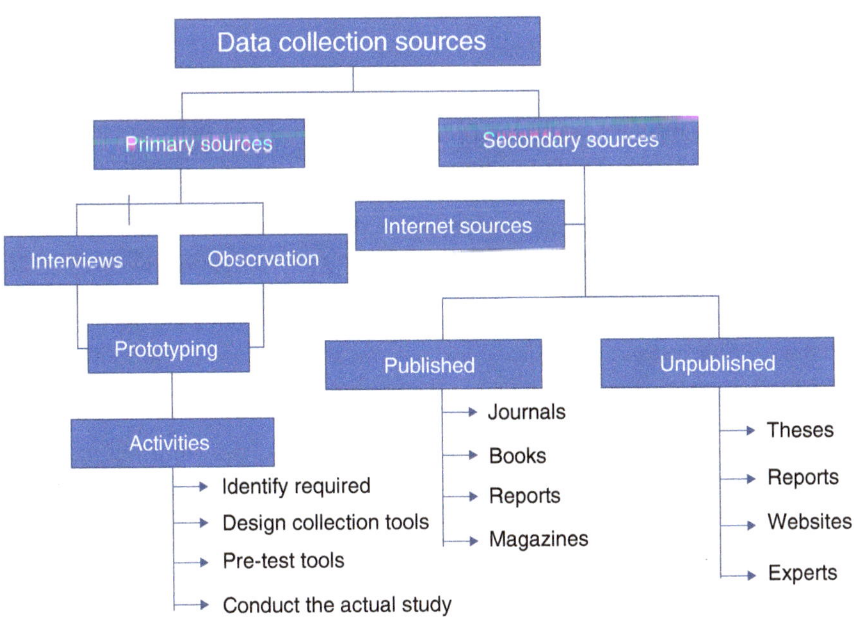


**Fig. 16.1** Data collection methods and sources

subjects and by the researcher to make them part of the DR process using personal, in-depth interviews); co-design (prototyping activities); and dialogical reasoning (listening to the data collected to identify and report any surprises between the research findings and theoretical pre-conceptions). Memorandums and information sheets were used to record the responses prior to actual data analysis (coding) by summarizing the main points derived from individual interview responses. This enabled the researcher to reflect, comprehend, and gain insight into the research phenomenon (Boore, 2018). The use of various methods enables the researcher to interact with participants and to listen to their opinions, views, experiences, and propositions to analyze and unearth various issues regarding maternal health information services in the research context. Semi-structure in-depth interviews, focus group interviews, participant observation, and co-design methods (service prototyping) were used (Salgado et al., 2017) to facilitate interaction and discussions between the researcher and research participants to discover relevant design requirements for the proposed new service.

Multiple methods of inquiry were employed by the researcher including empathizing with users, defining and redefining the problem, ideation, and service prototyping. New designs relevant to the needs within the constraints of the research context were created whereby the aim was to achieve the desired outcome as opposed to the scientific merits of the methods (Kaushik & Walsh, 2019). Secondary data on the other hand was used to complement primary data, thus working toward greater insight as well as scientific understanding (Jones, 2010).

## *Data Analysis*

This study used thematic analysis (TA) method to facilitate inductive coding of data (Elliott, 2018). According to Braun and Clarke (2014), TA technique enables theoretical freedom, hence a flexible and useful research tool capable of providing a rich and detailed, yet complex account of the data. Saldaña's streamlined codes-to-theory model for qualitative inquiry was used to identify themes and the patterns emerging from the data (Saldaña, 2015). This makes the TA method well suited to the varying needs and requirements of a wide range of qualitative research projects, including health research. The data collected was therefore subjected to a process of identifying codes from the data and categorizing concepts to derive themes and concepts to inform the research (Fig. 16.2).

Primary data was collected from five sources, namely, women using maternal services; community health workers (CHW); health professionals; management; and ICT experts. The interview questions were aligned to the research questions and represented the theoretical concepts of the study. The first research question deals with the current situation and needs of the different stakeholder groups within the context of use. The theoretical concepts associated with this question are information stakeholders, information needs, information practices, and information management challenges. The second research question deals with the co-design

| | A | B | C | D | E | F | G |
|---|---|---|---|---|---|---|---|
| 1 | **Source** | **TT** | **Theoretical themes** | **Initial codes** | **Categories** | **Data Themes** | **IQ** |
| 2 | ICTP | T4 | Information management challenges | A lot of bureaucracy | Management | Health system | 9 |
| 3 | CHW | T5 | Stakeholder perceptions | A mobile phone is not heavy and it saves time, but it require airtime and data bundles particularly for a smartphone | Resource provision | Service provider | 16 |
| 4 | HP | T3 | Information practices | A referral report/letter, phone call, partograph | Information sharing methods | Health communication | 10 |
| 5 | ICTP | T6 | Design considerations | Ability for scalability | Scalability | System quality | 4a |
| 6 | ICTP | T5 | Stakeholder perceptions | Active participation | Active participation | User-centered design | 3 |
| 7 | HP | T3 | Information practices | Admit new clients, medical tests, physical examinations, patient prescriptions, treat, baby deliveries, monitor patient progress, referrals, discharge clients and prepare handover report. | Routine activity documentation | Information services | 7 |
| 8 | AP | T5 | Stakeholder perceptions | Adoption of ICT in this facility is just medium that is, 50 % | ICT adoption | Service provider | 13 |
| 9 | ICTP | T6 | Design considerations | Affordable and Quality infrastructure | Infrastructure | System quality | 2 |
| 10 | ICTP | T5 | Stakeholder perceptions | Affordable quality health care service | Cost effective service | Quality health service | 2 |
| 11 | HP | T3 | Information practices | ANC booklet | Information sharing methods | Health communication | 10 |
| 12 | HP | T1 | Information stakeholders | ANC, PNC, Lab, Pharmacy, X-Ray, Out Patient Department, Theatre | Hospital departments | Stakeholders | 1 |

**Fig. 16.2** An extract from the coding sheet

process and the theoretical concepts associated with this question are stakeholder perceptions and design considerations. From the four sources, 265 unique codes were derived. The codes were then cleaned to deal with synonyms and potential mistakes in allocating the codes. The cleaned codes were then grouped through a categorization process into 166 categories. The categories were then grouped into 20 related themes that formed the data themes. The themes were linked to the five sources. The insights gained from the thematic analysis were used during the co-design process.

The source refers to the five groups from whom the data was collected. TT refers to the theory theme aligned to the research questions. The initial codes, categories, and data themes represent the coding process and IQ is linked to the interview question.

## Application of Co-design-Oriented SDR Strategy in Health Care

The concept of design can promote the development of new competences, and hence it can help an organization(s) to discover new technologies and business opportunities through exploring new modes of cooperation either at departmental, company, or industry level (Pekkala & Ylirisku, 2017). Discovery of new opportunities can then help to translate and shape the organization in diverse ways including in language and culture. Initial product concepts in innovation such as stories, new vocabulary, and prototypes provide a starting point in the product development process whereby the concept components change during the development. The new concept components can help in the initiation and reinforcement of a new course of development and design of a concept, which enables the commitment to development. This section highlights the design process this study used to design, develop, and test the proposed technology-enabled person-centered health and wellbeing information service (cloud-based PHR service) to improve maternal health information service. Particularly, IS development requires a thorough investigation of activities involved, context, and target group(s) concerned to gain insight into creating tangible artifacts, and to support intangible experiences as well as individual emotional needs (Huang et al., 2013).

The general perception is that engaging co-design research enhances meaningful involvement of end users in the research process, which may help to address the problem of wastage of research resources (Slattery et al., 2020). In this study, the co-design process commenced with contextual inquiry involving in-depth interviews and participant observation (Salgado et al., 2017). A model-oriented analysis and system development techniques were used to demonstrate knowledge and also to elicit ideas from participants to inform the design process as graphic presentations helps to deliver complex emerging concepts (Huang et al., 2013). Use case, data flow, and service system architectural design diagram were used to illustrate service requirements, which helped to visualize, construct, and document the proposed service system at a more conceptual level, hence conceptual knowledge (Group Object Management, 2010). This was necessary to facilitate the recognition of user experiences, motivation, and value to designing a service.

To produce user-centric services, stakeholders' insights were incorporated into service development whereby co-design-oriented SDR tools and techniques were used. This assisted the designer to capture customer activities, propositions, motives, and general user experience to produce the desired service (Patrício et al., 2011). Users' experience and opinions were used to capture and represent the new service requirements through their activities and the context (Patrício et al., 2011). More importantly, the user experience-based approach was used to enhance human-computer interaction whereby concepts and notation related to software engineering and SDR were used to create user-centric services (Huang et al., 2013). This models the service concept as the actual value offered to the customer in terms of various resultant service offerings. In this study, various co-design-oriented SDR methods, tools, and techniques were employed as illustrated in subsequent sub-sections.

## Design and Implementation of the Service-Oriented Information System Based on the Kenyan Context

Based on findings from field research data and a literature review, a cloud-based person-centered health and wellbeing information architecture (CPHIA) framework shown in Fig. 16.3 was developed to illustrate appropriate e-architecture model as

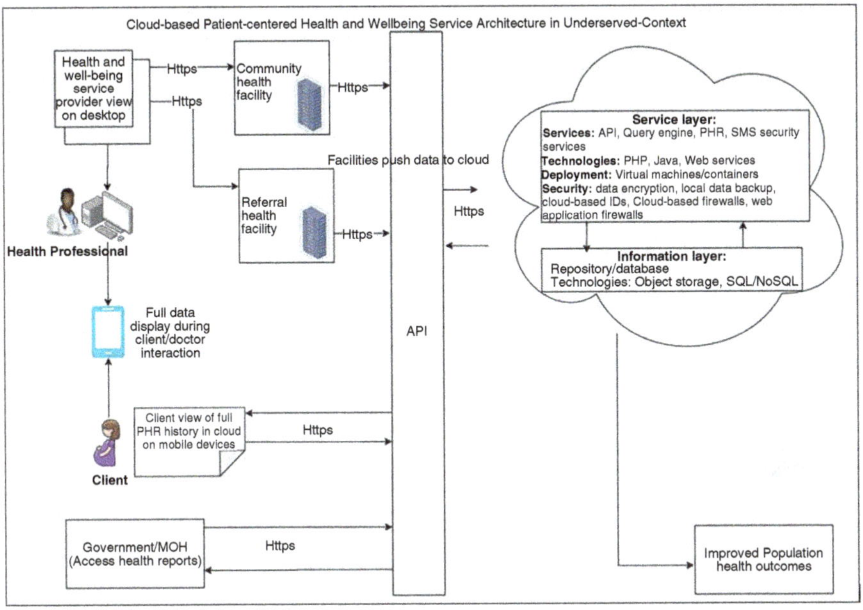


**Fig. 16.3** Cloud-based Patient-centered Health and Wellbeing Information Architecture (CPHIA) for maternal health care

the proposed PHR service in the research context. The CPHIA model, therefore, serves as a blueprint for determining information flows, selecting appropriate ICTs, creating, integrating, storing, and accessing health data according to various stakeholders' needs. For purposes of this study, a service prototype for the proposed PHR service was designed and implemented through four distinct components, namely, the relevant ICT infrastructure (CPHIA model), the service layer (cloud-based PHR service), the information layer (information architecture relevant to the user interfaces), and the *user layer* (user interfaces and service system functionality relevant to the various touch points).

Each of the aforementioned service components is illustrated in subsequent sections using various co-design-oriented SDR methods including service prototyping (service journey, case scenario, focus group discussions), sketches, business process modeling (use case and data modeling), and wireframes. This helps to illustrate a combination of material, tools, and techniques used in the research process.

## *The Service Layer*

This is basically the service prototyping stage of the proposed PHR service. Figure 16.4 is a maternity user journey the study adopted from the BOLD project

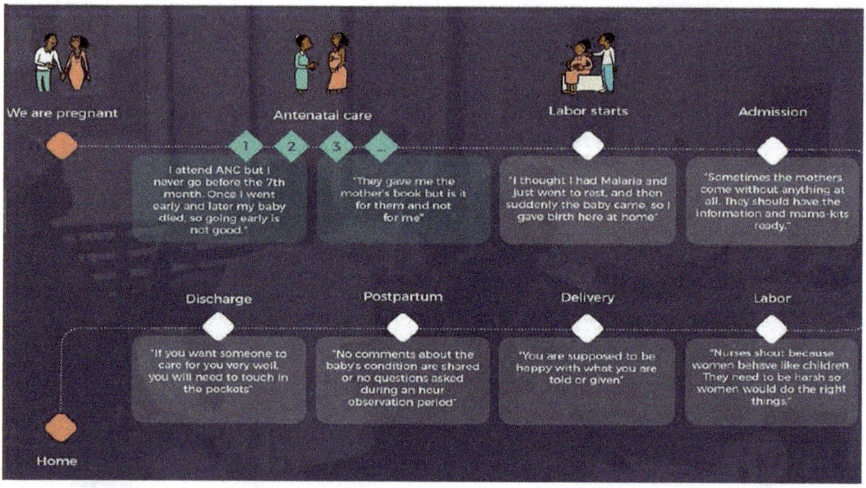


**Fig. 16.4** Touch points in the maternity pathway/user journey. (Source: Salgado et al., 2017:62)

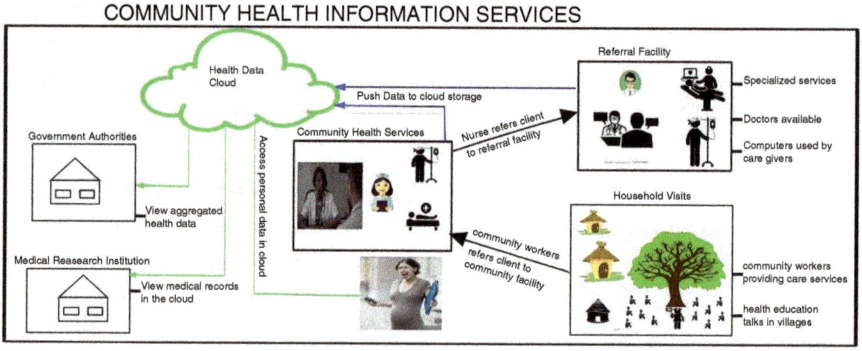


**Fig. 16.5** Case scenario for community health information service in Nairobi County

by Salgado et al. to create the service prototype, which was informed by the service touch points illustrated in the maternity pathway (Salgado et al., 2017). These include pregnancy → antenatal care → labor → admission → delivery → postpartum → discharge.

A case scenario of maternal health information services (Fig. 16.5) illustrates the nature of desired community health information services. The tool helped to capture the various stakeholders of maternal health care, interactions and information flows among different stakeholders, that is, clients, CHWs, health institutions (community and referral facilities), health practitioners (midwives and doctors), government authorities, and medical research institutions. It also illustrates the key technologies including cloud computing (CC), mobile, and Web technologies.

The study employed service prototyping techniques and design principles to co-create practical tools in the quest to improve maternal health information service in the research context (Salgado et al., 2017). Besides in-depth interviews, focus group discussions, and participant observation, more data were obtained using various design methods and tools such as the maternity service journey and Mother & Child booklet, that is, antenatal (ANC) booklet. These tools helped to identify possible barriers that women encountered while seeking maternity services from public health facilities. Also, design features for the service prototype, that is, maternal profile, medical and surgical history, clinical notes, physical examination, and ANC profile, were extracted from ANC booklet.

Focus group discussions were used to generate paper-based prototypes as well as role-playing situations to facilitate understanding of health and emotional needs of the users using maternity pathway. Sticky notes were used to gather ideas from participants during focus group discussion sessions, which were then displayed on a white board for comparison. This was done with reference to touchpoints in the aforementioned maternity user journey by Salgado et al. (2017).

Potential users of maternity services including women, CHWs, and health practitioners had the opportunity to express their different views/opinions and criticisms to generate the initial concepts as shown in Fig. 16.6. This helped to gauge the applicability of the desired change in the study context. This was done alongside mapping of user information practices, user information needs, and information behavior as input for the design. Particularly, common ideas regarding menu screen design and features for the new prototype service (cloud-based PHR service) were identified to help in tracking and monitoring of personal health data. This facilitated the identification and clarification of important themes as input for design requirements, and subsequent development of tools to provide better maternal information services.

**Fig. 16.6** Women discussing design requirements for desired PHR service

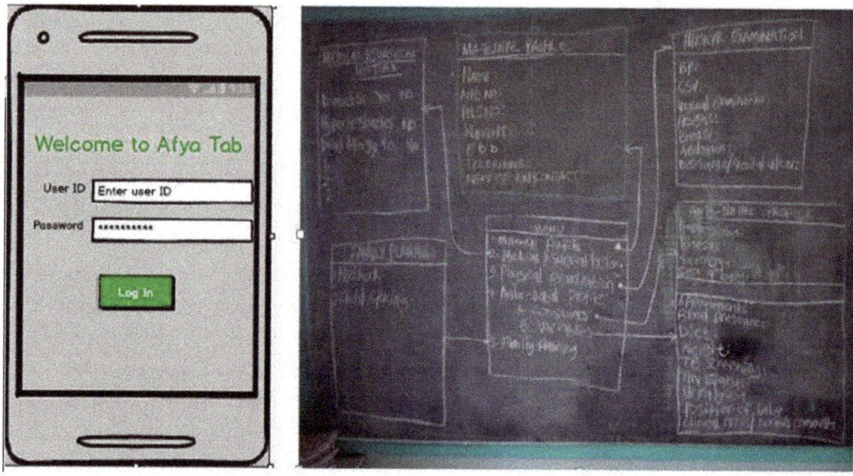


**Fig. 16.7** Mobile login screen and application flow sketch

During the co-design process, the designer/researcher constantly checked on service desirability, the functionality, and usability through monitoring participants' reactions. Participants' feedback was integrated into the prototype during co-design sessions until the final design solution was achieved. This involved close inspection of mobile screen layouts, menu screen, font size, and the functionality by both the designer and users.

The new proposed service will be accessed via both mobile device (for clients) and desktop computers (for health practitioners). Figures 16.7 and 16.8 are wireframes representing user's login window, application flow sketch, and mobile screen menus of the intended prototype service for clients (women).

## *Information Layer*

This component marks the design stage of the new PHR service, and it involves the system analysis and design process illustrating information flows and distribution of the business process modeling in relation to the business (maternal health care) functions. This helps discover the how, when, and what factors that would favor successful information flow within the proposed information service/system. Figure 16.9 is a business use case model developed for the project, showing the various actors, their roles, and interactions within the service ecosystem.

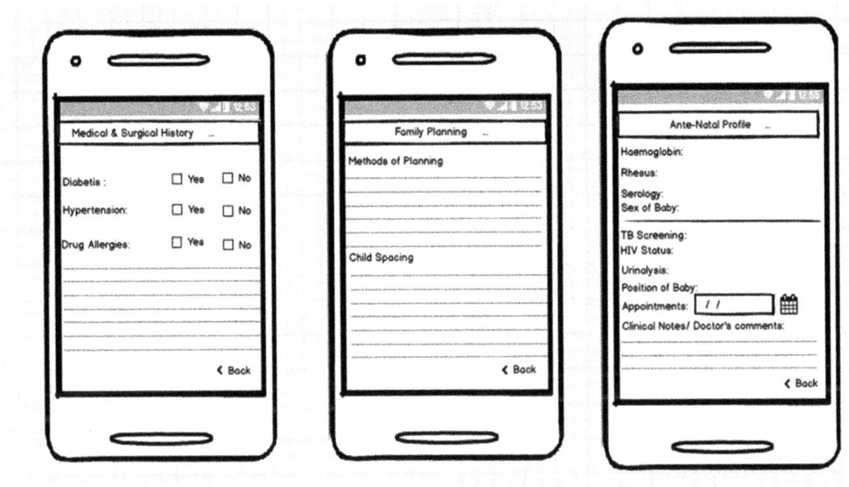


**Fig. 16.8** Mobile application screens for main menu, maternal profile, and physical examination

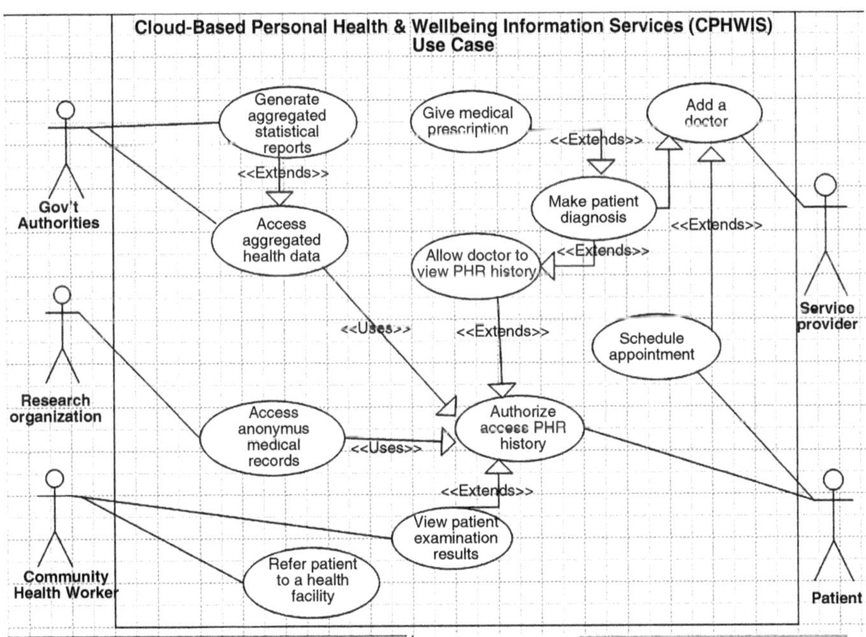


**Fig. 16.9** Use case diagram for proposed cloud-based PHR service

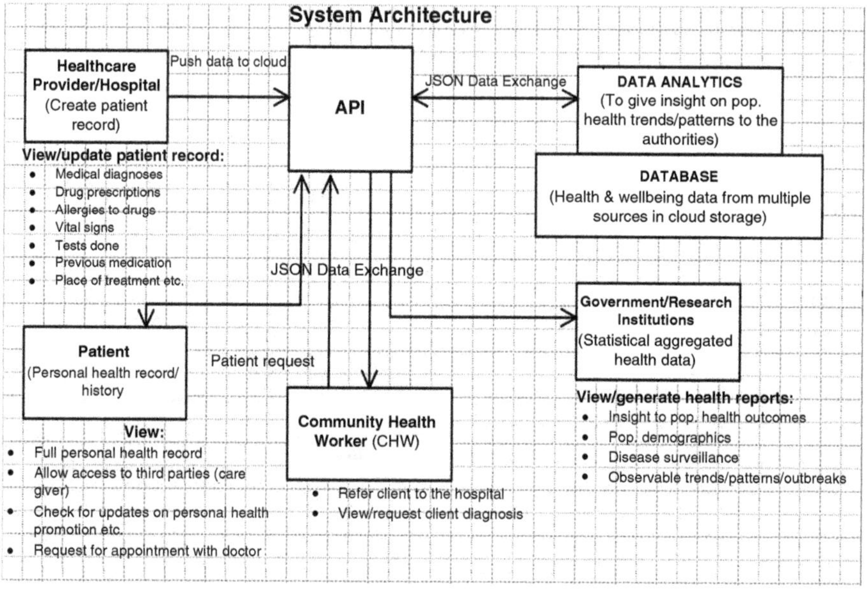


**Fig. 16.10** Design and implementation architecture for proposed cloud-based PHR services

## Use Case Diagram

## Architectural Design Diagram

The business architectural design model shown in Fig. 16.10 depicts the various components of the prototype service system, including data storage in the cloud, application programming interface (API) interface, the various entities, query engine, and service interactions. Patient health data are collected from different sources such as electronic health records (EHRs) from various hospitals/clinics and home monitoring devices which will be transmitted and updated in the personal health record (PHR) in the cloud. This may include patient demographics, allergies, vital signs, prescriptions, medications, immunizations, laboratory tests, radiology reports, blood sugar level, blood pressure, and heart activity, among others (Heart et al., 2017; Alyami et al., 2017).

## Functioning of the Proposed Service

Potential data security risks of the proposed PHR service will be addressed by the cloud service provider using key management techniques based on adequate cryptographic mechanisms on the data in cloud storage. Access control will be managed through role-based access with server authentication services on users' devices (Nkosi & Mekuria, 2010). Various users will be able to query cloud health data based on structured Query Language (SQL) database to enable users to generate various reports. The query engine will have an API key and a data exchange format (JSON). Care activity and service interactions will be supported by various electronic service touch points (human-to-computer interaction) through user interfaces. Different actors will interact via the service system platform during care activity, for example, patient versus CHW/V, patient versus doctor, department versus department, hospital versus hospital, and vice versa throughout the maternity partway/journey (Lee, 2017). Such activities may include maternal health care during clinic appointments/visits, during labor, admission, delivery, postpartum, and discharge from hospital. Distribution and access to electronic health information will involve such activities including patient-doctor consultations, patient access to view personal health data, doctors viewing and updating patient records, renewal of patient prescriptions, and communication (message alert and email) (Caligtan & Dykes, 2011). Also, government authorities and medical research institutions may want to view aggregated health data or anonymous medical records to generate reports.

## *The User Layer*

This section marks the implementation stage of the new PHR service which describes both functional and non-functional requirements, delivery of the potential prototype service and actual user interfaces for various actors/stakeholders as a proof- of-concept, and how they will interact with the new PHR service. Specifically, the proposed PHR service (AfyaTab app) will comprise of two main parts, namely, the web application (for service providers, i.e., community, referral facilities) and the government/MOH) and a mobile application for clients. The service requirements will comprise of user requirements, system requirements, and software requirements (ELkafrawy & Khalaf, 2016) as characteristics that must be fulfilled to meet users' expectations, hence the need for service quality in order to serve its purpose well (Chung & Do Prado Leite, 2009).

### Functional Requirements

These are requirements that describe service functionality, that is, what the system service must do/perform for users to accomplish their tasks. Based on various users' roles, the proposed prototype service system will be expected to enable users to perform the following tasks: clients should be allowed to have control over access to personal data in the cloud, and to authorize health professionals to view their full PHR history in the cloud; health care providers (health facilities and health professionals) should create, store, and update client records in local storage without seeking client consent; health care providers (health facilities and health professionals) should freely send client data in the local server to the central server in the cloud; government/MOH is the super admin for controlling service delivery – they can access and view anonymous medical records, view aggregated health data, and generate statistics in cloud storage for disease surveillance, decision-making, and research purposes.

### Non-functional Requirements

These refer to conditions on the services offered by the prototype service system that are not specifically related to system functionality or its execution such as timing requirements and limitations associated with the development process and standards (ELkafrawy & Khalaf, 2016). They place restrictions while the product is being developed which specify external constraints that the service should satisfy. The proposed service system will seek to satisfy but not limited to the following: interface requirements, usability requirements, security requirements, reliability requirements, performance requirements, and interoperability requirements.

## Development of a Potential Prototype Service System

Based on the analysis performed on the project variables, namely, time, cost, and project scope, the study adopted both the Rapid Application Development (RAD) model and the Structured System Analysis and Design (SSAD) approach. RAD involved prototyping with repetitive development hence an incremental software development process to emphasize a short development life cycle and encouraged constant feedback from the users during software development (Khan et al., 2012). Three specific development tools were required to create the proposed cloud PHR service (AfyaTab app), namely, user interface tools, database tools, and client/provider interfaces.

### *Web-Based Mobile App Interfaces*

The mobile user interface is implemented using the Java language while the web app (AfyaTab) is implemented using HTML5, JAVASCRIPT, JQUERY, and AJAX languages together with a style sheet, CSS. For data tools, Golang programming language was used to facilitate the front-end data. Deployment is done on Digital Ocean cloud-based system, which has capacity for scalable clusters, pooled bandwidth, and easy use of CC platform of virtual servers and easy deployment. Clients (women/family members) will mainly use a mobile device to access and track personal health data. Below are the menu screens illustrating actual mobile-based client user interfaces as shown in Figs. 16.11 and 16.12.

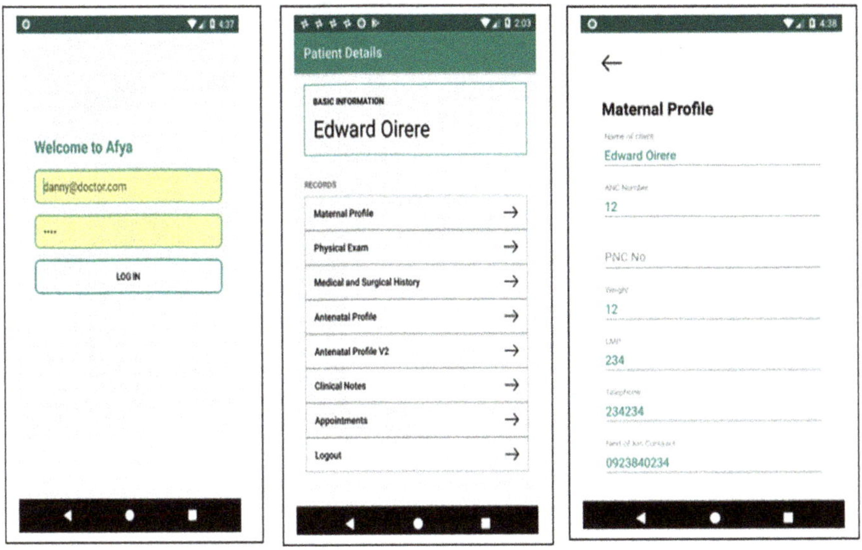


Fig. 16.11 Client user's interfaces for log in to client and maternal profile

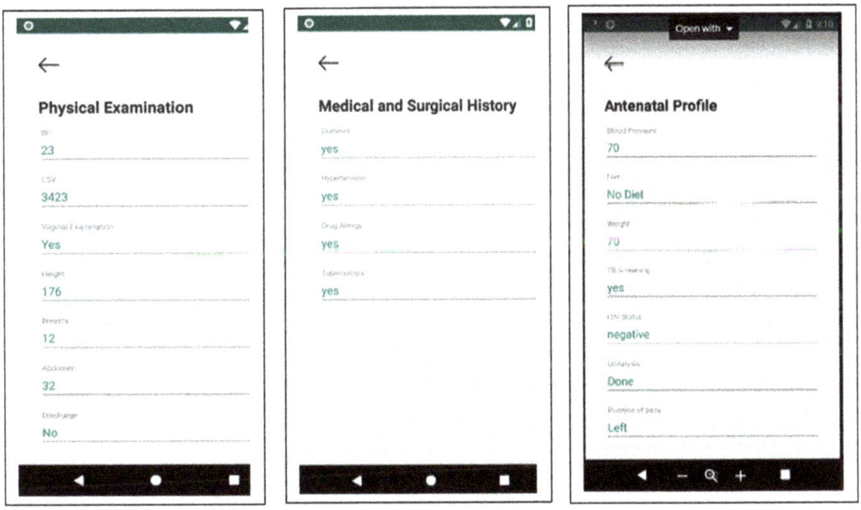


Fig. 16.12 User interfaces for physical examination, medical history, and ANC

## *Web-Based Desktop App Interfaces*

The service providers (hospitals/health institutions, health practitioners, and government/ministry of health (MOH) officials) will use the desktop app to access, view, and update records and generate reports from the system (Fig. 16.13).

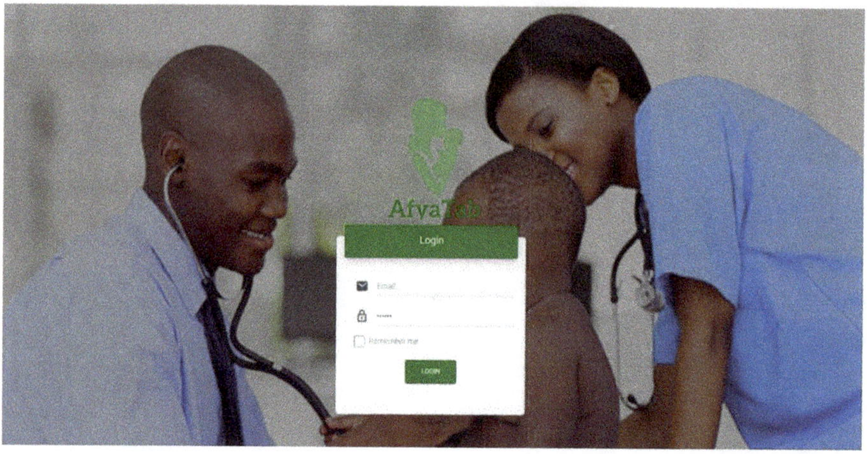


**Fig. 16.13** Health care provider's login window for cloud PHR service

Figures 16.11–16.13 are window screen images for various interfaces for health care providers.

## Lessons Learnt from Application of SDR in the Case Study

The following are some lessons learnt from the application of co-design-oriented SDR strategy in the case study:

SD is both a human-centered and iterative approach which facilitates creation of service innovation (Salgado et al., 2017). Being a participatory approach, various maternal health stakeholders were engaged using appropriate SD methods and tools including a service blueprint and case scenario to collect design data hence, the recognition of stakeholders' role (Nyatuka & de la Harpe, 2019). The co-design activities and process enabled the voice of the people to be heard, that is, stakeholders' views and opinions and human-centered solutions. The co-design process allows collaboration between the researcher and humans (stakeholders) to co-create services (Goldkuhl, 2012). While seeking to determine an appropriate design solution to the problem, the researcher and stakeholders collaboratively designed the solution(s) to create new realities, which were based on their needs and experiences within the constraints of the study context (Goldkuhl, 2012).

The approach is sensitive to the context of research, and thus it enhances the understanding of the primary needs of users to deliver solutions with context-specific design (Gregor et al., 2020; Nyatuka & de la Harpe, 2019). Particularly, SD made it possible for the researcher to employ both interpretive and pragmatic approaches to perform a contextual inquiry, which made it possible to identify values and opportunities through stakeholders' active participation, which were then incorporated in the design of a solution. The SD strategy, therefore, delivers evidence-based solutions that are meaningful in their context of use.

Also, it represents SDR as a multidisciplinary approach and service innovation as a multidimensional phenomenon (Joly et al., 2019). In this study, the researcher needed holistic understanding across the dimensions of the health service provider's services, the digital environment, and organizational properties (Nylén & Holmström, 2015) including. Particularly, the service dimensions that were of interest to the researcher included the context, roles, processes, and outcomes as critical considerations toward creating the desired situation (Malmberg et al., 2019). SDR approach therefore enabled building holistic health care service system that involved various design methods and techniques, stakeholders, products, interactions, experiences, and services within the specific health care environment (Pamedytyte & Akoglu, 2019).

SDR, therefore, is already regarded as an extremely essential and valuable approach for modern industries within service innovation meant to improve the service experience and quality while accelerating the service development process. More importantly, it gives a voice to customer views and opinions in the design and development process through participatory design, a merit which conventional methods do not embrace. This allows room for creation of context-specific and meaningful solutions, and thus guaranteeing maximum impact on organizations and communities. Individuals' lived experiences provide better insight into a study phenomenon through empirical scrutiny, thus enabling verifiable study conclusions and contributing to the improvement of the situation(s). Apparently, there is a paucity of empirical evidence to help in the understanding of the complex nature of technology use in health care services from a design thinking perspective (Nyatuka & de La Harpe, 2019). There is a lack of literature to guide the execution of multidisciplinary-oriented research toward improving health care service design particularly in underserved contexts such as Africa. This study will, therefore, go a long way to bridge this gap in terms of theoretical, discipline, and methodological knowledge to health practice, hence contributing to existing body of knowledge.

## The Nature of Service Innovation(s) Created

It was observed that SD strategy indeed plays a critical role in the creation of service innovations including in health care industry where various theories, methods, and techniques are combined to produce new service innovations, and hence regarded as a catalyst for organizational transformation (Simon, 1996; Malmberg et al., 2019). SD basically emphasizes designing new and/or improving existing services to make them more useful, efficient, and effective for organizations (Järvinen, 2007; Miah et al., 2013; Moritz, 2005). The main focus in this study, therefore, was to improve existing maternal health information services to make them better for women in their context (Schuler & Namioka, 1993). According to Sternberg et al. (2003), there are three main types of innovations, namely, incremental (continuous linear improvement of value acquired by the customer); disruptive (continuous improvement of value majorly linked to the needs of customers rather than capabilities of competitors); and radical (discontinuous, i.e., with or without a predecessor; essential non-linear improvement obtained by the customer respectively). In this case study, the innovation(s) created were partly incremental and disruptive since the aim was to transform the existing services toward a desired situation. For example, the paper-based maternal information services (Mother & Child Book) were transformed to an electronic information service (cloud PHR service). In order to meet the needs of stakeholders, some of the methods and technologies initially used to access information were improved while completely new ones were introduced, for example, changing from using a basic mobile phone to a smart phone with more advanced data features. Besides developing a mobile-based app for women, a desktop app was developed for health practitioners.

## Conclusion

Health care makes critical contribution toward economic development on a global scale, hence the need for health care service innovations. Particularly, the deficient HISs in underserved context require high-level innovation in order to leverage the benefits of ICT as a catalyst for health care reform. This will assist to effectively manage opportunities, challenges, and emerging trends in the modern health care space. The application of the service concept in creating health care innovations, and particularly the co-design-oriented SDR approach, should not only be viewed as a component of the design process but also as a means to "concretize" the nature of the service being created. This approach ensures integration between the how and what of SDR while mediating between the needs of various stakeholders through active participation, and thus creating person-centered, context-specific, meaningful, and sustainable innovations, and therefore, a catalyst for organizational transformation especially in underserved contexts. The use of SD strategy in this study reveals huge potential to create sustainable eHealth innovations to promote health and wellbeing, and thus improved population health outcomes. Specifically, in this case study, we created maternal health information services (cloud-based PHR service) comprising a web-based mobile app and windows desktop app to meet diverse stakeholder (customer) needs. The co-design process generated a service innovation comprising of characteristics for both incremental and disruptive innovations as the intent was to improve maternal health information services toward a desired situation.

## Study Limitations and Future Research

The main focus of this study was in government health facilities that provide maternity services as opposed to private health establishments, which are part of the health sector where a sizeable number of citizens equally receive health care services. Also, the findings of the study are based on maternal health care stakeholders' perceptions and opinions in the Kenyan context. The relevance of the study outside the study context should, therefore, be empirically verified. Furthermore, the study does not involve the in-use situation of the proposed new PHR service. As part of future research, therefore, the proposed solution needs to be further developed into a functional service and then implemented in the study context (particular situation). In addition, future work will discuss in general how this work can help in improving application of SD for emerging technologies product development such as artificial intelligence (AI), Internet of Things (IoT), robotics, automotive, and aircraft.

## References


Alamri, A. (2012). Cloud based e-health multimedia framework for heterogeneous network. In *Multimedia and expo workshops (ICMEW), 2012 IEEE international conference* (pp. 447–452). IEEE.

Alyami, M. A., Almotairi, M., Aikins, L., Yataco, A. R., & Song, Y. T. (2017). Managing personal health records using meta-data and cloud storage. In *2017 IEEE/ACIS 16th international conference on computer and information science (ICIS)* (pp. 265–271). IEEE.

Amoako, G., & Rivett, U. (2015). Towards the development of sustainable ICT projects in Africa–A review and synthesis of evaluation frameworks. In *The 2015 ACIST conference*. ICIST.

Blomkvist, J., Holmlid, S., & Segelström, F. (2010). This is service design research: Yesterday, today and tomorrow. In M. Schneider & J. Stickdorn (Eds.), *This is service design thinking* (pp. 308–315). BIS.

Bloom, D., Canning, D., & Sevilla, J. (2004). The effect of health on economic growth: A production function approach. *World Development, 32*(1), 1–13.

Boore, C. (2018). *A framework for e-Health implementation in the health care sector in Kenya: A Ground Theory approach* [PhD Thesis, University of Nairobi]. [Online] Available at: http://erepository.uonbi.ac.ke/bitstream/handle/11295/104378/Boore%2c%20Caroline.pdf?sequence=2&isAllowed=y. Accessed 20 Jan 2019.

Braun, V., & Clarke, V. (2014). What can "thematic analysis" offer health and wellbeing researchers? *International Journal of Qualitative Studies on Health and Well-Being, 9*, 1–2.

Caligtan, C., & Dykes, P. (2011). Electronic health records and personal health records. *Seminars in Oncology Nursing, 27*(3), 218–228.

Chung, L., & Do Prado Leite, J. (2009). On non-functional requirements in software engineering. In *Conceptual modeling: Foundations and applications* (pp. 363–379). Springer.

Clark, H., & Wu, H. (2016). *The sustainable development goals: 17 goals to transform our world. Furthering the work of the United Nations.* [Online] Available at: https://www.un-ilibrary.org/united-nations/furthering-the-work-of-the-united-nations_b528170b-en. Accessed 12 July 2019.

De Mello Freire, K., & Damazio, V. (2018). Promoting organisational change through new meanings: Designing the rhetoric and the aesthetics of service offerings. *Strategic Design Research Journal, 11*(3), 247–253.

Edvardsson, B., Gustafsson, A., Pinho, N., Beirão, G., Patrício, L., & Fisk, R. P. (2014). Understanding value co-creation in complex services with many actors. *Journal of Service Management, 25*(4), 470–493.

Elkafrawy, P., & Khalaf, M. S. (2016). Problems of system requirements in the new systems and its solutions. *International Journal of Computer Application (2250-1797), 6*(5), 1–15.

Elliott, V. (2018). Thinking about the coding process in qualitative data analysis. *The Qualitative Report, 23*(11), 2850–2861.
Girgis, A., Durcinoska, I., Levesque, J. V., Gerges, M., Sandell, T., Arnold, A., & Delaney, G. P. (2017). eHealth system for collecting and utilising patient reported outcome measures for personalised treatment and care (PROMPT-Care) among cancer patients: Mixed methods approach to evaluate feasibility and acceptability. *Journal of Medical Internet Research, 19*(10), e330.
Goldkuhl, G. (2012). Pragmatism vs. interpretivism in qualitative information systems research. *European Journal of Information Systems, 21*(2), 135–146.
Goldstein, S., Johnston, R., Duffy, J., & Rao, J. (2002). The service concept: The missing link in service design research? *Journal of Operations Management, 20*(2), 121–134.
Gregor, S., Kruse, L., & Seidel, S. (2020). The anatomy of a design principle. *Journal of the Association for Information Systems, 21*(6), 1622–1652.
Grisot, M., & Lindroth, T. (2019). *Bounded combinability of digital technologies for patient-centred care*. AIS.
Groeneveld, B., Dekkers, T., Boon, B., & D'Olivo, P. (2018). Challenges for design researchers in health care. *Design for Health, 2*(2), 305–326.
Group Object Management. (2010). *OMG unified modeling language, superstructure, version 2.3*. [Online] Available at: https://www.omg.org/spec/UML/2.3/About-UML/. Accessed 13 Apr 2020.
Guest, G., MacQueen, K., & Namey, E. E. (2014). Writing up thematic analyses. In *Applied thematic analysis* (pp. 3–40). Sage.
Heart, T., Ben-Assuli, O., & Shabtai, I. (2017). A review of PHR, EMR and EHR integration: A more personalised health care and public health policy. *Health Policy and Technology, 6*(1), 20–25.
Huang, K. H., Teixeira, J. G., Patrício, L., Constantine, L., & Nóbrega, L. A. (2013). Model-driven service design toolkit: From co-design to implementation. In *The 5th international congress IASDR 2013*. IASDR.
Hurley, E., Trischler, J., & Dietrich, T. (2018). Exploring the application of co-design to transformative service research. *Journal of Services Marketing, 32*, 715–727.
Järvinen, P. (2007). Action research is similar to design science. *Quality & Quantity, 41*(1), 37–54.
Johnson, G., Scholes, K., & Whittington, R. (2008). *Exploring corporate strategy: Text & cases*. Pearson Education.
Joly, M., Teixeira, J., Patrício, L., & Sangiorgi, D. (2019). Leveraging service design as a multidisciplinary approach to service innovation. *Journal of Service Management, 30*(6), 681–715.
Jones, C. (2010). Archival data: Advantages and disadvantages for research in psychology. *Social and Personality Psychology Compass, 4*(11), 1008–1017.
Karuri, J., Waiganjo, P., Daniel, O., & Manya, A. (2014). DHIS2: The tool to improve health data demand and use in Kenya. *Journal of Health Informatics in Developing Countries, 8*(1), 38–60.
Kaur, J. (2012). The role of litigation in ensuring women's reproductive rights: An analysis of the Shanti Devi judgement in India. *Reproductive Health Matters, 20*(39), 21–30.
Kaushik, V., & Walsh, C. (2019). Pragmatism as a research paradigm and its implications for social work research. *Social Sciences, 8*(9), 255.
Khan, A., Qureshi, M., & Khan, U. (2012). A comprehensive study of commonly practiced heavy & light weight software methodologies. *International Journal of Computer Science Issues, 8*(4), 441–450.
Klein, H., & Myers, M. (1999). A set of principles for conducting and evaluating interpretive field studies in information systems. *MIS Quarterly, 23*, 67–93.
Korpela, M., Ikävalko, P., Luukkonen, I., Martikainen, S., Palmén, M., Tiihonen, T., Toivanen, M., & Vainikainen, V. (2013). How to co-develop services, work, and information systems in health care: The daisy approach. *Studies in Health Technology and Informatics, 194*, 126–132.
Kumar, R. (2011). *Research methodology: A step-by-step guide for beginners* (3rd ed.). Sage.

Kurtmollaiev, S., Fjuk, A., Pedersen, P. E., Clatworthy, S., & Kvale, K. (2018). Organisational transformation through service design: The institutional logics perspective. *Journal of Service Research, 21*(1), 59–74.

Lade, V., Jaitpal, S., & Chitnis, A. (2014). A review of cloud based health care system. *International Journal of Enhanced Research in Science Technology & Engineering, 3*(10), 22–26.

Lee, E. (2017). Identifying key components of services in health care in the context of out-patient in Norway. In *10th international conference on health informatics* (pp. 354–361). SCITEPRESS – Science and Technology Publications. https://doi.org/10.5220/0006170803540361

Malmberg, L., Rodrigues, V., Lännerström, L., Wetter-Edman, K., Vink, J., & Holmlid, S. (2019). Service design as a transformational driver toward person-centred care in health care. In *Service design and service thinking in health care and hospital manage* (pp. 1–18). Springer.

Manya, A., Braa, J., Øverland, L. H., Titlestad, O. H., Mumo, J., & Nzioka, C. (2012). National roll out of District Health Information Software (DHIS 2) in Kenya, 2011–Central server and cloud based infrastructure. In *IST-Africa 2012 conference proceedings*. International Information Management Corporation (IIMC).

Manya, A., Nielsen, P., & Pundo, R. (2016). Cloud computing as a catalyst for integrated health information systems in developing countries. *Journal of the Association for Information Systems, 7*(3), 1–12.

Marshall, M. (1996). Sampling for qualitative research. *Family Practice, 13*(6), 522–526.

Miah, S. (2017). Large-scale innovations and approaches for community health care support in developing nations. In *Handbook of large-scale distributed computing in smart health care* (pp. 359–375). Springer.

Miah, S., Kerr, D., & Hellens, L. (2013). A collective artefact design of decision support systems: Design science research perspective. *Information Technology & People, 27*(3–4), 1–22.

Ministry of Health. (2016). *Kenya National eHealth Policy 2016-2030. Towards attainment of the highest standard of health through adoption and use of ICT*. Ministry of Health.

Moritz, S. (2005). *Service design: Practical access to an evolving field*. Köln International School of Design.

Nabyonga-Orem, J. (2017). Monitoring Sustainable Development Goal 3: How ready are the health information systems in low-income and middle-income countries? *BMJ Global Health, 2*(4), e000433.

Nkosi, M., & Mekuria, F. (2010). Cloud computing for enhanced mobile health applications. In *2010 IEEE second international conference on cloud computing technology and science*. IEEE.

Nyatuka, D. R., & De la Harpe, R. (2019). Evaluating mHealth interventions in an underserved context using service design strategy: A case of Kenya. In *Proceedings of the third international conference on medical and health informatics*. ACM.

Nyatuka, D. R., & De La Harpe, R. (2021). Service design as a catalyst for patient-centered eHealth innovation: An architectural design framework for cloud-based maternal health information service in underserved setting. *International Journal of Information System Modeling and Design (IJISMD), 12*(3), 62–85.

Nylén, D., & Holmström, J. (2015). Digital innovation strategy: A framework for diagnosing and improving digital product and service innovation. *Business Horizons, 58*(1), 57–67.

Ochieng, C., & Odhiambo, A. (2019). Barriers to formal health care seeking during pregnancy, childbirth and postnatal period: A qualitative study in Siaya County in rural Kenya. *BMC Pregnancy and Childbirth, 19*(1), 339.

Ogden, K., Barr, J., & Greenfield. (2017). Determining requirements for patient-centred care: A participatory concept mapping study. *BMC Health Services Research, 17*(1), 780.

Ostrom, A. L., Parasuraman, A., Bowen, D. E., Patrício, L., & Voss, C. A. (2015). Service research priorities in a rapidly changing context. *Journal of Service Research, 18*(2), 127–159.

Pamedytyte, R., & Akoglu, C. (2019). Value creation through service design in a health care environment. *The Design Journal, 22*(1), 655–668.

Patrício, L., Fisk, R., Falcão e Cunha, J., & Constantine, L. (2011). Multilevel service design: From customer value constellation to service experience blueprinting. *Journal of Service Research, 14*(2), 180–200.

Patrício, L., Sangiorgi, D., Mahr, D., Čaić, M., Kalantari, S., & Sundar, S. (2020). Leveraging service design for health care transformation: Toward people-centred, integrated, and technology-enabled health care systems. *Journal of Service Management, 31*(5), 889–909.

Pekkala, J., & Ylirisku, S. (2017). The role of design concepts in the development of digitalised industrial services. *The Design Journal, 20*(sup1), S2813–S2822.

Radojicic, M., Jeremic, V., & Savic, G. (2020). Going beyond health efficiency: What really matters? *The International Journal of Health Planning and Management, 35*(1), 318–338.

Saldaña, J. (2015). *The coding manual for qualitative researchers* (3rd ed.). Sage.

Salgado, M., Wendland, M., Rodriguez, D., Bohren, M. A., Oladapo, O. T., Ojelade, O. A., Mugerwa, K., & Fawole, B. (2017). Using a service design model to develop the "Passport to Safer Birth" in Nigeria and Uganda. *International Journal of Gynaecology and Obstetrics, 139*, 56–66.

Sangiorgi, D., Patrício, L., & Fisk, R. P. (2017). Designing for interdependence, participation and emergence in complex service systems. In S. Sangiorgi & A. Prendiville (Eds.), *Designing for service: Key issues and new directions* (pp. 72–86). Blooms.

Saravanan, M., Kumar, R., & Manoj, S. (2016). A survey on knowledge discovery of health care dataset using graph based approach. *Research Journal of Pharmaceutical, Biological and Chemical Sciences, 7*, 8–13.

Schuler, D., & Namioka, A. (1993). *Participatory design: Principles and practices*. CRC Press.

Simon, H. (1996). *The sciences of the artificial* (3rd ed.). MIT Press.

Slattery, P., Saeri, A., & Bragge, P. (2020). Research co-design in health: A rapid overview of reviews. *Health Research Policy and Systems, 18*(1), 17.

Sternberg, R. J., Pretz, J. E., & Kaufman, J. C. (2003). Types of innovations. In *The international handbook on innovation* (pp. 158–169). Pergamon.

Sun, Q. (2020). Towards a new agenda for service design research. *The Design Journal, 23*(1), 49–70.

Teixeira, J., Patrício, L., & Tuunanen, T. (2019). Advancing service design research with design science research. *Journal of Service Management, 30*(5), 577–592.

Thobias, J., & Kiwanuka, A. (2018). Design and implementation of an m-health data model for improving health information access for reproductive and child health services in low resource settings using a participatory action research approach. *BMC Medical Informatics and Decision Making, 18*(1), 45.

Thorbjørnsen, H., Pedersen, P., & Nysveen, H. (2009). Categorising networked services. *European Journal of Marketing, 43*(3–4), 371–397.

Tindall, R. M., Ferris, M., Townsend, M., Boschert, G., & Moylan, S. (2021). A first-hand experience of co-design in mental health service design: Opportunities, challenges, and lessons. *International Journal of Mental Health Nursing, 30*(6), 1693–1702.

Townsend, A., Adam, S., Birch, P., & Friedman, J. (2013). Paternalism and the ACMG recommendations on genomic incidental findings: Patients seen but not heard. *Genetics in Medicine, 15*(9), 751–752.

Trischler, J., Pervan, S., Kelly, S., & Scott, D. (2018). The value of codesign: The effect of customer involvement in service design teams. *Journal of Service Research, 21*(1), 75–100.

Tsekleves, E., & Cooper, R. (2017). Emerging trends and the way forward in design in health care: An expert's perspective. *The Design Journal, 20*(1), 2258–2272.

Wetter-Edman, K., Vink, J., & Blomkvist, J. (2018). Staging aesthetic disruption through design methods for service innovation. *Design Studies, 55*, 5–26.

World Health Organization. (2015). *WHO global strategy on integrated people-centred health services 2016-2026. Placing people and communities at the centre of health services*. WHO.

World Health Organization. (2016). *Framework on integrated people-centred health services. Sixty-ninth World Health Assembly, A69/39*. WHO.

Ziraba, A. K., Mills, S., Madise, N., Saliku, T., & Fotso, J. C. (2009). The state of emergency obstetric care services in Nairobi informal settlements and environs: Results from a maternity health facility survey. *BMC Health Services Research, 9*(1), 1–8.

# Index





U. Z. A. Hamid, M. Suoheimo (eds.), *Service Design for Emerging Technologies Product Development*, Springer Series in Design and Innovation 29,
https://doi.org/10.1007/978-3-031-29306-1

Printed in the USA
CPSIA information can be obtained
at www.ICGtesting.com
LVHW062337060823
754286LV00005BA/340

9 783031 293054